# SURFACE AND COLLOID SCIENCE

## Volume 11

## ADVISORY BOARD

A Continuation Order Plan is available for this series. A continuation order will bring delivery of each new volume immediately upon publication. Volumes are billed only upon actual shipment. For further information please contact the publisher.

# SURFACE AND COLLOID SCIENCE

## Volume 11

### EXPERIMENTAL METHODS

Edited by

## ROBERT J. GOOD

*State University of New York at Buffalo*
*Amherst, New York*

and

## ROBERT R. STROMBERG

*Bureau of Medical Devices*
*Food and Drug Administration*
*Silver Spring, Maryland*

PLENUM PRESS · NEW YORK AND LONDON

The Library of Congress cataloged the first Plenum Press edition of this title as follows:

Main entry under title:

Surface and colloid science.

Vol. 10-     published by Plenum Press, New York.
Includes bibliographies.
1. Surface chemistry—Collected works. 2. Colloids—Collected works. I. Matijević, Egon,
1922-     ed.

QD506.S78                          541'.345                          67-29459
ISBN 978-1-4615-7971-7        ISBN 978-1-4615-7969-4 (eBook)
DOI 10.1007/978-1-4615-7969-4

This book was edited by Robert R. Stromberg in his private capacity.
No official support or endorsement by the Food and Drug Administration
is intended or should be inferred.

Library of Congress Catalog Card Number 67-29459
ISBN 978-1-4615-7971-7


© 1979 Plenum Press, New York
Softcover reprint of the hardcover 1st edition 1979

A Division of Plenum Publishing Corporation
227 West 17th Street, New York, N.Y. 10011

# Preface

Surface science and colloid science are preeminently experimental subjects. They constitute complementary aspects of a field which has been notably active since World War II; there is every reason to expect that the level of activity will continue to rise in the coming decades, so it is timely to review certain experimental methods of surface and colloid science as they exist, and to evaluate and refine those methods. This volume, and others that will follow, are principally concerned with experimental methods.

The working scientist needs access to the latest techniques, of course. He also needs to learn of the potentialities of recently developed techniques which he may not have been aware of. Equally important, or perhaps even more so, he needs to learn of the pitfalls of existing methods. One might say, wistfully, that it would be nice to be able to pick up somebody's description of a new piece of apparatus, to go into the laboratory, to build it, and to have it work, the first time!

There is, however, a serious problem of the interaction between the experiment *per se* and the theory for which the experiment is designed. Very often, this interaction renders problematic the interpretation of "direct" observations. An example, from experience of the senior editor of this volume, is the question of contact angle hysteresis. (See Chapters 1 and 2.) There is reason to believe that experimenters have, for years, been "throwing away" important and easily accessible information about the surfaces they study, by neglecting to measure retreating angles as well as advancing angles. The interaction between theory and experiment, in regard to hysteresis, is a curious one: If a solid surface is flat and homogeneous, and if sufficient time is allowed in the measurement, theory says there should be no hysteresis. The theory of hysteresis indicates that it is a phenomenon with several causes; and it may be a measure both of the nonideality of the solid and of rate effects that are difficult to control. So there is a natural temptation for an experimenter to ignore hysteresis, and to assume that he has prepared his surfaces carefully enough that they are "ideal."

Considerations such as these determine some components of the editorial policy for this volume. It was evident to us that chapters on experimental methods would have to contain a modicum of what could best be called the theory of the phenomenon, in addition to the theory of the experiment *per se*. It soon became apparent that the relative proportions of these two types of theories would vary widely from one subject of research to another.

We have selected contributors who are truly authoritative in their fields, and have asked them to make their chapters as "critical" as possible. We feel that, to a major extent, they have succeeded. The proof of this contention will, of course, be in the utilization of these methods by future experimenters. At the very least, this work will be a considerable help in making these methodologies accessible to working scientists.

The editing of the present volume was started as part of an earlier series, the first volume of which was *Techniques of Surface and Colloid Chemistry and Physics*, Volume I, edited by Good, Stromberg, and Patrick, Marcel Dekker, Inc., 1972. We have, since 1972, combined forces with Professor Egon Matijević, editor of the series, *Surface and Colloid Science*, and moved our series to a new publisher. For sake of generality, we have preserved Professor Matijević's series title. We believe that the combined strength of the two efforts, with the support of Plenum Publishing Corporation, will make it possible to produce a new series which will be appreciably stronger than the sum of two separate works.

It is intended that the division of labor in editing will be such that Dr. Matijević will edit volumes in the series that deal primarily with results, both theoretical and experimental, in the various areas of surface and colloid science, while we (Good and Stromberg) will edit volumes that deal primarily with experimental methods. As noted above, the distinction is not always a clear-cut one, but we expect to be able to make it in a logical fashion.

ROBERT GOOD
ROBERT STROMBERG

# Contents

## 3.  Pendant Drop Technique for Measuring Liquid Boundary Tensions

*Durga S. Ambwani and Tomlinson Fort, Jr.*

# 4. Electrophoresis of Particles in Suspension
*Arthur M. James*

5.  Methods of Producing Ultrahigh Vacuums and
    Measuring Ultralow Pressures
*J. P. Hobson*

6.  Electron Probe Microanalysis
*Gudrun A. Hutchins*

## 7. Research Techniques in Detergency
### Anthony M. Schwartz

# Contact Angles and the Surface Free Energy of Solids

## Robert J. Good

### 1. Introduction

This chapter, and the following one by Neumann and Good, deal with the measurement of contact angles in three-phase systems. The contact angle is, intrinsically, a macroscopic property, and one that should be amenable to a phenomenological treatment, e.g., to measurement without regard to its thermodynamic or microscopic interpretation. But inevitably, the desire for a thermodynamic and molecular interpretation arises. In addition, some fundamental questions about solid surfaces come up, and the problem of hysteresis must be given explicit consideration. So there is a need to go beyond phenomenology.

This chapter deals with a number of fundamental questions. While it covers topics in which results of particular experiments could be reviewed and analyzed, to do that would be to depart from the philosophy of this volume. So this chapter is written more with a view to the experimental measurements *per se* than to the numbers that result from those measurements.

A contact angle on a solid can be defined, and is physically meaningful, only when a solid is such that a unique tangent plane to the solid surface can be constructed. (We will return to this important matter below.) A fluid fluid interface approaches the solid surface in such a way that the observable tangent to the fluid–fluid interface meets the solid (tangent) plane in a line, which is referred to as the three-phase line. In a plane perpendicular to that

---

*Robert J. Good* • Department of Chemical Engineering, State University of New York at Buffalo, Buffalo, New York 14260.

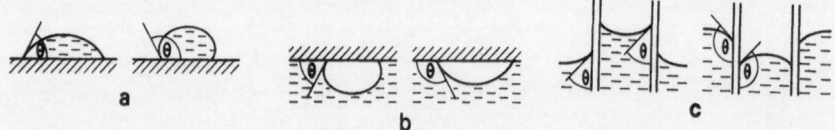

Figure 1. Some configurations in which contact angles greater than zero are observed : (a) sessile drops on a solid surface, (b) captive bubbles in a liquid underneath a solid surface, (c) capillary rise or depression in a cylindrical tube. (The figure can also be taken to represent capillary rise or depression between two parallel plates.)

line, there are two angles, whose sum is, of course, 180°. See Figure 1, which illustrates some configurations in which nonzero contact angles are observed. In the limiting case of zero contact angle, the three-phase line is the line at which the fluid–fluid interface becomes tangent to the plane of the solid surface. (If the approach is asymptotic, no such line exists.)

If, in contact with a solid, the second phase is liquid and the third is gas, the contact angle $\theta$ is conventionally defined to be measured through the liquid. When neither fluid is gaseous, there is no unique way of choosing between an angle $\theta$ measured through one of the liquids, and its supplement, $180° - \theta$. Then it is necessary for the observer to make an arbitrary decision and to identify his convention when reporting results.

We will not be concerned here with systems containing three fluid–fluid interfaces.

## 2. Solid Surfaces

### 2.1. Requirement of Effective Flatness

As noted above, the term "contact angle" can only be employed with regard to systems in which the solid phase is effectively flat, on the scale of observation. This is an important restriction. The surface may contain macroscopic singularities such as the intersection of two planes, or visible asperities or depressions, such that, for example, the radius of curvature at a ridge is small compared to ordinary laboratory dimensions. The angles that a fluid–fluid interface makes with the solid *at a singularity* are directly dependent on the macroscopic geometry of the solid. Conditions can be devised such that uniquely determined angles would be observable. But those angles—and the conditions under which they are found—depend so strongly on the local geometry of the solid and on the volume of liquid, that such measurements usually are of little general scientific usefulness.

The term "effectively flat" is used above for two reasons. First, a surface that is flat on a molecular scale is almost unattainable. (Carefully cleaved mica is probably the only exception.) Second, the topography of the solid, on the scale below that of observation by the naked eye or with low-power magnification, affects the macroscopically observable contact angle, and it is often possible to draw useful inferences about the microtopography from contact angle measurements. The distinction between macroscopic and microscopic is, of course, arbitrary. Low-power magnification is so conveniently available, using a hand lens or a simple combination of lenses, that it seems appropriate to set the limit of "macro" somewhere in the range of a resolution of lines separated by about 0.02–0.1 mm.

The methodology of measuring contact angles is established on the basis of the geometry of the surface, and of the uses to which the measurements are put. The uses are of two kinds. First, as is well known, the contact angle directly governs a number of important processes, such as wetting, penetration of porous solids by liquids, fluid–fluid displacement (as in certain processes of detergency and in secondary oil recovery) and adhesion. Second, the contact angle may be be used to characterize the solid surface itself. Thus, by means of a theory of Good and Girifalco[1] and Driedger, Neumann, and Sell[2] (see below) contact angle data may be used to deduce a fundamental property, $\gamma_s$, of a solid surface.

## 2.2. Surface Free Energy of a Solid

The physical significance of $\gamma_s$ has been discussed by Good[3]. $\gamma_s$, as obtained from contact angle studies, is one-half the free energy of cohesion, $\Delta G^c$, of the solid with respect to the plane that constitutes the surface.

$$\gamma_s = G^\sigma = \Delta G^c/2 \qquad (1)$$

$$= \left(\frac{\partial G}{\partial A}\right)_{T,P,n_i} \qquad (2)$$

where $G$ is Gibbs free energy and $A$ is area. For the purposes of measurement, there is no operational need to distinguish between Helmholtz and Gibbs surface free energy. See Good[4] for a discussion of this distinction, or its absence. See Good *et al.*[5–7] for more extensive thermodynamic discussion. The terms "surface free energy" and "surface tension" are both used in the literature, for $\gamma_s$. There are mechanical reasons for preferring the former.

Contact angles have been used recently to detect phase transitions in polymers[8] and molecular orientation of a polymer due to irreversible stretching in the solid.[9] These applications are consistent with Good's interpretation of $\gamma_s$.[3]

As a quantitative measure of surface properties, $\gamma_s$ is related to an important qualitative criterion for classifying solids. Thus, solids may be characterized as "high-energy" or "low-energy" on the basis of whether or not liquids with relatively high surface tension, such as water, spread with zero contact angle on the solid. Hard solids are generally high-energy solids; however, the surface energy characterization is not uniquely related to mechanical properties. Hardness and shear strength are functions of the concentration of defects such as dislocations in the bulk, and there is not a one-to-one correspondence between strength and the property of having high surface energy. This qualitative limitation of the correlation of bulk mechanical and surface properties applies to polymers also; for polymers, rheological behavior plays a major part in strength, but there is no direct relation between viscoelastic properties and surface properties.

An "ideal" solid surface would be homogeneous as to molecular composition, and, for a crystal, it would be a crystallographic plane; for a glassy solid, the ideal surface would be smooth at the molecular level. In either case, such a surface with appreciable area is exceedingly difficult to prepare. Essentially all available real surfaces (other than carefully cleaved mica) are appreciably rough, and the electron microscope reveals most surfaces to be exceedingly complex in their topography. Nearly all available surfaces are heterogeneous to an appreciable extent, and surfaces that have not been very carefully prepared and preserved usually are seriously heterogeneous, or become contaminated in a nonuniform manner.

### 2.3. Heterogeneous Surfaces

The heterogeneous nature of common surfaces may arise from at least three distinct causes. The first is, that matter has different chemical composition. For example, micrographs of steel reveal striations due to lamellae of $Fe_3C$ in a matrix of $\alpha$-Fe. (See any elementary textbook on metallurgy.) A second component of a solid surface may be an essential component of the material—as in the case of steel—or it may be an impurity. A surface impurity may be present as a distinct phase or as an adsorbed film that may not be identifiable as a phase; it may be present as discrete molecules or atoms, uniformly or randomly distributed, or in clumps. With respect to contact angle and surface energy, the types of heterogeneity that are of interest are those in which patches or bands exist, with local variations of $\gamma_s$ over the surface. It is noteworthy that some liquids, when used as "probes" to measure surface energy by contact angle methods, are much more sensitive to heterogeneity than others. Hydrogen-bonding liquids are particularly sensitive to polar contaminants on a nonpolar solid.

The second cause of heterogeneous surfaces is the presence of different crystallographic faces on a chemically homogeneous solid. The difference in density of atom packing in different planes, e.g., a (100) plane versus a (111) plane, leads to different energies of cohesion and hence to different values of $\gamma_s$ for a single solid. In addition, there may be different kinds of molecular groups exposed in different planes. Adam and Jessop[10] observed that the contact angle of water on crystalline stearic acid that had been cut through with a knife was variable, and lower than on a stearic acid surface formed by solidifying in contact with air. They attributed this result to the exposure of the edges of sheets of carboxyl groups in the cut crystal, that were buried, in the case of a crystal solidified in air.

A third cause of heterogeneity is the existence of grain boundaries, crystal edges and corners, and steps or ledges. Dislocations give rise to high-energy sites where they intersect a surface. A solid surface may be in a dynamic equilibrium condition, particularly if the temperature is high. (For metals, "high temperature" is considered to be above one-half the melting point on the absolute scale.) A surface that is in dynamic equilibrium and has a steady-state distribution of adatoms, steps, ledges, etc., would be considered uniform; but a heterogeneous surface is, in general, not in a condition of thermodynamic equilibrium. It is, of course, not always safe to assume that the structure of a solid surface remains the same when the solid is in a vacuum as when it is in contact with a liquid, even if no chemical reaction or chemisorption occurs.

The energy density of line defects, such as the intersection of a grain boundary with an external surface, has the dimensionality of energy per unit length, as opposed to $\gamma_s$ having dimensions of energy per unit area. For point defects, the energy density has dimensions energy per point. The contribution of point and line defect energies to $\gamma_s$ can be treated in terms of the defect energies only if the number of such defects per unit area is known. As a measure of $\gamma_s$, the contact angle of a liquid does not directly distinguish the structural features (such as defects) that cause the average $\gamma_s$ to have any particular value, when there is a distribution of $\gamma_s$ values over the surface.

The aspects of heterogeneity discussed above are basically the same as those considered by workers on gas adsorption on powders. See, for instance, the recent book *Clean Surfaces*, edited by Goldfinger, particularly the papers by Ross and Hinchen[11a] and by Sparnaay.[11b] Up to the present, there has been no direct comparison of heterogeneity effects in powders (studied by gas adsorption) with those in macroscopic surfaces (studied by contact angles.) The disparity between the high surface area requirement for convenient gas adsorption measurements, versus the low specific surface area of a macroscopic sample suitable for contact angles studies, has precluded such a comparison until recently.

We should also mention the study by White, of the effects of organic contaminants on the contact angle of water on various solids.[11c] White found evidence that contamination, such as by vapors of dioctyl phthalate or mineral oil (Nujol), was an adsorption process and "not just a case of material 'falling out' onto a surface." He showed that some surfaces, such as oxidized Nichrome, were more rapidly and far more drastically affected (rendered hydrophobic) than others, such as mica or glass.

## 2.4. Rough Surfaces

Roughness, as a departure from ideality in a surface, cannot be sharply separated from heterogeneity. Configurations such as ledges, crystal edges, corners, etc. can be considered both as line or point defects—high-energy sites—and as parts of ridges, hills, etc., i.e., as topographic features. For our present purposes, we can imagine a rough surface that is energetically homogeneous. Indeed, it is not difficult to prepare real surfaces where the microscale roughness has a far greater effect on contact angle than does heterogeneity.

The simplest parameter for describing roughness is the roughness ratio

$$r \equiv A/a \tag{3}$$

where $A$ is the true area and $a$ is the projected area on a plane parallel to the apparent surface.[7;12a,b] There is no agreement as to formal descriptions that are any more detailed than the roughness ratio. Certain idealized configurations have been analyzed by Johnson and Dettre[13,14] and by Eick et al.[6] There is no question that the distribution of hills and valleys is an important parameter. A surface with isolated, random hills will behave differently from one with isolated, random depressions; and isolated ridges on a flat surface will behave differently from isolated valleys. The degree to which ridges and valleys are parallel, and organized into ranges, will influence behavior. And complex surfaces, such as crosshatched ridges or valleys or organized arrays of hills or depressions, will have characteristic patterns of contact angle behavior.

Two further departures from ideality of a solid surface are (i) elastic distortion and (ii) swelling of the solid by the liquid that wets the surface.[15]

## 3. Thermodynamic Theory of Equilibrium Contact Angles on Ideal Solids

The intent of this and the following chapter is to describe methods of determining contact angles without prejudice toward one interpretation or another. But the variety and complexity of solid surfaces is such that some theoretical discussion is essential. Such a discussion will serve to show how the different kinds of nonideality affect the measurement.

On a smooth, homogeneous, rigid, isotropic solid surface, the equilibrium contact angle of a pure liquid is a unique quantity.[5] Young's equation[16] is obeyed.

$$\gamma_{sv} - \gamma_{sl} = \gamma_{lv} \cos \theta_e \tag{4}$$

where $\gamma_{sv}$ is the solid–vapor interfacial free energy, $\gamma_{sl}$ the solid–liquid interfacial free energy, $\gamma_{lv}$ the liquid–vapor interfacial tension, and $\theta_e$ the equilibrium contact angle. [See Good[7] and Johnson[17] for rigorous thermodynamic derivation of equation (4), and Neumann[18] for a discussion of the thermodynamic status of contact angles.] The term $\gamma_{lv} \cos \theta$ has, in the past, been referred to as the "wetting tension" or "adhesion tension." In terms of $\gamma_s$, Young's equation is written

$$\gamma_s = \gamma_{lv} \cos \theta_e + \gamma_{sl} + \pi_e \tag{5}$$

$$\pi_e \equiv \gamma_s - \gamma_{sv} \tag{6}$$

It is an active question in current research, whether the film pressure $\pi_e$ is negligible on smooth, homogeneous, low-energy surfaces with liquids for which $\theta_e > 0$.[19-21] On nonhomogeneous surfaces and on rough solids, it is to be expected that $\pi_e$ will often be appreciable, even when the average surface energy is low. When $\theta = 0$, $\pi_e$ must be greater than zero. We need not be concerned, in a paper on methods, with whether or not $\pi_e$ is negligible, except as regards the question of maintaining saturated vapor in contact with the solid. For interpretation in terms of $\gamma_s$ values, the $\pi_e$ question is of some importance, however.

After Young's equation, the most fundamental equation for the treatment of contact angle data is[1,4]

$$\gamma_{sl} = \gamma_s + \gamma_l - 2\Phi(\gamma_s\gamma_l)^{1/2} \tag{7}$$

where $\Phi$ is an interaction parameter that will be discussed below. Combining this with Young's equation[1,4,22,23] to eliminate $\gamma_{sl}$, the following equations are obtained:

$$\cos \theta_e = -1 + 2\Phi_{sl}\left(\frac{\gamma_s}{\gamma_{lv}}\right)^{1/2} - \frac{\pi_e}{\gamma_{lv}} \tag{8}$$

$$\simeq -1 + 2\Phi_{sl}\left(\frac{\gamma_s}{\gamma_{lv}}\right)^{1/2} \tag{9}$$

$$\gamma_s = \frac{[\gamma_{lv}(1 + \cos \theta_e) + \pi_e]^2}{4\Phi^2\gamma_{lv}} \tag{10}$$

$$\simeq \frac{\gamma_{lv}(1 + \cos \theta_e)^2}{4\Phi^2} \tag{11}$$

$$\gamma_{sl} \cong \gamma_{lv}\left[\frac{(1 + \cos \theta_e)^2}{4\Phi_{sl}^2} - \cos \theta_e\right] \tag{12}$$

$\pi_e$ may be neglected provided $\pi_e \ll \gamma_{lv}$. $\Phi_{sl}$ is a function of the molecular properties of the liquid and the solid, and it can easily be computed from readily accessible data. For nonpolar liquids on nonpolar solids, $\Phi \approx 1$.

More generally, $\Phi_{sl}$ lies between about 0.5 and 1.0 for common systems in which contact angles are observed. Equation (13) is the expression for $\Phi_{sl}$[4]

$$\Phi_{sl} = \frac{\frac{3}{4}\alpha_s\alpha_l[2I_sI_l/(I_s+I_l)] + [(\alpha_s\mu_l^2 + \alpha_l\mu_s^2)/2] + \mu_s^2\mu_l^2/3kT}{[(3\alpha_s^2I_s/4 + \alpha_s\mu_s^2 + \mu_s^4/3kT)(3\alpha_l^2I_l/4 + \alpha_l\mu_l^2 + \mu_l^4/3kT)]^{1/2}} \quad (13)$$

where $\alpha$ is the molecular polarizability, $\mu$ the dipole moment, and $I$ the ionization energy of the molecules. If either component is polymeric, these molecular properties are the properties of groups, e.g., the $(CH_2CHCl)$ group in polyvinyl chloride. If the component is a copolymer, an appropriate weighting must be given for the relative concentrations of the different kinds of groups that are considered as point centers of force. If preferred orientations of unsymmetrical groups are present, that must be taken into account in an *a priori* computation of $\Phi$.

$\Phi$, as calculated by equation (10), is related to the fractional polarities[24] of the two components. For component $i$,

$$D_i^p \equiv \frac{C_i^p}{C_i^L + C_i^i + C_i^p} = \frac{C_i^p}{C_i} \quad (14)$$

$$D_i^L \equiv \frac{C_i^L}{C_i^L + C_i^i + C_i^p} = \frac{C_i^L}{C_i} \quad (15)$$

$$D_i^i \equiv \frac{C_i^i}{C_i^L + C_i^i + C_i^p} = \frac{C_i^i}{C_i} \quad (16)$$

$$C_i^p \equiv \frac{\mu_i^4}{3kT} \quad (17a)$$

$$C_i^L \equiv \tfrac{3}{4}\alpha_i^2 I_i \quad (18a)$$

$$C_i^i \equiv \alpha_i\mu_i^2 \quad (19a)$$

$$C_i \equiv C_i^L + C_i^i + C_i^p \quad (20a)$$

where $D_i^p$, $D_i^L$, and $D_i^i$ are, respectively, the fractional polar contribution, the fractional London (dispersion) force contribution, and the fractional induction force contribution, to the intermolecular forces.† If these three kinds of force account for all the intermolecular forces in the system, then $D_i^p + D_i^L + D_i^i = 1$.

The force components for unlike substances can likewise be written in the form:

$$C_{sl}^p = \frac{\mu_s^2 \mu_l^2}{3kT} \qquad (17b)$$

$$C_{sl}^L = \tfrac{3}{4} \alpha_s \alpha_l \frac{2I_s I_l}{I_s + I_l} \qquad (18b)$$

$$C_{sl}^i = \frac{\alpha_s \mu_l^2 + \alpha_l \mu_s^2}{2} \qquad (19b)$$

$$C_{sl} = C_{sl}^L + C_{sl}^i + C_{sl}^p \qquad (20b)$$

Combining equations (13)–(20), we obtain

$$\Phi_{sl} = \frac{C_{sl}}{(C_s C_l)^{1/2}} \qquad (21)$$

For the polar component, equations (17a) and (17b) are valid provided the dipoles are not too large. (About 2–2.5 Debyes is the limit of validity.[4,23]) Therefore, we can write

$$D_{sl}^p = (D_s^p D_l^p)^{1/2} \qquad (22)$$

---

† There is a small, formal difference between equations (13) and (18)–(20), and the corresponding equations in Good and Elbing.[23] There, the dipole–dipole terms were multiplied by a semiempirical coefficient $B^2$ (with $B = 0.66$) and the induction terms were multiplied by $B$; and also, the numerical coefficient of the dipole–dipole terms was $\tfrac{2}{3}$ instead of $\tfrac{1}{3}$. This revision of the theory has been discussed in Good.[4] Briefly, it arose because of a contribution by Israelachvili,[25] who showed that the Keesom energy of interaction for two unhindered molecules

$$U_{12} = \frac{2\mu_1^2 \mu_2^2}{3kTr^6}$$

must be replaced by a free energy term

$$G_{12} = \frac{\mu_1^2 \mu_2^2}{3kTr^6}$$

The term $B$, which was employed by Good and Elbing,[23] was introduced because without it, the dipole–dipole component of the interaction coefficient, $\Phi$, was too large and a satisfactory fit to the data was not obtained. The revision indicated here, and in Good,[4] has very nearly the same quantitative consequence as the employment of the coefficient $B$ together with the original Keesom expression. It has the advantage of having no empirically adjustable coefficient.

This will break down if one of the dipoles is large enough that its rotation is restricted, in the pure compound, and the other is not, or if motion is restricted for any reason. We can also write, provided $I_s$ and $I_l$ are not very different (and this is frequently true),

$$D_{sl}^L \simeq (D_s^L D_l^L)^{1/2} \tag{23}$$

Also, if $\alpha_s \mu_s^2$ and $\alpha_l \mu_l^2$ are not too different,

$$D_{sl}^i \simeq (D_s^i D_l^i)^{1/2} \tag{24}$$

The approximation, equation (24), gives little trouble, because the induction term is relatively small compared to the other two terms in equations (20a) and (20b). Finally, we obtain the expression for $\Phi$ in terms of the fractional polarities of the two components:

$$\Phi_{sl} \simeq (D_s^L D_l^L)^{1/2} + (D_s^i D_l^i)^{1/2} + (D_s^p D_l^p)^{1/2} \tag{25}$$

This expression shows that, as Gardon has pointed out,[24] $\Phi$ is close to unity when there is a close match between the force components (expressed on a fractional basis) of the two substances, and is less than unity when there is a mismatch. Some further remarks on the subject of $\Phi$ will be found in Section 5 and in the Appendix.


## 4. Contact Angles on Nonideal Surfaces, Dynamic Effects, and Hysteresis

### 4.1. Nature of Hysteresis

On practically all surfaces, practically all liquids that form nonzero contact angles exhibit hysteresis. In liquid–liquid–solid systems, hysteresis is generally much more pronounced than in liquid–solid–gas systems. There are two ways of handling hysteresis. One is to develop simple methods by which reproducible data can be obtained in spite of it, and to report a single angle for any liquid on a particular solid. The second is to exploit the phenomenon, recognizing that it furnishes additional information about the solid. The amount of information coming out of a contact angle study can often be increased very markedly in this way. [It is sometimes suggested that hysteresis can be eliminated or reduced by employing very small drops. There is no experimental evidence for this; indeed, the contrary appears to be true. (See Section 4.3.) Theory indicates that hysteresis should persist with decreasing drop size, provided the drops are considerably larger than the scale of the solid's heterogeneity and roughness.]

When a liquid front is caused to advance over a solid, the contact angle is observed to depend on the rate of advance.[26] This rate dependence is observed to be greater when the system is liquid–liquid–solid than when it is liquid–vapor–solid. When the advance is stopped, the contact angle usually changes by a small amount for a short time (e.g., a few seconds to a quarter of a minute) after which—in the absence of vibration and evaporation—it stays effectively constant. When a liquid front is caused to retreat across a solid, the contact angle is also dependent on rate, and $\theta$ is lower than the angle in the advancing case; it may be zero. When the retreat is stopped, a constant angle is attained in a short time.

Hysteresis is ordinarily defined as the difference between the angle observed after an advance, $\theta_a$, and that observed after retreat $\theta_r$.

$$H \equiv \theta_a - \theta_r \qquad (26)$$

The common terms, "advancing angle" for $\theta_a$ and "retreating angle" for $\theta_r$, are misleading in that they give the impression that measurement is made while advance or retreat is in progress. The two common terms are probably too entrenched to be changed, and a convenient replacement is not obvious. Once the difference between a quasi-static system and dynamic system is recognized, this need not cause confusion, if experimenters will in every case report their procedure, e.g., the time elapsed after forming a drop or after advancing or withdrawing a liquid front. Since in dynamic systems the contact angle is rate-dependent, it may be desirable to use a different symbol; thus,

$$\theta_a^* \equiv \theta_a \text{ (rate)} \qquad (27a)$$

$$\theta_r^* \equiv \theta_r \text{ (rate)} \qquad (27b)$$

$$H^* \equiv H \text{ (rate)} \qquad (27c)$$

Recently, measurements have been made of $\theta_a^*$, $\theta_r^*$, and $H^*$ as a function of rate[27] for organic liquids on Teflon TFE. It was found that $\theta_a^* > \theta_a$, $\theta_r^* < \theta_r$, and $H^* > H$ for liquids such as decane, over the entire range of speeds employed, which were as slow as 0.01 cm/min. Attempts were made to extrapolate $\theta_a^*$ and $\theta_r^*$ to zero rate. No satisfactory extrapolation was found such that the limiting hysteresis, as rate approached zero, was equal to the quasi-static hysteresis observed after motion had been stopped.

In his pioneering work on contact angles, Alblett[29] found, using the rotating cylinder method, that above a limiting rate, the angle was constant, independent of rate. This result appears to be incompatible with modern observations, in which no such limiting angles were observed.[27] The inconsistency has not as yet been resolved.

The theory of hysteresis, and of rate effects, is currently in a state of development. It is believed that there are at least five more-or-less independent causes of hysteresis, some of which explicitly involve the rate of motion[30-32] (and hence may not be true hysteresis) and some of which predict, as well, hysteresis that is independent of rate.[5-7,13,14]

It is commonly accepted that, on a patchwise heterogeneous surface, the (quasi-static) advancing angle $\theta_a$ is related to—and may be equal to—the equilibrium angle that would be observed on a homogeneous, flat surface composed of the lower-energy component. Let us assume that a flat surface consists of two types of areas. On type 1, let the equilibrium angle be $\theta_{e1}$, and on type 2, let it be $\theta_{e2}$, with $\theta_{e1} > \theta_{e2}$. Then, on the composite surface,

$$\theta_a \simeq \theta_{e1} \tag{28a}$$

Similarly, if $\theta_r > 0$, the retreating angle may be considered to characterize the higher-energy component:

$$\theta_r \simeq \theta_{e2} \tag{28b}$$

Neumann and Good[5] have shown for a patchy surface that thermodynamic states or configurations of the system may exist that are accessible only when the liquid front is advancing or has previously advanced. Likewise, there are states that are accessible only if the liquid front is retreating or has previously retreated. Physically, this is a sufficient condition for the existence of hysteresis. (Good[7] and Johnson and Dettre[13,14] had previously proved that this condition, as regards accessible states, held for the geometry of concentric rings.)

On rough surfaces with asperities in the form of ridges, Eick *et al.*[6] and Johnson and Dettre[13] have shown that true hysteresis is to be expected. The values of $\theta_a$ and $\theta_r$ will depend on the slopes of the faces of the ridges, as well as on $\gamma_s$.

It is, in principle, possible to prepare rough or heterogeneous surfaces on which roughness or heterogeneity will not contribute to hysteresis—uniform ridges, all running parallel to the direction of liquid front motion, or parallel bands of the two types of solid similarly oriented. When the ridges or bands are perpendicular to the motion, maximum hysteresis will be observed. A surface with random roughness will be equivalent to a surface with random, intersecting systems of ridges. It is to be expected that, as regards hysteresis, such a random surface will be intermediate between a system of ridges perpendicular to the liquid motion and a system parallel to the motion.

Cassie[33] has proposed that the equilibrium angle on a patchy, heterogeneous surface should be the area-weighted average:

$$\cos \theta_C = \sigma_1 \cos \theta_{e1} + \sigma_2 \cos \theta_{e2} \tag{29}$$

where $\sigma_1$ is the fractional area of type 1 and $\sigma_2$ is that of type 2. Only when hysteresis is absent—as is possible in the case of parallel bands of $\theta_{e1}$ and $\theta_{e2}$ surfaces parallel to liquid motion—can $\theta_C$ be observed experimentally.

It has been shown with the aid of the electron microscope[34] that the retreating angle is more sensitive to roughness effects than is the advancing angle; and the same may be true regarding patchwise heterogeneity or heterogeneity arising from surface defects. This result gives an explanation why it has been much more popular to measure $\theta_a$ rather than $\theta_r$. It is possible to get reproducible results more easily. The explanation also leads to a prescription for evaluating data: The lower the hysteresis, the more confidence may be placed in $\theta_a$ as characterizing the low-energy component of the solid (or the solid as a whole) and $\theta_r$ as characterizing the higher-energy component. See Neumann[18] for a more detailed discussion of contact angle thermodynamics as related to heterogeneous and rough surfaces. For example, it was pointed out[18] that it is useful to identify $\theta_Y$ (the Young contact angle) as the angle with respect to which Young's equation is obeyed on a smooth, homogeneous surface of specified composition and structure. $\theta_Y$ is to be distinguished from $\theta_e$ in that an equilibrium angle, $\theta_e$, may exist on a heterogeneous or rough surface, and one cannot expect obedience to Young's equation, in terms of $\theta_e$, for such a surface.[5,6]

Diffusion, swelling, reorientation, and fluid mechanical effects are much harder to analyze theoretically, as causes of hysteresis, than the effects due to heterogeneity or roughness. This is so because the rate of the limiting process may be comparable to the rate of motion of the liquid front across the solid.[30] These rate-limiting processes themselves are, of course, matters of great practical and theoretical interest; and contact angle measurements can in principle be used to study them. Hysteresis due to reorientation is of particular interest. The orientation or configuration of molecules or groups at the solid–vapor interface will usually be the same as that at the solid–liquid interface, if the liquid–solid interaction is small. But this is not always true, and there is no logical necessity that it be so. The reorientation can be a dissipative process, and so can cause hysteresis. While Young's equation remains valid in systems where reorientation occurs, equations (7)–(12) either will not be valid or will only be approximations. This is because the physical basis for these equations includes the postulate that the molecular configuration of the solid is not affected by the presence or absence of the liquid. See below, Section 5.

It has very recently been found that hysteresis increases with decreasing drop size.[35] This effect apears to be part of a larger anomaly in the field of contact angles: the effect of drop (or bubble) size (i.e., of curvature of the three-phase line) on both advancing and receding contact angles.

## 4.2. Test for Contact Angle Equilibrium and Surface Quality

Neumann et al.[2,36] have suggested a graphical method, to detect cases where $\pi_e > 0$. This is to plot $\gamma_{lv} \cos \theta$ versus $\gamma_{lv}$, and fit an equation of the quadratic form

$$\gamma_{lv} \cos \theta = a\gamma_{lv}^2 + b\gamma_{lv} + c$$

to the data. This curve terminates at its intersection with the 45° line, $\gamma_{lv} \cos \theta = \gamma_{lv}$. If points close to the intersection fall below the quadratic line, that is a sign that $\pi_e$ is appreciable. A related method that is easier to carry out objectively is as follows: According to equation (11), a graph of $\gamma_{lv}(1 + \cos \theta)^2 / 4\Phi^2$ versus $\gamma_{lv}$ should be a horizontal straight line. If points for the lowest members of a series, which have the highest vapor pressure (e.g., the lowest *n*-alkanes on Teflon), fall below the best horizontal line through the rest of the data, that is an indication of nonzero $\pi_e$. (See below.) The observation that points other than those for liquids with high vapor pressures fall appreciably far from the best horizontal line may indicate either that nonequilibrium contact angles were observed or that incorrect values of $\Phi$ were employed.

If the surface is rough, the macroscopically observed $\theta_a$ will be greater than the microscopic $\theta_e$ by an amount that is not related to values of $\theta_e$ and $\gamma_s$, and $\theta_r$ will, similarly, be less than $\theta_e$. This will cause a trend, up or down, in the value of $\gamma_s$ computed using equation (11) that will give evidence that equilibrium contact angles are not being observed. A similar argument can be made with respect to other sources of hysteresis, such as those that involve fluid dynamics.[30,31]

A drawback, here, is that $\Phi_{sl}$ must be known if this method is to be used. Now, in order to compute $\Phi_{sl}$, one must know quite a bit about the composition, structure, and molecular properties of the solid and the liquid. However, a semiempirical equation [which may be derived from equation (9) with the aid of the linear experimental correlation that has been observed between $\Phi$ and $\gamma_{sl}$[2,46]] may be used with good effect:

$$\cos \theta_Y = \frac{(0.015\gamma_{sv} - 2.00)(\gamma_{lv}\gamma_{sv})^{1/2} + \gamma_{lv}}{\gamma_{lv}[0.015(\gamma_{lv}\gamma_{sv})^{1/2} - 1]} \tag{30}$$

The method of testing as to whether experimental $\theta$ values are $\theta_Y$'s is to measure the contact angle for a number of liquids on a single solid and calculate $\gamma_s$ in each case using equation (30). [See Driedger et al.[2] and Neumann et al.[36] for details of the computation of $\gamma_s$ from equation (30).] If the value of $\gamma_s$ obtained from $\theta$ measurements with different liquids is independent of the liquid used, the contact angles are probably closely approximate to Young angles, and $\pi_e$ is probably negligible for those liquids. This procedure may be carried out both with $\theta_a$ and $\theta_r$. Thus, from the $\theta_a$ data, values of $\gamma_{sl}$ (i.e., for the lower-energy component) can be deduced, and from $\theta_r$ data, values of $\gamma_{s2}$ (corresponding to the higher-energy component) can be obtained.

Deviations from constancy of $\gamma_s$ as estimated by this method, that are associated with roughness, have been observed to be the most serious with high-energy liquids such as water on low-energy solids. Deviations found with low-energy liquids, such as the lower alkanes of fluorocarbons, appear to be due to adsorption such that $\pi_e/\gamma_l$ cannot be neglected. Good and Yu[37] have very recently found evidence in contact angle data, with the aid of equations (8)–(11), that for $n$-heptane on Teflon TFE, $\pi_e$ is about $0.9 \text{ ergs/cm}^2$. For all higher $n$-alkanes on Teflon TFE, $\pi_e$ was too small to detect.

Since, as already noted, roughness on the scale that affects contact angle hysteresis can be observed with the electron microscope,[34] it is recommended that scanning electron microscopy, or at least high-power optical microscopy, be employed wherever possible to characterize surfaces whose contact angles are being measured. This allows the investigator to make a direct estimate of the "quality" of the surface with regard to roughness. This is particularly important when contact angle studies are made for the purpose of determining the surface energies of solids. On the other hand, when the purpose of the study is phenomenological, e.g., to evaluate the practical wettability of a surface, microscopic examination of the solid is less urgent.

## 4.3. Effect of Curvature of the Three-Phase Line

There is an anomaly in contact angle measurements which has not been resolved, up to the present, although it was first reported over 40 years ago.[35,38–41] This is a change in the measured contact angle with curvature of the three-phase line. It has been observed with captive bubbles[41] as well as with sessile drops. Recently, it was observed by Koo at Buffalo[35] and by Johnson and Dettre[42]; in both cases, it was found to be present with advancing angles and, more seriously, with retreating angles. In a typical experiment,[35] $\theta_a$ for water on a sample of Teflon TFE was observed to be 117° for drops larger than 2.5 mm diameter, but a decrease in $\theta_a$ set in with decreasing drop size, so that for drops 1 mm in diameter, $\theta_a$ was below 110°. The retreating angle was about 95° for drops larger than 5 mm diameter, and declined to below 80° for drops smaller than 1.5 mm.

Leja and Poling[41] suggested that the size effect was due to the influence of gravity; their analysis, however, is in error. Johnson[42] has found that the effect appears to be related to the presence of polar impurities.

If a line tension were invoked to explain the effect, it would have to be negative—which is physically unreasonable. However, an explanation can be suggested which phenomenologically resembles a negative line tension for a narrow range of drop sizes. If patchwise heterogeneities are postulated, the liquid surface near the solid, and also the three-phase line, will be contorted so that the line length will be greater than that which would be expected with a homogeneous solid. The maximum length of the three-phase line will depend on the size of the patches, while a true negative line tension would mean that there was no limit to the elongation of the line. The drop shape that would be expected for such a system, and hence the measurable contact angle, would be obtained by minimizing the entire energy of the drop, including the contortions of the surface in the region where it approached the solid surfaces of the two types. An approach of this type was used in reference 5.

If this model is valid, it would be expected that a similar drop size effect would be observed with rough surfaces.

The implications of these observations, with regard to the tactics of measuring contact angles, will be discussed in Chapter 2.

## 5. Treatment of Contact Angle Data Obtained with a Series of Liquids on a Solid

### 5.1. Series of Pure Liquids

We will review here some methods that have come into wide use in treating contact angle data. The first is the Fox and Zisman method,[43,44] in which for a particular solid, the contact angles are measured for a series of liquids, and $\cos \theta$ is plotted against $\gamma_{lv}$. A straight line is drawn (and if necessary extrapolated) to determine the value of $\gamma_{lv}$ for which $\cos \theta = 1$. This value of $\gamma_{lv}$ is called $\gamma_c$, the critical surface tension for wetting. Formally, this procedure corresponds to the use of the equation

$$\cos \theta = 1 + b(\gamma_{lv} - \gamma_c) \tag{31}$$

Good and Yu[4,37] have shown that equation (31) corresponds to the first two terms in a Taylor series expansion of equation (9):

$$\cos \theta = 1 - \frac{\gamma_{lv} - \Phi_{ls}^2 \gamma_s}{\Phi_{ls}^2 \gamma_s} + \frac{3}{4} \left[ \frac{\gamma_{lv} - \Phi_{ls}^2 \gamma_s}{\Phi_{ls}^2 \gamma_s} \right]^2 - \cdots \tag{32}$$

The coefficient $\Phi_{ls}$ in equation (32) is the parameter corresponding to the particular liquid–solid combination for which the contact angle is $\theta$. For a homologous series of liquids on a solid, $\Phi_{ls}$ can be expected to be constant, or else to exhibit a trend such that $\Phi_c$ is the limiting value of $\Phi_{ls}$ for that liquid for which $\theta = 0$ and $\gamma_{lv} = \gamma_c$. If an extrapolation is necessary, $\Phi_c$ is the value corresponding to a hypothetical liquid for which $\gamma = \gamma_c$.

If $\Phi$ is a constant independent of $\gamma_{lv}$ (as is the case for the alkanes on a solid such as a fluorocarbon), then from equation (11),

$$\Phi_c^2 \gamma_s = \gamma_c \qquad (33)$$

$$\cos \theta = 1 - \frac{\gamma_{lv} - \gamma_c}{\gamma_c} + \frac{3}{4}\left[\frac{\gamma_{lv} - \gamma_c}{\gamma_c}\right]^2 - \cdots \qquad (34)$$

This series converges for $\gamma_{lv} \le 2\gamma_c$. The first two terms on the right can be identified with equation (31). Therefore,

$$b = -1/\gamma_c \qquad (35)$$

It has been reported[45] that $b$ lies in the range of $-0.03$ to $-0.04$, which is just what would be expected for solids whose $\gamma_c$ is in the neighborhood of 30 ergs/cm$^2$. See also Kaelble and Uy,[46] who have reached a conclusion that is related to this one as regards the limiting slope when $\gamma_{lv}$ reaches $\gamma_c$.

A recommendation can be made on the basis of the comparison of equations (31) and (34): Linear extrapolation of contact angle data to estimate $\gamma_c$ should not be employed when the extrapolation is appreciable. For example, elementary computation shows that if the lowest $\gamma_{lv}$ for which a value of $\theta$ is available is larger than $1.25\gamma_c$, then the linear extrapolation will be in error, yielding a value of $\gamma_c$ about 10% too low. When any appreciable extrapolation is necessary, it should be made using the form of equation (9), i.e., by plotting $\cos \theta$ versus $\gamma_{lv}^{-1/2}$. This technique may be described as the Fox–Zisman method modified by Good and Girifalco. It has recently been shown[4] that this technique resolves a serious ambiguity in the interpretation of contact angle data on polystyrene, which had appeared to have two $\gamma_c$ values.[47] It is now clear that for this solid, $\gamma_c \cong 43$ ergs/cm$^2$, and that there is no inconsistency between data obtained with non-hydrogen-bonding liquids and with hydrogen-bonding liquids.

Zisman[48] has pointed out that for a solid such as nylon with different series of liquids, different surface conformations may exist, and that accordingly, different values of $\gamma_c$ may be found. Gardon[49] observed, with polyvinyl alcohol, drastically different contact angle behavior as between hydrogen-bonding liquids and non-hydrogen-bonding liquids. Baszkin and TerMinassian-Saraga[50] have found that polyethylene with a partially oxidized surface exhibits its full hydrophilic character only if the solid is heated, making the polar groups accessible to water. On the other hand, as noted above, polystyrene does not change its surface conformation, and the appearance of two $\gamma_c$ values was an artifact. It is the advantage of the plot of

$\cos \theta$ versus $\gamma_{lv}^{-1/2}$ that it eliminates the artifact of a long extrapolation and so allows a more rigorous interpretation of data than does the linear plot.

A tactic that often has considerable advantage over the $\gamma_{lv}^{-1/2}$ plot is based on equation (11). The quantity, $\gamma_{lv}(1 + \cos \theta)^2/4\Phi_{sl}^2$, or else $\gamma_{lv}(1 + \cos \theta)^2/4$, is plotted against $\gamma_{lv}$, and the best *horizontal* straight line is drawn. This approach has the methodological advantage that the straight line does not have an empirically adjustable slope. It is, of course, equivalent to, but more convenient than, making a Fox–Zisman type of graph with the slope $b$ forced to be equal to $-1/\gamma_c$.

The second method of handling data is that of Fowkes.[51-55] It is based on the Fox–Zisman–Good–Girifalco method, but it attempts to draw more information out of the data. Good[4] has recently reviewed the relationship between Fowkes' method and the FZGG method.

As applied to contact angles, Fowkes' method may be considered as starting with the contact angle data for saturated hydrocarbons on a solid such as Teflon TFE. Originally, Fowkes[51-53] used equation (9) in the modified form,

$$\cos \theta = -1 + 2(\gamma_s^L/\gamma_{lv}^*)^{1/2} \qquad (36)$$

The asterisk on $\gamma_{lv}^*$ denotes that only liquids in a homologous series, that are totally nonpolar are used. $\gamma_s^L$ is the London (or dispersion) component of the surface free energy of the solid.

$$\gamma_s = \gamma_s^L + \gamma_s^p + \gamma_s^i + \gamma_s^H + \cdots \qquad (37)$$

The superscripts $L$, $p$, and $i$ are as already defined; $H$ refers to hydrogen bonding. Fowkes has identified as many as seven components of $\gamma_s$. In later work,[52] Fowkes modified his treatment, to recognize that, with hydrocarbons on fluorocarbon solids, an interaction parameter other than unity should be employed:

$$\Phi_{\text{FC–HC}} = \frac{(I_s I_l)^{1/2}}{\frac{1}{2}(I_s + I_l)} = 0.91 \qquad (38)$$

[The form of equation (38) can be obtained from equation (13), since the dipole moment of an alkane is zero and that of a $CF_2$ group is small enough that the dipolar terms for the fluorocarbon versus hydrocarbon systems can be neglected.] Then equation (36) is replaced by

$$\cos \theta = -1 + 2\Phi_{sl}(\gamma_s^L/\gamma_{lv}^*)^{1/2} \qquad (39)$$

Equation (39) is identical with equation (9) for a nonpolar solid, since for such a system $\gamma_s = \gamma_s^L$. The graphical treatment of data that Fowkes used for these systems was identical with that employed by Good and Girifalco.[1]

Fowkes next proposed that the contact angle that a completely nonpolar liquid forms on a polar solid is insensitive to the dipoles in the solid. The expression $\mu_s^2\mu_l^2/3kT$ in equations (13) and (17) equals zero, since $\mu_l = 0$. The induction term in equation (13) is small, and other interactions, such as those due to hydrogen bonding, etc., are negligible or zero. The recipe for data handling is, then, to plot $\cos\theta$ versus $\gamma_l^{-1/2}$ for the hydrocarbons, extrapolate to the line $\cos\theta = 1$, and identify the intercept as the $\gamma_s^L$ of the solid. Since for hydrocarbons on solids other than fluorocarbons, $I_s$ is within about 1 eV of $I_l$, the expression for $\Phi$ corresponding to equation (38) yields $\Phi \approx 1$, so the use of equation (36) is not a bad approximation. Parenthetically, the graphical method based on equation (11) described above to determine $\gamma_s^L$ has considerable advantage over the $1/\gamma_l^{1/2}$ plot, because the latter, which ignores $\Phi$, necessarily allows the slope to be an adjustable parameter.

A further step along Fowkes' line of reasoning was taken by Owens and Wendt[56] and by Kaelble[57,58]. The two treatments are equivalent; we will follow that of Kaelble. Kaelble[57] employed the work of adhesion, using the well-known expression

$$W_{sl}^a = -\Delta G_{sl}^a = 2\gamma_{lv}(1 + \cos\theta) \tag{40}$$

He then treated the contact angle data for two liquids ($i$ and $j$) on the same solid, $\theta_i$ being the contact angle for the liquid with surface tension $\gamma_{li}$ and $\theta_j$ that for the liquid with surface tension $\gamma_{lj}$. The simultaneous equations were set up:

$$\begin{aligned}
\gamma_{li}(1 + \cos\theta_i) &= (\gamma_{li}^L\gamma_s^L)^{1/2} + (\gamma_{li}^p\gamma_s^p)^{1/2} \\
\gamma_{lj}(1 + \cos\theta_j) &= (\gamma_{lj}^L\gamma_s^L)^{1/2} + (\gamma_{lj}^p\gamma_s^p)^{1/2}
\end{aligned} \tag{41}$$

Here, the liquid surface tensions have been divided up into two components:

$$\gamma_{lv} = \gamma_l^L + \gamma_l^p \tag{42}$$

with the aid of liquid–liquid interfacial tension data for organic liquids versus water. Tables of values of $\gamma_l^L$ and $\gamma_l^p$ have been given in the references cited.† See the above and the papers by Fowkes, for the method of deriving $\gamma_l^L$ and $\gamma_l^p$ values. See also the paper by Good and Elbing[23] for the theoretical method of estimating these terms.

---

† In the tables of data of Kaelble, Fowkes, and others, the symbol $\gamma^d$ was used. The present author has discussed elsewhere[4] why this is an unsuitable nomenclature. Briefly, the term *dispersion* arose over a century ago, to mean the variation of the refractive index of a material body, such as glass, with wavelength—and consequent power of a prism to *disperse* white light into a rainbow spectrum. The use of the dispersion of refractive index was one of the techniques which London[59] suggested for evaluating an important coefficient in his theory. It has been almost totally supplanted by the use of the ionization energy $I$, as in equations (13) and (18), above.

Kaelble solved the equations[57] by a determinant method:

$$\Delta \equiv \begin{vmatrix} (\gamma_{li}^{L})^{1/2} & (\gamma_{li}^{P})^{1/2} \\ (\gamma_{lj}^{L})^{1/2} & (\gamma_{lj}^{P})^{1/2} \end{vmatrix} \tag{43}$$

$$\gamma_{s}^{L} = \frac{\begin{vmatrix} (1 + \cos \theta_{i}) & (\gamma_{li}^{P})^{1/2} \\ (1 + \cos \theta_{j}) & (\gamma_{li}^{P})^{1/2} \end{vmatrix}^{2}}{\Delta^{2}} \tag{44}$$

$$\gamma_{s}^{P} = \frac{\begin{vmatrix} (\gamma_{li}^{L})^{1/2} & (1 + \cos \theta_{i}) \\ (\gamma_{lj}^{L})^{1/2} & (1 + \cos \theta_{j}) \end{vmatrix}^{2}}{\Delta^{2}} \tag{45}$$

See Kaelble[57] for further discussion of how contact angle data with, say, seven liquids, may be combined, two at a time, to yield "best" values of $\gamma_{s}^{L}$ and $\gamma_{s}^{P}$. Kaelble has carried this treatment out for twenty-four different solids.

We may also refer the reader to the very recent paper by Sherriff,[60] who treated the contact angle data for all the liquids employed with a given solid (say, $m$ liquids) as an overdetermined set of $m$ equations in two unknowns, $m > 2$. Sherriff used a computer technique to keep the handling of data in this fashion from being too tedious.

A criticism of the Kaelble method may be raised, that it considers only the $\gamma_{s}^{L}$ and $\gamma_{s}^{P}$ components of $\gamma_{s}$. As a consequence, Kaelble[57] reached conclusions such as that for polystyrene, $\gamma_{s}^{P} = 2.17 \pm 0.68$ ergs/cm$^{2}$; for polyethylene, $\gamma_{s}^{P} = 2.21 \pm 1.10$; and for paraffin, $\gamma_{s}^{P} = 1.40 \pm 0.47$. On the basis of the physical nature of $\gamma^{P}$,[4,23] the surfaces of these three solids should have too low a degree of polarity to detect, as being different from zero, by contact angles.

Part of the anomaly in these results is, no doubt, due to errors in the measured values of $\theta$, and to errors in the assignment of $\gamma^{L}$ and $\gamma^{P}$ values to the different liquids. But a more basic reason for the anomaly is the attribution of the surface free energy to only two force components. Thus, for polystyrene at least, there is a component of liquid–solid interaction that should be attributed to an induction force term in the potential energy. The $\pi$ electron system of an aromatic ring is relatively highly polarizable, and there is a particularly strong interaction with hydroxyl groups, which has been estimated at 2.4 kcal/mole.[23] Kaelble's result, which indicates that a polystyrene surface is about 5% polar, is clearly a misinterpretation of an induction effect. Indeed, this contribution to the surface behavior of polystyrene should not be considered as a separable component of $\gamma_{s}$ or of the free energy of cohesion. It is a property that is latent in the surface structure, which becomes manifest only in the presence of polar molecules in the other phase. (Compare heat of vaporization, which is not considered to be a distinguishable *component* of the enthalpy of a gas.)

Actually, it is very fruitful, in terms of physical concepts, to consider the induction force as being a cross term, as between the London dispersion force and the dipole–dipole attraction of permanent dipoles. See Good.[4]

A more general criticism of the Fowkes–Kaelble–Dann–Owens and Wendt technique is that it tempts the investigator to overinterpretation of the data. A number of different sorts of interactions get lumped into $\gamma^L$ and $\gamma^p$—not only induction force, but also hydrogen bonding (which is known to be partially covalent) and other weak covalent interactions.

We may summarize the above criticisms, in a constructive fashion: The division of $\gamma_s$ into two components, $\gamma_s^L$ and $\gamma_s^p$, is *phenomenologically* acceptable to the extent that it is carried out by a specifiable procedure. The interpretations of the numbers that come out of this procedure are qualitatively reasonable.[23] Quantitatively, problems often arise, as with polystyrene. On the basis of the discussion, above, we recommend that results as to relative or fractional polarity of a surface be considered to be physically meaningful only to one significant figure. For example, if $\gamma_s^p$ is less than 5% of $\gamma_s$, it probably should be considered not significantly different from zero.

For a nonpolar solid, this limitation does not arise, because there is no operational distinction between $\gamma_s^L$ and $\gamma_s$. Indeed, there is no need for the concept of a $\gamma^L$ or $\gamma^d$ for such a solid at all.†

### 5.2. Mixtures of Liquids

In the Fox–Zisman method of handling contact angle data, and in the Good–Girifalco method, as well as in the Fowkes version of the latter, *the only data that may be employed are measurements made with pure liquids.* The following is an expansion of the discussion in Good[4] on this subject.

Misleading conclusions can come out of contact angle studies, using mixtures of liquids in order to vary $\gamma_b$, if the data are employed to estimate $\gamma_c$ values or if equations (8)–(12) or (36)–(45) are used to estimate $\gamma_s$ or $\gamma_s^L$, etc. Mixtures such as ethanol and water, or formamide and 2-ethoxyethanol, have been suggested[61] for this purpose. It is often found in practice that consistent results are not obtained, between one series of liquids and another, or between series of mixed liquids and series of pure liquids. This is not surprising. First, there is at present no way to estimate $\Phi$ for a three-component system; and if $\Phi$ differs appreciably from naively expected values, inconsistent results may be anticipated. Moreover, it is a very well-known fact that components from a binary solution will not adsorb

---

† The principle referred to is known as Occham's razor (William of Occham, 1280–1349). It states that the number of theoretical entities or mental constructs should not be increased any more than necessary.

equally at the three different interfaces. Adsorption isotherms are never linear over wide ranges of composition of mixtures, and the adsorption at a solid–liquid interface will be seriously different from that at a solid–vapor interface. Hence, the film pressure term $\pi_e$ for the solid–vapor interface, equation (6), may not be zero, and the corresponding term for the solid–liquid interface will certainly be quite appreciably different from zero.

To put this argument in a different form, equation (7) is valid only in the absence of adsorption of either component of the liquid at any of the three interfaces. Without equation (7), there is no theoretical basis for the Fox–Zisman "$\gamma_c$" method. And the widely variable results show that there is no empirical basis for it, either, with respect to the use of mixtures of different liquids.

Dann[62,63] has used mixtures in a way that attempts to escape from this trouble. He measured the contact angles of solutions such as ethanol–water or ethylene glycol–2-ethoxyethanol on a completely nonpolar solid, and from them he deduced values of $\gamma_{lv}^p$ as $(\gamma_{lv} - \gamma_{lv}^L)$. He then used these data, in combination with values of the contact angle of those solutions on a polar solid, to attempt to deduce values of $\gamma_s^p$ and $\gamma_s^L$. He obtained results that were not highly consistent for a given solid with various solutions.[63]

Now, such a procedure would be valid only if one could write an expression such as

$$\gamma_{sl} = \gamma_s^L + \gamma_l^L - 2(\gamma_s^L \gamma_l^L)^{1/2} \quad .$$

for a binary liquid, using the actual surface tension of the *solution* for $\gamma_l$. This cannot be done unless the liquid is an ideal solution[64] and the liquid layers at both the liquid–vapor and solid–liquid interfaces are of the same composition as the bulk.[15,23] These conditions are very far from being met in systems such as ethanol–water. Therefore, this method does not avoid the troubles already mentioned.

Experimentally, it is observed, when mixed liquids are used, that the $\gamma_c$ or $\gamma_s^L$ deduced is quite appreciably lower than the value obtained with pure liquids. This is in accord with theoretical expectations; see above. For example, adsorption of the less-polar component at a nonpolar surface should lower $\gamma_{sl}$, and hence raise $\theta$, according to Young's equation. W. A. Zisman has warned against the incautious use of mixed liquids to determine $\gamma_c$ values.[48] It is important that his objection should be on record here, in addition to our own.

If mixed liquids are used, the investigator should employ at least two or three different series of liquids, and establish whether or not they yield consistent results for $\gamma_c$. (It is a matter for future research to establish whether this tactic will succeed.) If only one series is used, the results should be considered to be of no *theoretical* value.

To conclude this discussion, we should note that there is some empirical value in the measurement of the contact angle of liquid mixtures, the components of which have different degrees of hydrophilic character. For polymeric solids that have been chemically treated, e.g., by etching, chemical oxidation, or exposure to flame, to improve their adhesion behavior when a glue or coating is applied, there is a correlation between treatment level and contact angle data with such mixtures. (Physically, this correlation reflects the increase in adsorption of the more hydrophilic component of a mixture at the hydrophilic sites of the solid–liquid interface; see above.) And there is a correlation with adhesion behavior. But while these correlations may be good enough for use in process control, e.g., on a narrowly defined system, they are only of limited value when any appreciable change is made in a system, such as a change from a homopolymer to a copolymer, or a change in the treatment or in the formulation of the adhesive or coating.

## Appendix. Note on Recent Developments

### A.1. Microscopic Contact Angles

An issue was raised in 1976 by Jameson and Del Cerro,[65] who produced a mathematical derivation that led to a somewhat different relation from Young's equation, between the contact angle and the surface free energies, $\gamma_{sv}$, $\gamma_{lv}$, and $\gamma_{sl}$:

$$\gamma_{sv} - \gamma_{lv} \cos \theta - \gamma_{sl} = \tfrac{1}{2}\gamma_{lv} \cos \theta \sin^2 \theta \qquad (46)$$

Very recent papers by White,[66] Navascués and Berry,[67] Pethica,[69] and Davis and co-workers[68] have revealed the essential nature of the problem. Jameson and Del Cerro assumed that the geometric surface specifying the liquid–vapor interface could be extrapolated from the macroscopically observable contour, as a plane, all the way to the solid surface. White points out that this extrapolation is not rigorous, and indeed, a region with local curvature is to be expected, as illustrated in Figure 2.

Jameson and Del Cerro[65] assumed that the molecules within the liquid and the solid interacted according to a Lennard-Jones 6–12 potential. Davis *et al.*[68] have independently examined the microscopic statistical–mechanical problem, allowing the local liquid–vapor surface curvature and slope to be mathematically deduced consequences of the theory. They recovered the Jameson–Del Cerro result when they used the 6–12 potential. But when they replaced the repulsion term (in $r^{-12}$) with an exponential (which is physically more realistic than $r^{-12}$) they obtained a different relationship. These results show that equation (46) is model-dependent, and so its disagreement with thermodynamics (i.e., with Young's equation) is a

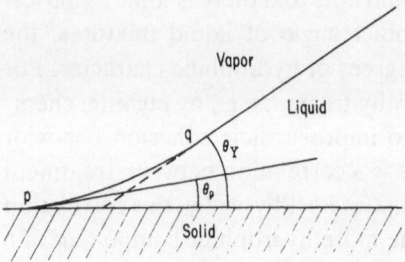

Figure 2. Possible microscopic configuration of a liquid surface in the vicinity of a solid. $\theta_Y$ is the contact angle formed by the macroscopically observable surface, extrapolated (dashed line) for the last few molecular diameters. $\theta_p$ is the macroscopic contact angle at point $p$. Note that this is a representation that ignores atoms and molecules and treats the liquid–vapor interface as being sharp. The actual dimensions in which this phenomenon occurs are, in all probability, such that the size of molecules of the liquid and the solid must be taken into account in any complete analysis. The distance from the solid to the point $q$ is of the order of 3–10 molecular diameters, and the thickness of the region between bulk liquid and bulk vapor in which the gradient of density is appreciable is of the order of 5 molecular diameters for a liquid that boils in the range of 50–200°C.

consequence of attraction and repulsion interactions in the dimensional regions that are not directly accessible to experiment.

Now, molecules of the liquid in the trihedral region, e.g., up to perhaps 10 or 20 Å or more from the solid, will be under the influence of the solid, as well as being attracted by the molecules in the subsurface layers of the liquid (see Figure 2). This means that a description of the liquid–vapor interface by the surface tension of the *bulk* will be a poor approximation. Indeed, an investigator will be seriously misled if he tries to use the naive concept of surface tension at all. The situation is somewhat similar to that discussed by Tolman[70] with regard to the effect of curvature on the "surface tension" of a small drop. The case of a liquid near a solid differs from that of a drop in that the surface molecules of a very small drop are under the influence of a total field that is weaker than the field in a flat liquid surface; but near a solid, if $\theta$ is less than 90°, the total field on surface molecules of the liquid is stronger than the field affecting surface molecules far from the solid.

A further restriction on all theoretical treatments of the local contact angle $\theta_p$ is noted in the caption of Figure 2; the nature of the liquid, as consisting of molecules, cannot be ignored. Hence, any mathematical treatment of the local contact angle that in any way employs a continuum assumption is appreciably lacking in reality.

If an attempt is made to measure $\gamma_{lv}$ in the region $pq$ (Figure 2), it is found that that operation cannot be performed. The area (i.e., the length $\overline{pq}$ times the perimeter $2\pi r$ for a circular drop) cannot be varied mechanically, so that a force associated only with an increase in that liquid–vapor area can be measured. If the area $2\pi r(\overline{pq})$ is increased, so is the area of solid–liquid interface under it. The force field that a molecule in the $pq$ region feels must be larger for elements of surface nearer the solid, so the force per unit length, which corresponds to $\gamma_{lv}$, must be anisotropic and must vary between $p$ and $q$, approaching $\gamma_{lv}$ as the slope in the $pq$ region approaches (asymptotically) the slope of the bulk surface.

In Figure 2, which illustrates the probable configuration of the liquid near a solid, the extrapolation of the liquid–vapor interface to the solid–vapor interface leads to an intersection at an angle equal to the Young angle $\theta_Y$. The continuation of the liquid–vapor interface to the trihedral line (at $p$) leads to a microscopic contact angle $\theta_p$. This latter angle is not physically observable with present-day instrumentation. It is, however, the angle that would obey an equation of the type proposed by Jameson and Del Cerro.

If in addition to there being a $\theta_p < \theta_Y$, as in Figure 2, there is an adsorbed film of molecules from the liquid, on the solid, the microscopic configuration becomes even less definite than that discussed so far. Suppose, for instance, that the equilibrium coverage of the solid is 0.1 of a monolayer[19] in regions far from the liquid. Then, as the terminus of the liquid is closely approached, it is possible that the fractional coverage may increase on account of the presence of the liquid. If we consider film thickness averaged over a considerable area, it is evident that 0.1 of a monolayer is effectively equivalent to a layer 0.1 of a molecular diameter thick. (Such a "thickness" is observable directly by ellipsometry.) And in a traverse that approaches the liquid upon the solid, there would be a gradual increase in the effective film thickness. Thus, even a zero value of $\theta_p$ is, in principle, compatible with a nonzero value of $\theta_Y$.

The fact that a $\theta_p$ may exist that is not equal to $\theta_Y$ does not vitiate the analysis above, e.g., equations (4)–(25). The Young equation is not directly relevant to the angle $\theta_p$, or vice versa. Young's equation is a thermodynamic relationship, concerned with the angle formed by the solid and the tangent to the liquid–vapor surface, the tangent being determined far enough from the solid (e.g., at least 10 or 20 Å), that the liquid–vapor surface tension is that of the bulk liquid. The apparent paradox of Jameson and Del Cerro[65] is resolved by establishing that $\theta_p$ is not necessarily equal to $\theta_Y$.

### A.2. Acid–Base Interactions at Solid Surfaces

Very recently, Fowkes[71] has revised the theory which we have expressed in equation (37). He points out that acid–base interactions across an interface (including hydrogen bonding) cannot be given a formulation that is as simple as that given for the London force. Fowkes and Mostafa[71] employ a treatment due to Drago *et al.*[72,73] who divided the Lewis acid–base component of heat of mixing into two terms. They designated one term $C$: a covalent component. The other term was designated $E$: an electrostatic component. For substances $i$ and $j$,

$$-\Delta H_m^{ab} = C_i C_j + E_i E_j \qquad (47)$$

Drago has published tables of empirical values of $C$ and $E$ for a large number of substances. Fowkes and Mostafa's expression for the work of adhesion, their equation (14), may be written

$$-\Delta G_{sl}^a = 2\frac{(I_s I_l)^{1/2}}{(I_s + I_l)/2}(\gamma_s \gamma_l)^{1/2} - f_s(C_s C_l + E_s E_l)n_a + W_a^p \quad (48)$$

Here, $W_a^p$ is the polar component of the work of adhesion; Fowkes states that it is usually small. The coefficient $f$ is "the constant (near unity) that converts enthalpy per unit area into surface free energy"; $n_a$ is the number of moles of acid–base pairs per unit area.

An examination of equation (48) shows that both the covalent and electrostatic terms are in the form of simple products. Therefore it is possible to make the following definitions, which are in the Pauling tradition of geometric mean representation of covalent interactions:

$$\gamma_s^{cv} = fC_s^2, \qquad \gamma_l^{cv} \equiv fC_l^2 \quad (49)$$

$$\gamma_s^e = fE_s^2, \qquad \gamma_l^e = fE_l^2 \quad (50)$$

We now can write

$$-fC_s C_l = (\gamma_s^{cv}\gamma_l^{cv})^{1/2} \quad (51)$$

$$-fE_s E_l = (\gamma_s^e \gamma_l^e)^{1/2} \quad (52)$$

Physically, $\gamma^{cv}$ is the component of surface free energy due to structures such as unshared electron pairs which can take part in (weak) covalent interactions, and $\gamma^e$ is the component due to electrostatic interactions. [We may also point out that equation (48) omits the induction term, $2(\gamma_s^i \gamma_l^i)^{1/2}$. The omission of this term leads to physically incorrect conclusions, such as the attribution of appreciable polar character to solids such as polystyrene.[57] Compare equations (13)–(25) above; see also reference 4.]

We may simplify the treatment of Fowkes and Mostafa by putting it into a form that is consistent with previous work on interfacial energies. We note, first, that the term $E_s E_l$ is unquestionably a polar interaction, and that $W_a^p$ can equally well be called an electrostatic interaction. $W_a^p$ is usually equated to $2(\gamma_s^p \gamma_l^p)^{1/2}$, and is identified with the attraction of unhindered dipoles for each other [see equation (17b) above]. Therefore it is reasonable to merge $W_a^p$ into the empirical term, $-fE_s E_l$ (rather than dismissing it as being "usually small"). This means merging the terms $\gamma^e$ and $\gamma^p$ into one. Since the former is often much larger, we will retain it to represent the merged quantities. Then we may rewrite equation (37) as follows:

$$\gamma_s = \gamma_s^l + \gamma_s^i + \gamma_s^{cv} + \gamma_s^e + \cdots \quad (53)$$

Also, we may represent the coefficient of the term $2(\gamma_s^L \gamma_l^L)^{1/2}$ in equation (48) as

$$\Phi_I = 2(I_s I_l)^{1/2}/(I_s + I_l) \tag{54}$$

Equation (48) now takes the form

$$-\Delta G_{sl}^a = 2\Phi_I (\gamma_s^L \gamma_l^L)^{1/2} + 2(\gamma_s^i \gamma_l^i)^{1/2} + 2(\gamma_s^{cv} \gamma_l^{cv})^{1/2} + 2(\gamma_s^e \gamma_l^e)^{1/2} + \cdots \tag{55}$$

Equations (53) and (55) have the advantage over previous formulations in that they do not misrepresent the hydrogen bond (or any other acid–base interaction) as a one-parameter property. Also, they allow for treatment of the cases in which strong polar interactions are present which are not properly described as contributing to hydrogen bonding. There is the further advantage that it should be possible to evaluate the new parameter $\gamma^{cv}$ independently, for example, with the aid of Drago's data.

ACKNOWLEDGMENT

The author thanks Dr. A. W. Neumann for discussion of this material, and the contribution of certain items from his experience.

## References

1. R. J. Good and L. A. Girifalco, *J. Phys. Chem.* **64**, 561, (1960).
2. O. Driedger, A. W. Neumann, and P.-J. Sell, *Kolloid-Z. u. Z. Polym.* **201**, 52 (1965); **204**, 101 (1965).
3. R. J. Good, in *Wetting, a Symposium, Bristol, England, 1966, SCI Monograph* No. 25, p. 328 (1967).
4. R. J. Good, *J. Colloid Interface Sci.* **59**, 398 (1977).
5. A. W. Neumann and R. J. Good, *J. Colloid Interface Sci.* **38**, 341 (1972).
6. J. D. Eick, R. J. Good, and A. W. Neumann, *J. Colloid Interface Sci.* **53**, 235 (1975).
7. R. J. Good, *J. Phys. Chem.* **74**, 5041 (1953).
8. W. Funke, G. E. H. Hellwig, and A. W. Neumann, *Angew. Makromolec. Chem.* **8**, 185 (1969).
9. R. J. Good, J. A. Kvikstad, and W. O. Bailey, *J. Colloid Interface Sci.* **35**, 314 (1971).
10. N. K. Adam and G. Jessop, *J. Chem. Soc.* **1925**, 1863 (1925).
11. G. Goldfinger (ed.), *Clean Surfaces*, Marcel Dekker, New York (1970): (a) S. Ross and J. J. Hinchen, p. 115;   (b) M. J. Sparnaay, p. 153;   (c) M. L. White, p. 361.
12a. R. N. Wenzel, *Ind. Eng. Chem.* **28**, 988 (1936).
12b. *J. Phys. Colloid. Chem.* **53**, 1466 (1949).
13. R. E. Johnson, Jr. and R. H. Dettre, *Advan. Chem. Ser.* **43**, 112, (1964).
14. R. E. Johnson, Jr., and R. H. Dettre, *J. Phys. Chem.* **68**, 1744 (1964).
15. R. J. Good and E. Diamanti-Kotsidas, 173rd National ACS Meeting, 1977; *J. Adhesion* (1978), in press.
16. T. Young, *Miscellaneous Works* (G. Peacock, ed.), Vol. 1, J. Murray, London (1855).
17. R. E. Johnson, Jr., *J. Phys. Chem.* **63**, 1655 (1959).

18. A. W. Neumann, *Advan. Colloid Interface Sci.* **4**, 105 (1975).
19. R. J. Good, *J. Colloid Interface Sci.* **52**, 308 (1975).
20. A. W. Adamson and I. Ling, *Advan. Chem. Ser.* **43**, 57 (1964).
21. M. E. Tadros, P. Hu, and A. W. Adamson, *J. Colloid Interface Sci.* **49**, 184 (1974).
22. R. J. Good, *Advan. Chem. Ser.* **43**, 74 (1964).
23. R. J. Good and E. Elbing, *Ind. Eng. Chem.* **62**(3), 54 (1970).
24. J. L. Gardon, *Encyclopedia of Polymer Science and Technology* (Mark, Gaylord, and Bikales, eds.), Vol. 3, p. 883, Interscience, New York (1965).
25. I. Israelachvili, *Quart. Rev. Biophys.* **6**, 341 (1976).
26. A. M. Schwartz, *Advan. Colloid Interface Sci.* **4**, 349 (1975).
27. R. J. Good and N.-W. Han, unpublished results.
28. N. K. Adam, *The Physics and Chemistry of Surfaces* (3rd ed.), Oxford University Press, Oxford, England (1941).
29. R. Alblett, *Phil. Mag.* **46**, 244 (1923).
30. R. S. Hansen and M. Miotto, *J. Amer. Chem. Soc.* **79**, 1765 (1957).
31. T. D. Blake and J. M. Haynes, *J. Colloid Interface Sci.* **30**, 421 (1969).
32. E. B. Dusan V and S. H. Davis, *J. Fluid Mechan.* **65**, 71 (1974).
33. A. B. D. Cassie, *Discuss. Faraday Soc.* **3**, 11 (1948).
34. A. W. Neumann, D. Renzow, H. Reumuth, and I.-E. Richter, *Fortschr. Kolloid u. Polym.* **55**, 49 (1971).
35. R. J. Good and M. L. Koo, unpublished work.
36. A. W. Neumann, R. J. Good, C. J. Hope, and M. Sejpal, *J. Colloid Interface Sci.* **49**, 291 (1974).
37. R. J. Good and K. H. Yu, unpublished results.
38. G. L. Mack, *J. Phys. Chem.* **40**, 159 (1936).
39. Ya. I. Frenkel, *Zh. Eksp. Teor. Fiziki* **18**, 659 (1948).
40. Ya. B. Aron and Ya. I. Frenkel, *Zh. Eksp. Teor. Fiziki* **20**, 453 (1950).
41. J. Leja and G. W. Poling, in *Proceedings of the International Mineral Processing Congress, London, 1960*, Institute of Mining and Metallurgy, London (1960).
42. R. E. Johnson, personal communication.
43. H. W. Fox and W. A. Zisman, *J. Colloid* **5**, 520 (1950).
44. W. A. Zisman, *Adv. Chem. Ser.* **43** 1 (1964).
45. A. W. Adamson, *The Physical Chemistry of Surfaces* (2nd ed.), Wiley Interscience, New York (1968).
46. D. H. Kaelble and K. C. Uy, *J. Adhesion* **2**, 50 (1970).
47. A. H. Ellison and W. A. Zisman, *J. Phys. Chem.* **58**, 503 (1954).
48. W. A. Zisman, *J. Paint Technol.* **44**, 42 (1972).
49. J. L. Gardon, *J. Phys. Chem.* **67**, 1935 (1963).
50. A. Baszkin and L. TerMinassian-Saraga, *J. Colloid Interface Sci*, **43**, 90 (1973).
51. F. M. Fowkes, *J. Phys. Chem.* **66**, 682 (1962).
52. F. M. Fowkes, *J. Phys. Chem.* **67**, 2538 (1963).
53. F. M. Fowkes, *Advan. Chem. Ser.* **43**, 99 (1964).
54. F. M. Fowkes, *J. Phys. Chem.* **72**, 3700 (1968).
55. F. M. Fowkes, *J. Adhesion* **4**, 155 (1972).
56. D. K. Owens and R. C. Wendt, *J. Appl. Polym. Sci.* **13**, 1741 (1969).
57. D. H. Kaelble, *J. Adhesion* **2**, 50 (1970).
58. D. H. Kaelble, *Physical Chemistry of Adhesion*, Chap. 5, Wiley Interscience, New York (1971).
59. F. London, *Trans. Faraday Soc.* **33**, 8 (1937).
60. M. Sherriff, *J. Adhesion* **7**, 257 (1976).
61. *ASTM D-2578-67*, *ASTM Standards*, Part 26, p. 370 (1968).

62. J. R. Dann, *J. Colloid Interface Sci.* **32**, 302 (1970).
63. J. R. Dann, *J. Colloid Interface Sci*, **32**, 320 (1970).
64. I. Prigogine and R. Defay, *J. Chem. Phys.* **46**, 367 (1949).
65. G. J. Jameson and M. G. Del Cerro, *J. Chem. Soc. Faraday Trans. I*, **72**, 883 (1976).
66. L. R. White, *J. Chem. Soc. Faraday Trans. I*, **73**, 390 (1977).
67. G. Navascués and M. V. Berry, *Mol. Phys.* **34**, 649 (1977).
68. H. T. Davis *et al.*, personal communication.
69. B. Pethica, *J. Colloid Interface Sci.* **62**, 567 (1977).
70. R. C. Tolman, *J. Chem. Phys.* **16**, 578 (1948); *J. Chem. Phys.* **17**, 118 (1949).
71. F. M. Fowkes and M. A. Mostafa, *Ind. Eng. Chem. Prod. Dev.* **17**, 3 (1978).
72. R. S. Drago, G. C. Vogel, and T. E. Needham, *J. Am. Chem. Soc.* **93**, 6014 (1971).
73. R. S. Drago, L. B. Parr, and C. S. Chamberlain, *J. Am. Chem. Soc.* **99**, 3203 (1977).

# 2

# Techniques of Measuring Contact Angles

## A. W. Neumann and R. J. Good

### 1. Introduction

The previous chapter was largely theoretical, in that it dealt with the interpretation of contact angle results in terms of solid surface energies. It also delved into the question of how the structure of a solid surface affects the contact angle that a liquid forms on the solid. The level of structure considered there included features that are not macroscopically observed, such as microheterogeneities, or minute peaks, pits, hills, and grooves in various geometries. Their existence may be inferred from certain observations, such as contact angle hysteresis, and sometimes they can be observed directly, e.g., with the optical or electron microscope.

In this chapter, a different question of structure is taken up. The choice of method for measuring contact angles depends quite directly on the gross geometry of the system. For example, the most convenient method for a flat plate is not usable at all for the inner surface of a capillary tube, and fine textile fibers and powders raise specific problems of measurement that are not present in other systems. The direct observational problems, the optics, the mechanics of manipulation, and the way the Laplace equation is involved in the measurement all vary widely from one kind of system (in terms of its configuration, shape, orientation, etc.) to another. An investigator who has complete freedom with respect to size, shape, and orientation of his samples can choose that method which suits his needs best, as regards accuracy, convenience, speed, cost, etc. But an investigator may need to

*A. W. Neumann* • Department of Mechanical Engineering, University of Toronto M5S1A4.  *R. J. Good* • Department of Chemical Engineering, State University of New York at Buffalo, Buffalo, New York 14260.

study a system in which there are strong constraints. His solid may be in the form of, say, a powder; and in general, a powder cannot be converted into a flat, solid plate without drastic changes in its surface structure. Drawn polymer fibers have highly oriented surfaces, and a film or thick plate made of the same polymer will not necessarily have the same surface properties. So the investigator must make do with methods of measurement that are often less accurate and much less convenient than those he would employ had he complete freedom of choice.

For the convenience of the reader, we have arranged the treatment below fairly systematically, in the order of increasing difficulty of accurate measurement for the different solid geometries.

## 2. Experimental Methods

### 2.1. Flat Plate

#### 2.1.1. Sessile Drop or Adhering Gas Bubble

*2.1.1.1. Direct Measurement of Angle from Drop Profile.* The method of the sessile drop or, alternatively, of the adhering gas bubble (Figure 1) is at present the most widely used technique. It is, in general, the most convenient method if high accuracy is not required, and it has the two great advantages of requiring only very small quantities of liquid and that samples as small as a few square millimeters can be used. It is commonly claimed that the accuracy of this method is $\pm 2°$. It would be better to call this a matter of precision; agreement between independent laboratories using this method has in the past been as poor as a $5°$ discrepancy. It is not clear, however, to what extent these disagreements are due to differences between samples. The methods described below should reduce the instrumental component of the discrepancies.

Major credit for popularizing the direct method belongs to Zisman and his co-workers,[1,2] who set up a convenient instrument. A somewhat refined instrument has been built by R. H. Ottewill.[3] The method has been described by other authors, often with only slight modifications. We will not attempt to describe all the variants, but we will discuss the basic equipment

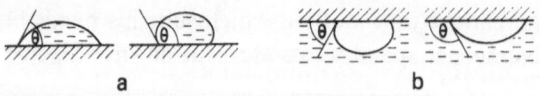

a                                         b

Figure 1. Examples of systems with nonzero contact angles; (a) sessile drops, (b) adhering bubbles.

together with some important precautions. Most of the optical and mechanical considerations are the same for the sessile drop and the attached bubble.

There are two obvious possibilities for the direct measurement. The contact angle may be either determined using a telescope equipped with a goniometer eyepiece, or a photograph may be made and the angle measured at leisure. The greater precision with measurements of a photograph is balanced by a decrease in convenience when dynamic effects are taken into account (see below). Because of dynamic effects, the reproducibility of individual drops is commonly poor enough that, for any high accuracy, a large number of photographs would be required, and many measurements on a single photograph will not constitute the equivalent of the same number of measurements on photographs of different drops.

There is a constraint on the direct method that has not been formally recognized up to the present. In the previous chapter, the drop-size effect was mentioned, together with a tentative theory of the effect. The variation of contact angle with drop size seems to be related to the existence of contact angle hysteresis. If hysteresis is small, e.g., less than 3°, the variation of $\theta_a$ and $\theta_r$ with drop size appears to be absent for drops that are of a convenient size for measurement. When hysteresis is appreciable, e.g., 20°, the contact angle exhibits a decrease with decreasing size, below about 2 mm diameter. The effect is more pronounced and sets in with larger-sized drops, with the retreating angle than with the advancing angle.

For best reproducibility of results, the same-size drops should be employed. We may also recommend that the drop-size effect be given explicit attention by experimenters in the future, since this appears to be a previously unrecognized cause of disagreements between laboratories.

The equipment for the sessile drop case consists of a horizontal stage on which the flat specimen is mounted, a micrometer pipette with small tip or other device for forming the drop, a source of illumination of the drop from behind, a light filter to minimize heating by the light source, and a telescope with a protractor eyepiece.

The telescope should be mounted so that it can be moved left or right relative to the stage on which the drop is placed. A traveling microscope micrometer cross-carriage provides this motion quite conveniently. (The Gaertner Scientific Co. micrometer slide no. M300, M301, or M340 is well suited to this purpose.) This mounting makes the study of contact angle hysteresis and the drop-size effect, as well as the control of drop diameter, relatively easy. A camera may be built into the telescope.

For the bubble method, the specimen is immersed in the liquid and either attached to the top or laid on the bottom of the chamber. The gas for the bubble must be supplied via a small-tipped tube, and the volume of the gas must be controlled.

In the manipulation of the equipment, the stage or the telemicroscope may be tilted slightly so that the front edge of the stage, which is out of focus, is out of the line of sight. This prevents the edge from causing a fuzzy appearance in the image of the drop edge. It can be shown geometrically that a tilt of one or two degrees causes negligible error in the measured contact angle. The tilt angle must, of course, be kept to the minimum necessary for clear viewing. Provision must be made to measure angles at both the left and right sides of the drop (or bubble) to investigate symmetry.

For sessile drops, it is preferable to use an enclosed chamber with at least the front wall of high-quality glass. The enclosure serves two purposes. The first is to protect from contamination due to dust, oil in the atmosphere, etc. The second is to make it possible to have the atmosphere around the drop saturated with the vapor of the liquid. This precaution is a particularly important one with volatile liquids.

In cases where equilibrium spreading pressure $\pi_e$ is appreciable, the observed contact angle may depend on whether or not equilibrium with respect to adsorption is reached. Fox and Zisman[1] found a significant difference in the contact angle of silicones (but not other liquids) on Teflon TFE, depending on whether or not careful precautions were taken to saturate the vapor. It has been shown recently by Good,[4] that for liquids other than the lowest-boiling members of a series, $\pi_e$ should be negligible on an ideal, low-energy solid. The nonideality of most practical surfaces is notorious, and so it is not, in principle, safe to assume that $\pi_e$ can be neglected, with an arbitrary solid, and particularly with one that has not been prepared with considerable attention to smoothness and homogeneity. (See Chapter 1 of this volume.)

A very practical reason for using an enclosure is that evaporation may cause the liquid front to retract, so that a retreating or an intermediate contact angle is observed unintentionally. Therefore, we must advise that (especially with liquids that boil below 100°C) it is usually worth the trouble to use a closed system to ensure saturation of the vapor if high-precision results are desired. With liquids that boil much above 100°C, in measurements at room temperature, this practical saturation requirement is not extremely important. For water, it is usually allowable to use an open stage if measurements are made quickly after drop formation. (The use of the attached-bubble method avoids most of the problems of vapor saturation.)

To summarize these arguments as to open versus closed stages, an open stage may be used (except with liquids that boil near room temperature) if an accuracy no better than perhaps 5° in $\theta$ will suffice for the purposes of the measurement, or if other, more serious causes of imprecision (such as roughness or heterogeneity) are known to be present.

The telemicroscope found in common commercial cathetometers has a magnification of about 5–20×. The optics of such instruments are essentially

Table 1. Components for Assembling a Suitable Telescope

| Component | Catalog number[a] |
|---|---|
| Telescope[b] | |
| Viewing telescope | 45 0100 |
| Auxiliary lenses | |
| 30 ×, 165–200 mm working distance | 44 9006 |
| 50 ×, 105–120 mm working distance | 44 9010 |
| Eyepiece—goniometer eyepiece, magnification 15 × mount[c] | 25 6510 |
| Tripod base | |
| Tripod with bubble level | 02 4311 |
| Precision cross-carriage ($\frac{1}{100}$ mm) | 02 3414 |
| Column with fine vertical adjustment ($\frac{1}{10}$ mm) | 02 3806 |
| Triangular rail base | |
| Triangular rail—1000 mm, with scale (other lengths available) | 02 2103 |
| Adjustable bracket supports (two) for level adjustment of rail | 02 2906 |
| Sliding carriage with precision cross-carriage ($\frac{1}{100}$ mm) | 02 3416 |
| and column with fine vertical adjustment ($\frac{1}{10}$ mm) | |
| Telescope stem—$\frac{3}{8}$-in. thread, 70 mm long | 02 4041 |
| Mounting ring—$\frac{3}{8}$-in. thread, 44-mm diameter | 45 0901 |

[a] As listed by Spindler and Hoyer, except for the eyepiece, which is by Leitz.
[b] Obtainable from Klinger Scientific Co., Jamaica, New York, distributors for Spindler and Hoyer Co., Göttingen, West Germany.
[c] Telescope should be mounted for horizontal and vertical travel. Mounting accessories listed are available from Spindler and Hoyer.

those of a low-power microscope with an objective lens that has a long focal length. It has been found, however, that the use of higher magnification is very advantageous (see below). High magnification is not at present available from most microscope or telemicroscope makers without the limitation of a prohibitively short working distance. However, it is possible to assemble a suitable telescope using the components in Table 1. Simpler and cheaper mountings than those in Table 1 for the telescope can also be employed.

If high magnification is used, it is particularly important that the measuring telescope be mounted so that it has the capability of horizontal travel in order for the observer to locate the edge of the drop and follow it as it is caused to advance or retreat. Even a small change in drop volume will move the edge of the drop out of the field of view at 50× or higher magnification.

The reason for employing high magnification is that, at low power, it is difficult to line up the cross hair tangent to the edge of the drop at the point where the drop edge terminates in the solid. The radius of curvature is typically of the order of 1 mm, and at 10×, an entire drop may well be in the field of view. See Figure 2, which illustrates the error that can easily be made. The 5° error in $\theta$, as drawn in Figure 2a, is large enough that it can be seen on

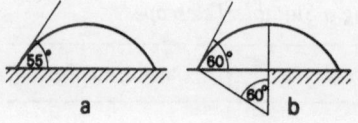

Figure 2.   Illustration of optical error in sessile drop measurement of contact angle; (a) tangent as lined up by eye by a novice observer—apparent angle $\theta = 55°$; (b) tangent drawn by construction, using arc extended past surface of solid—true angle $\theta = 60°$.

close inspection even without comparison to the true angle, drawn in Figure 2b; yet this error is easily made by an operator who is not aware of it. Errors of 2° can easily escape detection entirely.

The error can be regarded as the result of an optical illusion. The observer must imagine the curve that would be present if the drop edge continued past the solid surface, and this cannot be done with great accuracy, or even with great consistency, between one observer and another. This systematic error in measurements by an operator is probably responsible for much of the disagreement between operators and between laboratories, in contact angle measurements. We now will describe three methods by which this error may be eliminated.

The first method is to use higher-magnification optics, as already mentioned. The second is to set up a photographic system so that the drop can be photographed and measured at leisure. (This has the minor drawback of loss of convenience. Also, if stronger illumination is required, that will mean more heating of the drop.) If the drop is small enough that equation (1) holds (see below), then the measurement of height and base diameter yield the value of $\theta$; for a larger drop, the more complicated methods described in Section 2.3 are required. In measuring $\theta$ directly on a photograph, the optical illusion already mentioned will still exist, even though the measurement is made at leisure. So it is better, with a photograph, to locate the center of the circular arc that is tangent to the drop profile, draw a radial line, and obtain $\theta$ as the angle between that line and the normal to the surface (see Figure 2b).

The third method is to train the observer so that he reports the correct angle in spite of the optical illusion. One of us (R.J.G.) has recently developed a self-training technique for new workers that is particularly valuable when a low-power optical system is used. The method is to check the goniometer readings against the values of $\theta$ obtained from linear dimensions using spherical droplets of a high-boiling liquid. An elementary trigonometric construction shows that if $\Delta$ is the chord of a segment of a circle (i.e., the base of the silhouette of a spherical cap) and $h$ is the height, then[5]

$$\frac{2h}{\Delta} = \tan \frac{\theta}{2} \tag{1}$$

This equation can, of course, be used for direct measurement of contact angle, as noted above.

If a drop of liquid is used, it must be very small, so that the gravitational distortion is negligible. (But the use of very small drops is not recommended for *measurement* of contact angles, because of the drop-size effect noted above, page 33, and in Chapter 1, pages 13 and 15–16.)

A drop 0.5 mm in diameter of a high-boiling liquid is usually suitable; a flat piece of polished Teflon TFE block is a satisfactory substrate. (The Teflon should be polished on glass that has been sandblasted and ground on fine emery paper.) $n$-Dodecane (b.p. 214°) is a suitable liquid. It forms a contact angle of about 42° on Teflon TFE (see Section 5). The departure from sphericity of such a drop can be estimated with the aid of the tables of Bashforth and Adams; see the review by Padday on this subject.[6] For dodecane on Teflon, the error in $\theta$ that is made using equation (1) is less than 0.1° if the drop base is less than 0.8 mm in diameter. For mercury on glass ($\theta \approx 130$–$140°$), the error is less than 0.1° for a drop whose maximum diameter is 0.3 mm. For such drops, the shape parameter $\beta$, given by[5]

$$\beta \equiv b^2 \, \Delta\rho \, g / \gamma_{lv} \qquad (2)$$

is less than 0.05 for $\theta \approx 45°$ and less than 0.003 for $\theta \approx 135°$. The parameter $\beta$ and the departure of drops from spherical shape will be discussed at greater length below.

Equipment suitable for this self-training is currently available, e.g., a Gaertner coordinate cathetometer (Gaertner Instruments Co., Chicago, model no. 1236, fitted with a goniometer eyepiece, model no. M205).

Alternatively, it is possible to photograph a drop (or a sphere such as a ball bearing and to mask off the lower part of the silhouette so as to simulate a drop on a solid) using the optical system of the microscope and to measure both the angle and the linear dimensions in the photograph with a measuring projection microcrope such as a Nikon profile projector. The magnification in the photograph and the projection system must be such that the size of the image that is measured is approximately the same as that of the apparent drop image in the microscope when the contact angle is being observed directly. In this way, the optical illusion in setting the cross hairs tangent to the drop at the solid surface will be the same in the simulated measurement and a real measurement.

With this self-training method, it has been possible to obtain agreement between the direct goniometer measurement of $\theta$ and the value computed using equation (1) within 0.1°. Allowing for a degradation of the accuracy with which the observer makes the correction for the optical illusion after he has received the training, it should still be possible to attain an accuracy appreciably better than 0.5° in routine direct measurements. A warning should be noted, here, against the intuitively reasonable practice of training an observer by attempting to duplicate contact angle results that are to be found in the literature. See Section 5 for a discussion of standards. Not only is it, at present, impractical to prepare surfaces on which contact angle of any liquid is known and reproducible to better than 1°. It is also true that hysteresis effects, and the drop-size effect, are large enough that an arbitrary experimenter cannot be sure he is employing the same technique, in all its details, that another worker employed in taking the data that are reported in the literature.

On relatively rough or heterogeneous solid surfaces, particularly if the heterogeneity is due to the presence of adsorbed contaminants, refinement in the optical measurement of $\theta$ may be less important than refinements in the mode of forming the drop. A micrometer syringe should be used if possible.

The sessile drop method is not particularly well adapted to the quantitative measurement of the dependence of contact angle on the rate of advance or retreat, because a linear rate of change in drop volume does not correspond to a linear rate of motion of the drop front. (See below for more convenient methods.) It is possible, by use of a motor-driven syringe, to control the rate of addition of liquid or its removal, and to compute the rate of drop front motion, or to measure it directly. An appropriate rate for a sessile drop is of the order of 0.01–0.1 mm/min linear advance or retreat. If the motion is stopped, then it is best to specify some time after stopping, e.g., 10 sec or $\frac{1}{2}$ min. If these precautions are not taken, the imprecision of the measurement is increased appreciably. (An experienced operator usually finds himself employing such measures, often without being given detailed instruction.)

In some laboratories, tapping or other vibrating is employed with the expectation that it will lead to the contact angle approaching the equilibrium contact angle $\theta_e$. Such measures lead to less reproducible angles than those obtained by direct employment of the advancing angle, i.e., they lead to an angle somewhere between $\theta_a$ and $\theta_r$. Indeed, it is probably best to make an effort to insulate against vibrations.

We will now describe the important technique employed in Zisman's laboratory for applying a drop of liquid to a surface.[7] A drop of the liquid is picked up on a piece of platinum wire (about 8 cm long and 0.05–0.1 mm in diameter) that has been cleaned by heating to red heat in a Bunsen burner. (The use of a *fine* wire is important.) The wire is flicked gently, so that a drop of the liquid hangs at the tip of the wire. The drop is then brought slowly to the surface, so that it makes contact and flows off the wire, forming the sessile drop. Reproducibility to about ±2° in $\theta$ is claimed; it is not known how much of this fluctuation is attributable to the visual reading of the angle and how much to the mode of drop application.

Failure to follow the details of this application technique—particularly using a loop of wire so that a larger drop can be deposited—is likely to lead to results that are much less self-consistent and are inconsistent with those reported by workers who do follow the application technique of the Zisman laboratory conscientiously. The larger amount of kinetic energy carried by a larger drop as it flows to the surface, and the large deformation that occurs when the applicator is removed, are no doubt responsible for this loss of precision.

There is a perturbation of a liquid–vapor surface that some experimenters worry about needlessly. If liquid is being added to a drop from a syringe with the needle passing through the upper surface of the drop, there will be some capillary rise of liquid up the needle, and distortion of the surface. *This does not perturb the liquid in the region of the contact line with the solid.* The shape of the drop in the vicinity of the solid surface is completely controlled by the energetics of the three interfaces (solid–vapor, solid–liquid, and liquid–vapor) and the topography of the solid[8] and not by the shape of the surface far from the three-phase line.

If the needle enters the drop at a point very close to the solid, it may obscure the liquid profile. A more important perturbation occurs if appreciable hysteresis is present, and if the drop is not symmetrical about the needle, particularly if the distance from the needle to the surface is comparable to the diameter of the needle. Then, after advancement or retraction of the drop, the angles at the left and right sides of the drop may be unequal. This inequality is often quite easily observable, with very small drops. When this effect is present, then either the observation should be rejected, or else the measurement at the side farther from the needle should be considered as the more representative of the system.

If the needle is in as asymmetric location in the drop, and enters "behind" or "in front of" the drop, the drop profile will be symmetrical even though the drop itself does not have an axis of symmetry. This condition will lead to a spurious angle measurement, with $\theta_r > \theta_{measured} > \theta_a$. Probably the only way to eliminate this cause of scatter of the measurements is to refrain from using very small drops. Removing the needle from the drop does not help, because that makes it impossible to study hysteresis. The use of the smallest available needle, for adding liquid to and removing it from the drop, should mitigate this trouble to some extent.

A more important distortion occasionally arises in the direct measurement, on account of which a drop may appear to have a small rim around it when magnified. This is usually caused by heterogeneity or roughness. It is not related to the kinetic process of spreading. If, for instance, the surface is patchwise heterogeneous or contains bands of one species or the other extending perpendicular to the line of sight, then the edge of the drop may be contorted in order that Young's equation can be satisfied locally[9] (see Figure 3). If the width of the patches or bands is of the order of 0.01–1% of the drop diameter, the contortions will be of a magnitude that can produce the apparent rim. The portion of the drop surface (dashed line in Figure 3a) that comes down to the lower-energy

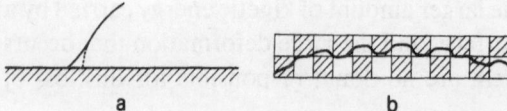

Figure 3. Edge of drop on a heterogeneous solid surface. (a) Schematic profile of the edge of a drop on a heterogeneous solid surface; the broken line indicates liquid surface coming down to meet the solid at angle characteristic of low energy component. (b) Top view of a ragged drop edge on an idealized heterogeneous solid surface; crosshatched areas indicate the solid surface with the higher equilibrium contact angle. If the drop were on a solid having the same surface composition but very much smaller patches, the liquid would meet the solid in a smooth arc that was in a position given approximately by a area-weighted mean location relative to the contorted line in (b). See Neumann and Good[9a] and Boruvka and Neumann.[9b]

patches of the surface (which are crosshatched in Figure 3b) is not visible in profile. An extrapolation of the main drop surface would fall somewhere between the solid line and the broken line. A microscopic view from the top would reveal the contorted nature of the three-phase line; this condition is illustrated for an idealized surface having a regular pattern of equal areas of two different kinds of surface in Figure 3b. Certain types of roughness of the right dimensions can also cause troubles of this kind.

The first remedy for this problem is to move the drop or to rotate the solid. The feature that looks like a rim may be absent in other regions of the periphery. The more general remedy is to prepare the solid surface in such a way that it is smoother and more homogeneous. That remedy may not be possible; for example, the experimenter may want to use the contact angle to characterize his *real* surface and not an ideal surface. In this case, the tangent required for estimating $\theta$ should be established for the curve of the main drop surface extrapolated to the solid. If the profile of the rim of the drop can be established, the angle of the tangent at the actual edge of the rim can also be measured. In general, however, the accuracy of this latter measurement will not be high.

Novices at contact angle measurements are sometimes overly concerned about one other effect, which is that the observer sees the reflection of the drop in the surface of the solid, which is usually smooth enough that its reflectivity is fairly high. Therefore, a drop with a low contact angle (e.g., 30°) may look like a football rather than a spheroidal cap. This effect may be avoided by two methods—either illumination from behind with well-collimated light, or having the telescope axis exactly in the plane of the solid surface. As already noted, the latter measure is undesirable for other reasons. However, the practical interference with measurement caused by this optical effect is usually negligible, since the line defined by the (pointed) tips of the football must lie in the plane of the solid surface.

The choice between the sessile drop and the adhering bubble may be made with the following considerations in mind: A sessile drop needs only very little liquid, whereas a bubble immersed in liquid will invariably need far more liquid. Complications due to solubility are usually dealt with more easily using a sessile drop. These complications may arise when the solid is swelled but not dissolved by the wetting liquid, or when adsorbed material in a film on the solid is appreciably soluble in the liquid. The advantage of the sessile drop is that, at least for advancing contact angles, the reading may be taken immediately after formation of the drop. The adhering bubble, on the other hand, has the advantage of minimizing contamination of the solid–vapor interface from sources such as airborne oil droplets, since the bubble is formed at an interface without contact with the general atmosphere. As already mentioned, saturation of vapor is more easily maintained with the bubble method.

It is somewhat easier to thermostat a system consisting of a solid immersed in a liquid than one consisting of a drop resting on a solid surface, so the use of the adhering bubble has advantages for the measurements of the temperature dependence of contact angles. However, the temperature coefficient of a contact angle is commonly small enough that, in routine measurements, thermostatting is not necessary. (See below, for a thermostatted system.)

Three commercial instruments for measuring contact angles by the direct method are available. One, employing an enclosed (but not sealed) chamber and integral microscope assembly, may be purchased from Kernco Instruments Co., Huntington Station, Long Island, New York.†

A disadvantage of the Kernco instrument is that it is quite difficult to hold a syringe tip stationary relative to the stage when the stage is moved—and the stage must be moved to bring the edge of the drop to the axis of the crosshairs. (The telemicroscope is not movable relative to the specimen chamber.) This limitation on the Kernco instrument makes accurate measurement of advancing and receding angles difficult, though not impossible.

---

† The instructions furnished with the Kernco instrument (catalog no. 15) are overly optimistic with respect to the measurement of $\theta$ by linear dimensions of the drop. The constraint that the drop must be a spherical cap if equation (1) is to be used is very severe; see above. See also the remarks, above, on the drop-size effect. In addition, the instructions contain an error. They recommend [in addition to equation (1)] the use of the following equation, when $\theta > 90°$:

$$\theta = 90° + \sin^{-1}(h/r - 1)$$

where $h$ is the height of the drop and $2r$ is the diameter of the drop base. The argument of the $\sin^{-1}$ function should be $(h/R - 1)$, where $R$ is the radius of curvature of the drop, i.e., half the maximum diameter for a spherical cap when $\theta > 90°$.

The bulletin also indicates that the $\sin^{-1}$ is an elliptic integral, which, of course, is incorrect. More important is the failure of the bulletin to stress that the equations cited are valid only for truly spherical drops.

These matters have been called to the attention of the Kernco organization, and they should be remedied in the next edition of the bulletin.

A second instrument, which employs an open sample stage that can be raised and lowered, on a small optical bench, is made by Ramé–Hart, Inc., Mountain Lakes, New Jersey. It is known as the NRL contact angle instrument, since it was designed in the laboratory of W. A. Zisman, at the Naval Research Laboratory, Washington, D.C. An optional component, which should be purchased, is an enclosed, controlled-environment chamber.

The Ramé–Hart instrument, model 100-00-00, has a stationary telescope. The position of the stage is controlled by graduated micrometer screws, so that the edge of a drop can be moved horizontally and vertically, to bring it to the axis of the telescope cross hairs. The optional microsyringe attachment for this instrument is mounted so that the needle may be held stationary relative to the stage. This is a valuable feature used for the measurement of advancing and receding angles. The horizontal micrometer drive enables the operator to follow an advancing or retreating drop front. It is, of course, also possible to use the micrometer screws to measure the height and width of a drop. A high-pressure chamber is available for use in studying liquid–liquid systems.

Ramé–Hart also makes a low-priced instrument, their Model 104 Wetometer. It employs a drop-dispensing syringe by which a known volume of liquid is deposited on the solid. The drop height is measured, and from it the contact angle can be computed. No provision is made by which advancing and receding angles can be measured. This instrument should be suitable for use in control analysis in many systems.

The third commercial instrument is the "Control Angle Analyzer" made by Imass, Inc.; their address is Box 134, Hingham, Massachusetts. This instrument employs projection optics; it has a magnification of 40×, which is high enough to minimize the optical illusion described above. Hence, the lining up of a crosshair with the tangent to the drop is somewhat easier than with the Ramé–Hart or Kernco instruments.

The Imass instrument, as it is presently constructed, has a stationary, transparent protractor and a movable sample stage. The angle is determined by laying an external ruler on the protractor. It should be possible to modify this instrument at only a modest increase in cost, to enable the operator to slide the protractor to left and right (and also to measure the drop diameter) and to determine the angle by rotating an accurately positioned cross hair or ruled line on a rotatable disk. Thus while this instrument is not suitable for measuring advancing and receding angles, it probably can be modified by the manufacturer so as to overcome this drawback.

For the sake of completeness, mention should also be made of a modification of the sessile-drop method due to McDougall and Ockrent,[10] which was advocated more recently by Gray.[11] Unlike the more conventional procedure where advancing and receding contact angles are produced by changing the drop volume, constant drop volume is employed, and advancing and receding contact angles are obtained by tilting the solid until the drop just begins to move. The contact angles obtained at the lowest and the highest point where the drop is in contact with the solid are then, respectively, the advancing and receding contact angles.

The reproducibility of direct angle measurements with sessile drops is usually given as about $\pm 2°$, as mentioned above. Use of the precautions listed should improve the reproducibility, in favorable cases, to better than $1°$. In view of the variability in precision of this method, as between specific designs of the laboratory setups, we can only urge that experimenters report the precision of their measurements, together with pertinent information such as the optical magnification employed. We should point out that, in the past, conclusions have been drawn from contact angle measurements which were not justified by the accuracy of the data.

*2.1.1.2. Interference Microscopy.* Interference microscopy provides a method of measuring contact angles in certain cases.[12] Several arrangements are possible, depending on the relative reflectivities of the solid and the liquid surface, respectively.

If the reflectivities at these two interfaces are comparable, fringes due to interfering beams reflected from the solid–liquid and the liquid–vapor interfaces may be observed conveniently. The principle of such a setup is illustrated in Figure 4. Destructive interference (dark fringes) will occur when the optical path difference between adjacent interfering beams is given by

$$t = \frac{\lambda}{2\mu} \tag{3}$$

where $\mu$ is the refractive index of the liquid and $\lambda$ the wavelength. From geometry (Figure 4), we have

$$\theta = \tan^{-1} \frac{t}{x} \tag{4}$$

where $x$ is the separation between dark fringes. Combining equations (3) and (4), we have

$$\theta = \tan^{-1} \frac{\lambda}{2\mu x} \tag{5}$$

If the solid substrate is opaque and less reflective than the liquid surface, Michelson optics (i.e., based on the Michelson interferometer) may be employed. In this case, the superposition of light reflected from an optical flat mirror with light reflected from the surface of the liquid drop is employed. See the very recent paper of Jameson and Del Cerro[13] on the use of interferometry when the underlying phase is transparent.

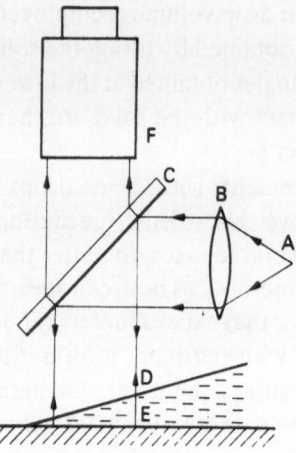

Figure. 4.   Optical arrangement for measuring contact angles by interference microscopy: (A) light source, (B) lens, (C) half-silvered glass mirror, (D) liquid–vapor interface, (E) substrate–liquid interface, (F) microscope.

The limitation of this method is imposed by the lateral magnification for resolving fringes. Large contact angles are hard to measure. The inconvenience of the method is serious, and if large quantities of liquid are available, other methods are to be preferred. But for very small drops of liquids that form low contact angles, it offers real advantages.

*2.1.1.3. Measurement of Contact Angles from Drop Dimensions*

A. Spherical Cap Drops. The contact angle may be determined from the dimensions of a single drop. For a spherical drop we have equation (1) connecting the contact angle with the base diameter $\Delta$ and height of the drop $h$.[5] There is also the equation

$$\frac{\Delta^3}{V} = \frac{24 \sin^3 \theta}{\pi (2 - 3 \cos \theta + \cos^3 \theta)} \tag{6}$$

connecting the angle $\theta$ with the base diameter and $V$, the volume of the drop. Care must be used with these methods. The size of the drop must be kept small so that the drop is indeed a spherical cap. For drops of liquid in air, under ordinary laboratory conditions, the constraint on drop size is quite severe. A drop that is small enough so that equation (1) or (6) is valid to $\pm 0.1°$ in $\theta$ may well be small enough so that $\Delta$ and $h$ or $V$ cannot be measured with a corresponding accuracy. This stringent limitation on drop size is relaxed somewhat for liquid–liquid systems. Whereas $\Delta\rho$ for an organic liquid in air is of the order of unity, it is of the order of 0.1 or smaller for a number of organic liquids in water, and also, the value of $\gamma$ may be appreciably higher, e.g., near 50 dyn/cm for many hydrocarbons versus water. For such systems, the sessile drops are excellent approximations to spherical caps for diameters as large as 5 mm, and the required linear dimensions to better than 0.1% accuracy should be easily measurable.

The limitations of this method are similar to those of the direct measurement of the contact angle at a sessile drop. The drop may not have perfect axial symmetry, and the contact angle may vary somewhat from point to point; the measurement loses significance when such variations occur. The measurement of two drop dimensions takes more time, of course, than one measurement; it can best be done after a photograph of the drop has been made. This further diminishes the suitability of the method for measuring advancing and receding angles. Perhaps even worse, it fosters the illusion that an equilibrium angle has been measured.

B. Nonspherical Drops. Measurements using drops that depart appreciably from spherical shape have the advantage that the surface tension of the wetting liquid $\gamma_{lv}$ can be determined simultaneously with the contact angle. These methods consist, essentially, of measuring parameters that determine the entire profile of an axially symmetrical drop. Most of these methods start with the Laplace equation describing the shape of fluid interfaces.[14] Bashforth and Adams[15] put the Laplace equation in dimensionless form for an axially symmetric drop in a gravitational field (Figure 5).

$$2 + \beta\frac{z}{b} = \left[\frac{1}{R/b} + \frac{\sin\phi}{x/b}\right] \tag{7}$$

Here, $R$ and $x/\sin\phi$ are the two principal radii of curvature in a plane perpendicular to the paper and in the plane of the paper, respectively, at the point $S$; $\phi$ in Figure 5 is the angle between a tangent and the horizontal. At the apex of the meniscus, $O$, the two radii are equal, and are designated $b$. Equation (2) defines $\beta$.

$$\beta \equiv b^2 \Delta\rho\, g/\gamma_{lv} \tag{2}$$

It is the parameter $\beta$ that defines the shape of the profile; $g$ is the acceleration due to gravity and $\Delta\rho$ is the density difference between phases. The larger the absolute value of $\beta$, the further the drop is from spherical shape.

Several graphical, curve-fitting techniques have been developed (see Padday[6] for details) that can be used in conjunction with the numerical integration of the Laplace equation by Bashforth and Adams (and by subsequent workers) to obtain $\gamma_{lv}$ and also to determine $\theta$. The most precise of these techniques appears to be the one by Smolders,[16,17] who used a number of coordinate points $(x, z)$ of the profile of the drop for his

Figure 5. Geometric construction for development of general method for calculating the contact angle from shape of a sessile drop.

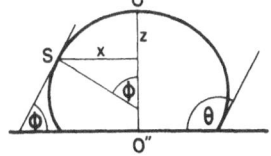

curve-fitting. If the surface tension of the liquid is known and if $\theta > 90°$, a perturbation solution of the Laplace equation derived by Ehrlich[18] may be used to determine the contact angle, provided the drop is not far from spherical. Input data are the maximum radius of the drop and the radius at the plane of contact of the drop with the solid surface. The accuracy of this calculation does not depend critically on the accuracy of the interfacial tension.

Another method of determining contact angles from the dimensions of sessile drops is by Staicopolus,[19] who obtained a digital computer solution of the Laplace equation, from which he developed approximate equations and nomographs permitting the computation of surface tensioi. and contact angle, provided the latter is larger than 90°. Parvatikar[20] has introduced a relation connecting $z/r$ and $V/r$ to the contact angle, which can be used in conjunction with the tables of Bashforth and Adams; here, $r$ is the maximum radius of the drop and $V$ its volume.

The main disadvantage of these and similar curve-fitting techniques[21] is the fact that they can be used to determine the contact angle only if $\theta > 90°$. This limitation is not an intrinsic one. A nonlinear regression procedure developed by Maze and Burnet[22] permits calculation of surface tensions and contact angles both above and below 90°. In this method, a calculated drop shape is made to fit the coordinate measurements taken over the drop surface by varying the two parameters $\beta$ and $b$ until a best set of values is obtained, i.e., when there is a minimum sum of squares of differences between measured and calculated coordinates. In order to start the calculation, reasonable estimates of $\beta$ and $b$ are needed. However, the range within which the values must fall is fairly broad. Convergence was obtained by Maze and Burnet for initial estimates of 0.5–5.0 times the true value of $\beta$ and of 0.7–2.2 times the true value of $b$. Thus, one would normally estimate the radius of curvature at the apex of the drop $b$ and then estimate $\beta$ from equation (7) using plausible values for $\gamma_{lv}$ and the density $\Delta\rho$. Working from a photograph, an estimate for $b$ inside the limits given above should present no problems. For cases where $\theta > 90°$, the authors give $\beta$ as a function of an eighth-order polynomial in the ratio $r/h$, where $h$ is the height of the apex of the drop above the equatorial plane, and the radius at the equatorial plane is $r$. Knowing $\beta$, $b$ may then be obtained from equation (2). The computer program to perform all the calculations is given in the paper by Maze and Burnet.

While the mathematical technique described above is a very usable procedure, the general curve-fitting method is not as easily handled as several other techniques. However, this method appears to be particularly valuable for measurement with liquid metals, for two reasons. First, the surface tensions and contact angles of such liquids are highly sensitive to contaminants, including gases, and so the value of $\gamma_{lv}$ for the particular drop studied is often needed anyway. Second, because of the need for operation at high temperature and in vacuum, it is often found that direct optical methods are not much more convenient than photographic methods. When an elaborate vacuum equipment is being built, the addition of photographic equipment is only a small incremental complication and so is worth the trouble. It should be noted, however, that metal–solid systems are just as subject to hysteresis as are any other systems.

A method that has a deceptively attractive appearance is based on the fact that, for a large enough drop, the height is independent of drop size. Then

$$\cos \theta = 1 - \frac{\Delta\rho \, gh^2}{2\gamma_{lv}} \tag{8}$$

where $h$ is the limiting height of the drop.[23,24] Only a single measurement, a length, is required. The pitfall of the method is in the requirement that the drop must be "large." Burdon[25] has discussed this requirement in some detail, in a description of methods of measuring surface tension. His conclusion (which can be directly carried over to contact angle measurements on systems in which the contact angle is 90° or higher) is as follows: "For water ... the simple formula is valid for drops *over one meter* in diameter ... and also for drops having a particular diameter slightly less than 2 cm. Some workers have, apparently fortuitously, used drops of the latter size." He points out that the error may easily be 10% or worse. For drops that are not of the huge size called for by Burdon's analysis, the methods already mentioned, employing tabulated functions or computer solutions based on the numerical integration of the Laplace equation, must be used. The restriction to drops having the specified, smaller diameter requires that the diameter be measured and adjusted. See Padday.[23]

### 2.1.2. Wilhelmy–Gravitational Method[26]

If a smooth, vertical plate is brought into contact with a liquid, as indicated in Figure 6a, the liquid will exert a downward force on the plate, which is given by

$$f = p\gamma_{lv} \cos \theta \tag{9}$$

where $p$ is the perimeter of the plate. If the depth of immersion is not zero, so that a volume $V$ of liquid is displaced (see Figure 6b),

$$f = p\gamma_{lv} \cos \theta - V \, \Delta\rho \, g \tag{10}$$

where $\Delta\rho$ is the difference in density between the two fluids and $V$ the displaced liquid volume. These two equations are general and rigorous.[7,26–28] To evaluate the contact angle, it is obvious that the surface tension $\gamma_{lv}$ must also be known.

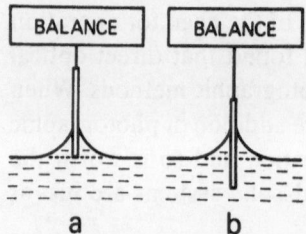

Figure 6. Vertical plate methods; plate extends perpendicular to plane of paper; (a) zero net depth of immersion, (b) finite depth of immersion.

If the plate is not effectively smooth, it will not be suitable for determination of the contact angle of the liquid on that solid. A well known, useful technique for obtaining a zero contact angle, in order to use the Wilhelmy method for measuring surface tension, is to sandblast a glass plate or roughen it with an abrasive. See the paper by Princen[29] for an analysis of the Wilhelmy plate with a highly roughened surface.

With the advent of electric balances, the Wilhelmy method has come to offer certain advantages over all other methods. For one, it makes it convenient to record changes of contact angles with time. Such changes may be due, for example, to changes in temperature or to adsorption at various interfaces.

Several experimental setups have been described in the literature.[27,30,31] The basic apparatus is shown schematically in Figure 7. The plate (1), the perimeter of which should be constant, is suspended by a thin rod (2) from the electrobalance (3). The downward force is recorded on the recorder (4). The liquid (6) is contained in a beaker (5), which is partially covered by a lid (11) to minimize evaporation. This lid may conveniently consist of two halves so that it may be put on after the plate has been

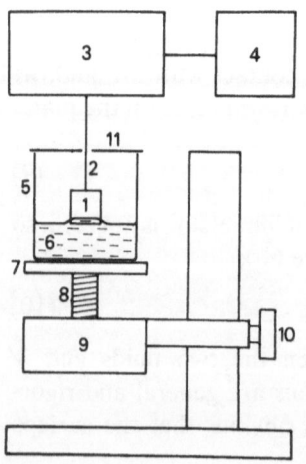

Figure 7. Apparatus for Wilhelmy technique: (1) measuring plate, (2) glass fiber or rod, (3) electrobalance, (4) recorder, (5) measuring cell, (6) liquid, (7) movable platform, (8) screw or gear mechanism to raise or lower the platform, (9) motor, (10) clamp and support, (11) lid.

suspended from the electrobalance. The table (7) carrying the beaker may be moved up and down by means of the screw (8) driven by the electric motor (9) or by hand in order to establish advancing and receding contact angles. Motion of the table (7) at speeds of the order of 1 mm/min appears to be convenient. To facilitate a rough positioning of the table (7) and the motor (9) it is convenient to mount them on a heavy vertical rod and to bring them into the desired position using a clamp (10). Vibration insulation, which will be discussed in the next section, may also be desirable. It is obvious that the motor and screw must not be sources of appreciable vibration.

Advancing and receding conditions are established as follows: After suspending the plate from the electrobalance, the table is raised from an initial low position to such a height that the surface of the liquid is approximately 1 mm below the lower edge of the plate. The table, together with the motor, are fixed in this position by means of the clamp. Then the table is raised slowly by means of the screw. After contact is established between the plate and the liquid, the motor is stopped immediately. The resulting configuration is one of advancing contact angle. The depth of immersion is zero, so that equation (9) may be used.

Receding conditions may be established by immersing the plate to some depth and subsequently withdrawing the plate through exactly the same distance. This may conveniently be done by timing the immersion and the withdrawal or by measuring the depth of immersion and withdrawal by means of a cathetometer. If the latter technique is used, a simpler setup without electric motor may be employed. To obtain the true receding contact angle, it is imperative to immerse the plate to such a depth that motion of the three-phase line with respect to the plate occurs. The upper limit for that height $h$ may be calculated from equation (11), which assumes zero contact angle.

$$h = \left(\frac{2\gamma_{lv}}{\Delta\rho\,g}\right)^{1/2} \tag{11}$$

This equation will be discussed in more detail below.

The apparatus described in Figure 7 may also be used to test the constancy of the contact angle over the length of the specimen. For this measurement, the perimeter must be constant along the plate. Prior to the establishment of contact between the plate and the liquid, the recorder indicates constant weight (line $AB$ in Figure 8). Immediately after contact is established, the recorder jumps from $B$ to $C$ due to the capillary rise of the liquid at the plate. If the motor is not stopped at this moment, the plate will be continuously immersed deeper into the liquid, so that the weight on the balance decreases again (straight line $CD$ in Figure 8). If the average contact angle along the line of contact does not remain constant during the

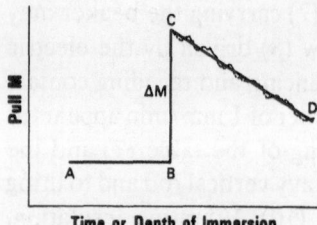

**Time or Depth of Immersion**

Figure 8. Weight of the measuring plate as a function of the depth of immersion in Wilhelmy method.

immersion or emersion, or if the surface of the liquid is vibrating, the chart line will not be straight but contorted, as indicated by the dotted line. In such cases, the point $C$ may be established by drawing an average straight line through the contorted line $CD$ and letting it intersect with the normal to $AB$ erected at $B$. Local deviations from the straight line $CD$ may be interpreted as deviations from the overall average contact angle. From the length of the line $BC$, which is equal to $\Delta M$ in grams, the force $f$ in equations (9) and (10) is obtained from

$$f = \Delta M g \qquad (12)$$

An electrobalance, such as that manufactured by the Cahn Instrument Co., Paramount, California, is calibrated to read out mass, not weight; its calibration changes with latitude and altitude. This is a consequence of the fact that the signal is a measure of the electric current in an electromagnet, which is needed to balance the gravitational pull on the object that is being weighed. It is for this reason that equation (12) is written including the gravitational acceleration $g$.

The preceding description makes some of the advantages of the Wilhelmy method obvious. The measurement of an angle (the contact angle $\theta$) is reduced to the measurement of a weight, which can be performed with much higher accuracy and objectivity than the direct reading of an angle with a goniometer. The reproducibility is therefore only limited by the reproducibility of the solid surface and the reliability of determining the perimeter $p$ of the plate. The perimeter may be established from the slope of line $CD$ (Figure 8) or by direct measurement. The method has the further advantage that it can be completely automated. Continuous recording is almost a necessity if the contact angle changes with time.

While the Wilhelmy method at first sight seems to be an ideal method for obtaining high-precision contact angles, it has several drawbacks that are not immediately obvious and that can restrict the usefulness of the method. The high sensitivity of an electrobalance employed in the Wilhelmy method can be exploited only if the perimeter of the plate is constant. Furthermore, the plate must have the same composition and morphology at all surfaces—front, back, and both edges. This condition may be difficult to meet, particularly if one wants to investigate films deposited in vacuum, or polished surfaces, or anistropic systems. In measurements that extend over appreciable time intervals, swelling or solution of the solid may become a problem. Swelling or solution may change the volume $V$ and mass of the displaced liquid [see equation (10)] in an uncontrollable manner. Also, adsorption of the vapor of the liquid at various parts of the gravimetric system other than the plate may change the recorder readings. This is particularly serious for measurements of the temperature dependence of contact angles. Lastly, on surfaces that are not completely uniform, the interface does not intersect the solid in a perfectly smooth curve of constant shape (cf. Figure 3b, regarding which it was noted that the edge of a spherical drop may not meet the solid surface in an arc of a circle). Since the simple Wilhelmy method does not have a provision for microscopic examination of the surface, such effects may escape notice. When these limitations are absent, the Wilhelmy method has the highest sensitivity of any method available, and values of $\cos \theta$ that are as accurate as the value of $\gamma_{lv}$ of the contacting liquid can be obtained.

All of these difficulties are virtually absent in another indirect method, which we now describe.

### 2.1.3. Capillary Rise at a Vertical Plate

In this method, the solid surface is also aligned vertically and brought into contact with the liquid. Instead of the capillary pull as in the Wilhelmy method, the capillary rise $h$ at the vertical surface is measured[27] (see Figure 9). This method has been found particularly effective for measuring contact angles as a function of rate of advance and retreat and for determining the temperature coefficient of $\theta$.

For an infinitely wide plate, a straightforward integration of the Laplace equation[8,27,32] yields

$$\sin \theta = 1 - \frac{\Delta \rho \, g h^2}{2\gamma_{lv}} \tag{13}$$

It is by rearranging this equation, and assuming $\theta$ to be zero that we obtain equation (11). For practical purposes, plates that are about 2 cm wide satisfy the theoretical requirement of "infinite" width. For such a plate (assumed to have a uniform surface), the line of contact is straight in the middle part of the plate for all liquids of moderate surface tension, including water. If we assume $g$, $\Delta\rho$, and $\gamma_{lv}$ to be known, the task of determining a contact angle is

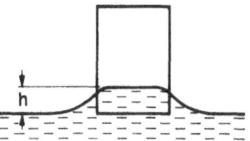

Figure 9.  Capillary rise at a vertical plate.

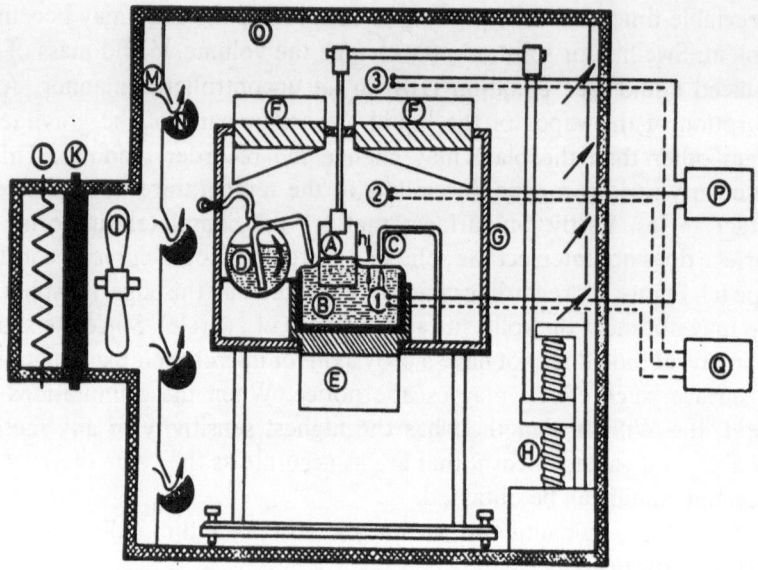

Figure 10.   Apparatus vertical plate technique in thermostat; (A) sample plate, (B) rectangular cell containing liquid, (C) pointer or glass fiber for use in determining elevation of flat liquid surface, (D) bottle supplying liquid to cell, (E) Peltier elements for heating or cooling liquid, (F) Peltier elements for heating or cooling air space surrounding liquid, (G) inner thermostat box, (H) screw for raising or lowering plate, (I) fan, (K) cooling element for climate chamber, (L) heating element, (M) slits that direct passage of air over the desiccant N, (O) enclosure for climate chamber, (P and Q) electronic temperature controller and recorder; (1, 2, 3) locations of temperature sensors, e.g.. thermocouples; $h$ indicates capillary rise; cathetometer and vibration insulation are not shown.

reduced to measuring a length (the capillary rise $h$), which may be determined optically with a cathetometer.

Since this method is broadly useful, and since it has proved to be particularly suitable for measuring the temperature dependence of contact angles,[33–38] it may be appropriate to describe the apparatus in detail, including the temperature control system. The essential parts of the equipment (Figure 10) are the following:

1.  The plate (A) (about $2 \times 3$ cm and as thin as is practical), which is aligned vertically; in certain cases it may be tilted (see below). It is attached to a support that can be raised or lowered by a screw (H). The screw may be driven by a variable-speed, reversible motor.

2.  The liquid (B) is contained in an optical-quality glass cell whose top is horizontal and that is about $10 \times 10$ cm or larger. This relatively large area is needed to ensure that the liquid surface is horizontal in the region where the reference height is measured. It also helps in the attainment of equilibrium between liquid and vapor. In principle, a smaller cell can be used if it

can be filled so that the liquid level is exactly at the top of the cell, with the only source of curvature being the capillary rise at the plate. However, if the liquid surface area is small, the lowering of the plate into the liquid will raise the liquid level appreciably, particularly if the plate is not extremely thin. This will interfere with the determination of the reference level from which the capillary rise $h$ is measured, so it is desirable to use a cell with the largest area that is practical. The plate dips into the liquid, which rises as indicated in Figure 9. If the contact angle is greater than 90° there is, of course, a capillary depression instead of a rise. The optical quality of the front of the cell is necessary in order for accurate observations to be made below the liquid level when there is a capillary depression.

3. A cathetometer, not shown in Figure 10.

4. A pointer (C) located in the same plane as the plate, for use in determining the level of the horizontal liquid surface. It is difficult to see the surface directly to determine its elevation, so a vertical fiber (which may be of glass) is mounted above and close to the surface. The location of the liquid surface can be determined by observing the tip of the pointer and its mirror image in the surface. Three precautions are necessary in this measurement:

a. The regions of the liquid surface that is curved, on account of nearness to the edge of the cell or to the plate, must not be used in observing the mirror image of the pointer. This trouble can be minimized by filling just as little as possible above the top of the cell, and by using a cell that is as long as possible in the direction perpendicular to the line of sight.

b. Some possible troubles due to curvature can be removed by tilting the telescope slightly downward, but this entails another error unless the pointer is at exactly the same distance from the telescope as the plate. In practice, it is possible to align the pointer and plate with sufficient accuracy, provided the experimenter is aware of the need for the alignment. The needed precision of alignment depends on the tilt of the telescope, as can be seen from an elementary construction.

c. The cross hair on the telescope must be as fine as possible, and errors due to backlash must be avoided. These are very elementary precautions, but worth mentioning because the need for them does not always arise in microscopic studies. In precision measurement with a traveling telemicroscope, accurate measurements can be made with a coarse cross hair, provided the travel of the microscope is always in the same direction. (This precaution, of course, also prevents errors due to backlash.) But the pointer and its mirror image are symmetrical about the liquid surface. If a cross hair is not of good quality, one might inadvertently use one side of the cross hair for the upper image and the other for the lower image. The error due to this technique is similar to the error due to backlash in the lead screw of the traveling microscope.

5. A pneumatic device (D) for transferring liquid into the cell to maintain the level of the liquid is a considerable convenience; for high precision work, it is a necessity. The bottle shown (D) is mounted on a

horizontal axis so that it can be rotated and the tip brought in contact with the liquid; a vertical bottle supported on a platform that can be raised or lowered can serve the same purpose.

The line of contact of the liquid on the solid appears as a sharp light–dark border when the specimen is illuminated with light from the direction of the cathetometer telescope (illumination from within the telescope is best). The contrast may not be great, but it is usually possible to observe the boundary quite distinctly.

The optical measurements are made with the cathetometer. Readings can conveniently be made with commercial equipment to precision of about $2 \times 10^{-4}$ cm. Since three position readings are required, the overall precision in the measurement of $h$ is no better than about $3 \times 10^{-4}$ cm. The corresponding uncertainty in the value of $\theta$ for a typical liquid (assuming $\gamma_{lv} = 30$, density $\Delta\rho = 1.00$, $\theta = 30°$) is about 0.1°.

In practice, it is only by special preparation of a surface that a straight enough meniscus line can be obtained, such that this accuracy can be attained. But it is easily seen that, if the effort is to be made to prepare surfaces good enough for thermodynamic measurements and if sufficient liquid is on hand, the effort involved in using this method is worthwhile, because it provides a better accuracy of measurement than any other method that is available. It is second only to the Wilhelmy method in precision, and is more broadly useful and convenient.

When the meniscus line is not exactly straight, the mean position of the line should be recorded. If the waviness of line is very regular, the average of the maximum and minimum values can be used. If it is irregular, one may estimate, by eye, a mean that is weighted for equal areas above and below the cross hair. This is an approximation, of course. If the exact contour were known, and if the fluctuations were known to be due to heterogeneity, it might be possible to use the thermodynamic method of Neumann and Good[9] to establish the best kind of weighted mean. Such a computation would, however, be too specialized to be worth the trouble of developing it in this chapter.

We might note at this point that, when the meniscus at a vertical plate is not a straight line, the perimeter of a drop on the same solid will not be perfectly circular. So it would be illusory to assume that the profile of a drop, as used in the telescope–goniometer method, would give any more valid a result than that obtained by the vertical plate method. The angle as observed with a goniometer could vary around the perimeter of the drop, and the drop would very likely exhibit the rim effect described above.

Vibration insulation is not shown in Figure 10. It is generally more important to insulate against vibrations with this method than with the sessile drop method. A convenient technique for a system that weighs 5 or 10 kg is to support the system on tennis balls that are filled with a silicone fluid to increase damping. Another kind of support uses small blocks (e.g., a few cubic centimeters) of a foamed, "lossy" polymer such as polyethylene or polypropylene. An appropriate location for antivibration padding is between the box (G) and its supports. The optimum mode of vibration insulation is strongly dependent on the mass, size, and shape of a system, as well as on the vibrations that are to be insulated against. Therefore, the investigator should experiment to find the optimum support for the size, weight, and shape of his equipment.

Dynamic advancing and receding contact angles can be measured by moving the plate up or down. It may be possible to choose a rate in a range such that the position of the line of contact is essentially independent of rate. The appropriate rate should be determined by the experimenter; about 1 mm/min is a good rate to start with. The motion can, of course, be stopped and the angle measured after a specified time. See above, Section 2.1.1.1.

If high-precision measurements are to be made, or measurements other than at room temperature, then for the purpose of the temperature control, the apparatus is doubly enclosed, with separate controls on the temperatures of three regions. First, the liquid can be heated and cooled from a battery of Peltier elements (E, Figure 10). In order to establish vapor and thermal equilibrium, the interior of the box (G) can be thermostatted with the Peltier batteries (F). Condensation at the walls of (G) would be a serious problem at temperatures different from ambient. Therefore, the box (G) is installed inside a climate chamber (O), which can be thermostatted by means of a resistance heater (L) and an external refrigerating unit with the cooling element (K). Heat exchange is enhanced by means of the ventilator (I). The hot or cold air enters through the slits (M) and is directed over troughs containing a desiccant such as calcium chloride. The temperature control is effected by resistance thermometers or thermocouples in conjunction with external electronic control units (P). The temperature is controlled in the three positions 1, 2, and 3. To avoid condensation, the temperature in G is always kept slightly (about 0.2°C) higher than the temperature in B, and the temperature outside B in the climate chamber somewhat higher still. Peltier batteries and the heating and cooling device of the climate chamber (O) are connected in such a way that each heating pulse is followed by a cooling pulse. By this means, the temperature may be kept constant in B to ±0.1°C, in G to ±0.2°C, and in O to ±1.5°C. This apparatus may be used in the temperature interval from −80°C to +80°C.

Actual measurements may be performed in the following way: The specimen (A) is installed vertically in its holder inside the box (G), and the liquid is transferred into the cell (B) and the bottle (D). The glass tip is adjusted so that it is at the same distance from the axis of the cathetometer as is the specimen. Then the whole apparatus is closed, and when temperature equilibrium is reached at the first temperature to be investigated, the cell (B) is filled slightly higher than level from the bottle (D). The specimen is lowered until it dips about 0.2 cm into the liquid. Since each reading for an advancing angle measurement should be taken at a region of the surface that has not been previously in contact with the liquid in that run, the plate is lowered very slowly, e.g., at a rate of the order of 0.01 cm/min, while a number of readings are taken at that particular temperature. The three-phase line usually remains stationary with respect to the telescope of the cathetometer, while the depth of immersion of the plate increases steadily. The procedure of slow, constant rate immersion has the additional advantage of decreasing the effects of small random vibrations. The procedure just described yields dynamic advancing contact angles. Receding contact angles may be obtained by withdrawing the plate A from the liquid in the same manner.

Measurements of the temperature dependence of contact angles may stretch over many hours, since proper time intervals have to be allowed for the establishment of thermal equilibrium. Therefore, disturbances from trace impurities adsorbed or absorbed by the liquid or the solid may become a problem. For nonpolar liquids such as the long-chain $n$-alkanes, an electrical purification *in situ* may be employed. A potential drop of 400 V d.c. has been used[37] between inert electrodes such as two platinum-plated copper blocks (not shown in Figure 10). The copper blocks are mounted inside the measuring cell (B) in such a way that they are completely immersed in the liquid when the cell is filled to the brim. This purification keeps the electrical conductivity at a low and constant value and has a very favorable effect on the reproducibility of the capillary rise measurements. The presence of the copper blocks also displaces unneeded liquid.

The optical alignment of the cathetometer needs special attention, and it is desirable to mention some further elementary precautions here. Since a high magnification by the telescope is desirable, the object field will normally be so small that it will be necessary to rotate the cathetometer about its vertical axis between readings of the line of contact and the meniscus of the undisturbed liquid. First of all, the horizontal cross hair of the eyepiece may possibly need adjustment. To test vertical alignment of the cathetometer beam and also the optical quality of the windows of the inner box (G) and of the outer climate chamber (O), the plate (A) is replaced by a second glass fiber. Determination of the position of the meniscus at these two locations by measuring the distance between the tips and their mirror images should give identical results.

As already noted, a slightly different arrangement is necessary when the contact angles to be measured are larger than 90°, i.e., when there is a capillary depression rather than a capillary rise. The cell (B) in Figure 10 is then filled slightly less than level, and the glass tip (C) should be under the liquid, supported from below and pointed upward, but again, close to the liquid surface. In other words, the tip and its mirror image are interchanged. Observations are then made through the wall of the cell (B) and the liquid.

The accuracy and reproducibility of this method is such that, for example, the solid–solid phase transition of cholesterol acetate at 40°C has been detected by means of the temperature dependence of the contact angle of water.[34] The discontinuity in the contact angle curve was about 0.3° of arc. For water on a siliconed glass plate, the deviations of individual points in the plot of contact angle versus temperature were found to be, on the average, about 0.1°,[36] in good agreement with limits of error estimated above. The precision of measurements obtained with this method is, thus, an order of magnitude better than that of direct methods such as the sessile drop–goniometer technique.

*2.1.3.1. Note on Filar Microscopic Measurement.* The theoretical limit of resolution of an optical system such as a microscope[39] is commonly considered to be

$$d = \frac{0.6\lambda}{n \sin(\alpha/2)} \tag{14}$$

where $d$ is the minimum distance between object points resolved as distinct points in the image, $\lambda$ the wavelength of light employed, $n$ the refractive index of the medium between object and objective lens, $\alpha$ the angle subtended by the objective lens at a central point of the object, and $n \sin (\alpha/2)$ the numerical aperture. For a telemicroscope with, say, a 10-cm object-to-objective lens distance and an objective, say, 2 cm in diameter, in air, and assuming $\lambda = 5000$ Å, the estimate of the limit of resolution is $3 \times 10^{-4}$ cm by equation (14).

This criterion, equation (14), is known as the Dawes limit. It was developed by astronomers, who were interested in the resolution of double stars considered as point sources. As a limitation, it is often surpassed by as much as a factor of 2 or 3 in metrology using a filar microscope,[40] and if photoelectric techniques are employed, a precision of the order of $10^{-5}$ cm is claimed for routine measurements. The reason this is possible is that the bringing of a cross-hair line of a microscope up to a sharp black–white edge or to another cross hair is a different and less demanding optical problem from that in astronomy, in which two point sources or bright disks are resolved.

One of the authors (R.J.G.), in instructing students in measurement for the purpose of the capillary rise determination, was surprised to find that he was regularly able to surpass the Dawes limit with low-cost equipment by a factor of 2 or better.

Michelson[41] has reported that, using a 6.5-mm focal length objective, a 6-mm eyepiece, and 200-mm object-to-objective distance, he could measure the diameter of a small object using a filar microscope to ±1.3 ×

$10^{-6}$ cm. This surpasses the Dawes limit by three orders of magnitude for his system.

### 2.1.4. Combination of the Capillary Rise at a Vertical Plate with the Wilhelmy Method

Despite the many advantages that indirect methods such as the capillary rise or the Wilhelmy method have over direct methods, they have the disadvantage that the surface tension of the liquid $\gamma_{lv}$ must be known. While this restriction does not usually present severe problems with most pure liquids, there may be a serious uncertainty for solutions of surface active substances. In such systems, adsorption of the surface active material at the various interfaces may change the surface tension and the contact angle simultaneously. The joint use of the capillary rise method and the Wilhelmy method makes it possible to determine contact angle and liquid surface tension at the same time.[28,42,43] Eliminating $\gamma_{lv}$ from equations (9) and (13) and with the aid of the identity $\sin^2 \theta + \cos^2 \theta = 1$, we obtain

$$\cos \theta = \frac{4 \, \Delta M \, \Delta \rho \, h^2 p}{4(\Delta M)^2 + p^2 (\Delta \rho)^2 h^4} \tag{15}$$

Measuring $\Delta M$ and $h$ as described above and knowing the density $\rho$ of the liquid and the perimeter $p$ of the plate, we can calculate the contact angle $\theta$ from equation (15) without explicit knowledge of the surface tension of the liquid. $\gamma_{lv}$ may then, of course, be calculated from equation (9) or (13). Alternatively, we may eliminate $\theta$ from equations (9) and (13) to obtain for the liquid surface tension

$$\gamma_{lv} = \left( \frac{\Delta M \, g}{p} \right)^2 \frac{1}{\Delta \rho \, g h^2} + \frac{\Delta \rho \, g h^2}{4} \tag{16}$$

The contact angle may then be calculated from either equation (9) or (13). Details, particularly referring to work with solutions of surface active substances, are given by Neumann et al.[42,43]

### 2.1.5. Tilting Plate Method

The tilting plate method, which was devised by Adam and Jessop,[44] has lost most of its former popularity. The principle of the method is illustrated in Figure 11. The plate, which should be about 2 cm wide, dips into the liquid, which will form a concave or convex meniscus near the plate. The plate is tilted until the meniscus becomes flat; the angle between the plate and the horizontal is then the contact angle. The straightness of the meniscus may be judged by viewing it over a sharp straight edge, e.g., a razor

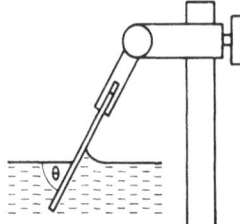

Figure 11. Tilting plate method.

blade, or by illuminating the meniscus and plate through a slit. The lines of light on the surface of the liquid and the plate will appear as two straight lines meeting each other in the three-phase line if the liquid meniscus is flat right up to the solid surface. Otherwise, the line of light on the liquid near the plate will appear to be curved.[44,45] The method was improved by Fowkes and Harkins[46] by providing precautions for keeping the surface of the liquid clean, as in film pressure investigations. Suspending the plate at both ends from two micrometers and dipping it into the liquid makes it possible to measure small contact angles, e.g., below 10°.[47] The precision and reproducibility of measurements taken with this method depend to a considerable extent on the means by which the straightness of the meniscus near the plate is detected. A reproducibility of 0.1° or better is claimed in some cases.[47] It is to be noted that contact angle measurements performed with this method correspond to the properties, not of a point, but of a small section of the three-phase line. In this respect, the method is similar to the method of sessile drops or bubbles.

The most important advantage of the tilting plate method is the simplicity of the apparatus, which may be constructed in a good workshop within a few days. No special optical equipment is needed. In this respect, the method is probably equalled only by the method of specular reflection from a sessile drop or meniscus, which is described in the next section.

There is a serious disadvantage to this method; measurements of advancing and retreating contact angles are made only with considerable difficulty, so the numbers obtained by the method may lie somewhere between $\theta_a$ and $\theta_r$. This indefiniteness is a very serious trouble when there is appreciable hysteresis—which is nearly always. Probably because of this defect, it is found in practice that it takes considerable skill to operate the apparatus successfully.

### 2.1.6. Reflection Method

Langmuir and Schaeffer[48] employed the specular reflection from a drop to measure the contact angle (see Figure 12). This method was refined by Fort and Patterson,[49] and combined with the tilting plate method by

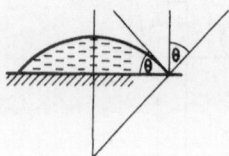

Figure 12.  Reflection from a drop. Geometry for limiting angle at which reflection can be seen.

Good and Ferry.[50] It was used by Good and Paschek[51] to measure the contact angle of mercury on glass under vacuum (see below).

The method employs a light source mounted on a beam, which is pivoted on an axis containing the three-phase line at which the contact angle is to be measured (see Figure 13). This may be at the edge of a drop, or a meniscus at a flat plate or rod, or on the inside of a tube. The observer sights along the beam to observe the specular reflection from the liquid surface. The measurement consists of pivoting the beam on its axis and determining the orientation at which the specular reflection disappears. For angles of incidence lower than $\theta$ (see Figure 12), there is no reflection. The location of the cutoff angle can be found to within 1°, usually; and the accuracy of the method is comparable to the accuracy of other direct measurements.

The apparatus of Fort and Patterson has a stage on which the specimen is placed; the stage may be raised or lowered to bring it to the pivot axis, and it may be leveled by screws. A small flashlight is mounted in a block at the end of a long, hinged arm, and a peephole is located close to the light source. The angle of elevation is measured with a protractor.

## 2.2. Contact Angle Measurement at a Capillary Tube

### 2.2.1. Wilhelmy Method

The derivation of equations (9) and (10) does not postulate that the surface be flat. Any shape of solid surface is allowable, provided that it is perpendicular to the liquid surface. Hence, the Wilhelmy method can be applied to plates, rods, wires, tubes, and capillaries. For the geometry of a tube, the perimeter $p$ is the sum of the inner and outer perimeters. The inner and outer diameters of a capillary may be determined by a measuring microscope. The procedure for obtaining contact angles in capillaries is identical with that described in Section 2.1.2 for thin plates.

### 2.2.2. Method of Capillary Rise

For vertical capillaries that are so narrow that the meniscus may be considered to be spherical, we have

$$\Delta P = \Delta \rho \, gh = \frac{2\gamma_{lv} \cos \theta}{r} \tag{17}$$

where $r$ is the capillary radius and $h$ the capillary rise. Both $r$ and $h$ may be determined optically. In cases where the radius is too small for accurate measurement with a microscope, $r$ may be determined by measuring the length of a column of mercury of known mass in the capillary. Advancing and receding angles may be obtained by raising or lowering the capillary at a slow rate.

The capillary rise method for measuring contact angles is restricted to small tubes at present. Relatively wide tubes, in which the meniscus is not spherical, have been used for the measurement of surface tension of pure liquids using tubes in which the contact angle is zero.[52] The general correction tables that must be used to account for deviation from sphericity have not been worked out as yet for systems with $\theta > 0$. While the computations required are not unduly difficult, the need does not seem to have been sufficiently great to stimulate anybody to carry them out.

A special case of this technique should also be mentioned. For a spherical meniscus, the Laplace equation yields equation (17). Instead of measuring the hydrostatic pressure $\Delta \rho\, gh$ necessary to compensate the Laplace pressure $\Delta P$ as in equation (17), we may determine the pressure needed to prevent the liquid from penetrating the capillary at all.[53] This is also the basis for a method of measuring contact angles on powdered solids that will be discussed below.

Methods that involve penetration of a liquid into a capillary are not in general recommended for solutions or for two-liquid-phase systems, either for measurements of surface tension or of contact angle. This is because adsorption on the capillary wall may reduce the concentration of one component inside the capillary, and transport of solute down the capillary from the bulk liquid is likely to be slow. This adsorption and transport

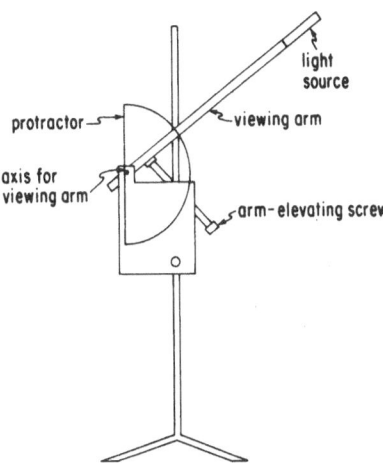

Figure 13. Apparatus for measuring contact angle by Langmuir–Schaeffer method.

limitation may cause the contact angle to change with time, so that the angles reported may have little resemblance to those measured on open surfaces.

### 2.2.3. Reflection Method

For relatively wide, transparent tubes, the reflection technique described by Good and Paschek[51] may be used. It works particularly well in the case of liquid metals, on account of their high reflectivity. See Figures 13 and 14. Good and Paschek measured the contact angle of mercury on glass and on fused quartz in vacuum and in atmospheres containing various degrees of saturation of water vapor. The configuration of a vertical tube was found to be a particularly convenient one for subjecting the solid surface to a variety of pretreatments.

The method as used had the limitation that angles near 90° were difficult to measure. This could be overcome by aligning the light source and viewing direction at a convenient angle to each other and adding (or subtracting) half that angle from the angle read off the protractor.

### 2.2.4. Rate of Penetration Method

The rate of penetration method is often employed for measuring the contact angle with powders and porous solids (see Section 2.4). The theory of that technique was developed by a generalization of the law that governs penetration into capillaries—the so-called Washburn equation.[54] As a practical method *with capillaries*, this technique does not seem to have any advantage over the capillary rise technique.

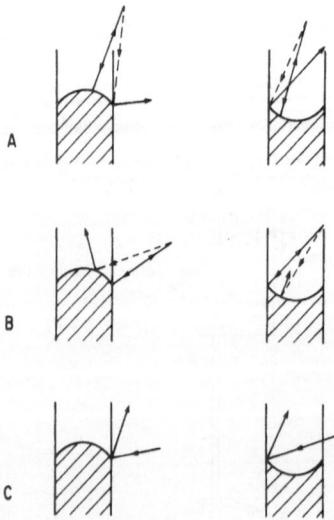

Figure 14. Adaptation by Good and Paschek of Langmuir–Schaefer method for a liquid in a vertical tube; (A) viewing angle larger than $\theta$, (B) viewing angle equal to $\theta$, (C) viewing angle less than $\theta$.

## 2.3. Elongated Solid Bodies—Cylinders, Rods, and Fibers

Contact angles on drawn fibers or natural fibers such as cellulose are likely to be anisotropic (cf. Good *et al.*[55] regarding stretched films). This anisotropy appears to be a subject that has not been investigated, up to the present, with fibers. It may be a matter of technological importance in textiles, nonwoven fabrics, etc., for example, with respect to wetting and drying of fiber mats and the forming of junctions bonded by adhesives. The anistropy is, in any case, a matter of fundamental scientific importance.

The direct method, described in Section 2.3.1.1, and the methods in Sections 2.3.1.4 and 2.3.1.5 measure the contact angle at a portion of the three-phase line that is perpendicular to the fiber axis. It is possible, or even very probable, that the angle measured at a portion of the three-phase line parallel to the axis (see Sections 2.3.1.2 and 2.3.1.3) would be appreciably different. Therefore, discrepancies may be expected between angles measured on fibers depending on the methods used, and these discrepancies will not be experimental artifacts.

### 2.3.1. Individual Cylinders (Fibers)

The basic principles of measurement with fibers are exactly the same as with flat plates and with capillaries; we can best describe such measurements by adapting the previously discussed methods to this geometry. Thus, for example, it is possible to derive by numerical integration of the Laplace equation of capillarity[56] a method using the capillary rise at a convex, vertical, cylindrical surface by direct analogy with the capillary rise at a vertical plate or inside a capillary tube. However, the rise $h$ on the outside of a cylinder decreases with decreasing cylinder radius and quickly becomes too small for convenient measurement.

*2.3.1.1. Direct Measurement—Profiles of Drops on Fibers.* A system for direct contact angle measurements at fiber surfaces was developed by Schwartz *et al.*[57,58] In their apparatus, the fiber is suspended horizontally in the field of a microscope; the supporting clamps are mounted so that the fiber can be rotated. A drop that is larger (but not a great deal larger) than the fiber diameter is deposited on the fiber, and the angle that the drop profile makes with the fiber is measured directly with the aid of a goniometer eyepiece.

The first precaution that must be taken is to focus the microscope on the edge of the fiber, not on the profile of the drop. The drop, which is brought to the fiber by a wire or rod or capillary tip, often assumes an asymmetric position. This fact can be ascertained by rotating the fiber and seeing whether the apparent contact angle changes. Rotation of the fiber is important also in case the fiber does not have a circular cross section and uniform surface composition. If the fiber surface is smooth and (macro-scopically) homogeneous, the asymmetry of the drop may be attributed to the contact angle hysteresis. The maximum and minimum angles the drop makes with the fiber may be taken as approximately the advancing and receding angles, respectively. A better approximation can be achieved, however, if a micromanipulator is used to hold the drop in position and to add liquid to or remove it from the drop. By appropriate manipulations, it should be possible to observe $\theta_a$ and $\theta_r$ directly.

Because of the small dimensions of the system, the accuracy and reproducibility of the measurement are somewhat poorer than with measurements at extended flat surfaces. This is especially true for angles below about $10°$.

An alternative method of measuring contact angles directly at a single filament was described by Bascom and Romans.[59] In this method, the drop is held in a small ring in the microscope field, and the fiber, which is suspended between support posts, is passed (advancing or retreating) through the center of the ring. If the ring is small enough and the fiber homogeneous, the liquid surface will be axially symmetrical about the fiber. Gravitational distortion from symmetry can be eliminated by mounting the microscope horizontally and the filament vertically. These measures diminish but do not remove the need to rotate the fiber, which was a requisite in the method of Schwartz and Minor.[57] Rotation, in this case, serves to aid in detecting inhomogeneity of the fiber surface.

With a thin fiber, there is little difficulty in ensuring that the focal plane of the microscope contains both the outermost edge of the fiber and the profile of the meniscus. With a thick rod sticking into a liquid with a horizontal surface, it is not quite so easy to ensure this condition. As already noted, a slightly improper focus can lead to an apparent contact angle that is considerably too high.[50] For a drop of liquid on such a rod, the apparent contact angle could easily be too low.

In addition to the difficulties of microscopic observation already noted, there is an error that could arise for certain values of the contact angle and with large ratios of drop size to fiber diameters. Roe[60] has computed the equilibrium shapes for a drop clinging to a fiber. In at least one case that he illustrates ($45°$ contact angle and drop diameter greater than five times the fiber diameter), the drop profile has a region of reverse curvature near the fiber surface. [The effect has some resemblance to the effect described in Section 2.1.1.1, in which a rim is seen at the base of a drop on a flat but heterogeneous surface (see Figure 3a). However, for a large drop on a small fiber, the cause of the observed effect is independent of heterogeneity and microscale roughness.] If this region of reverse curvature is not clearly visible in the microscope, it might be disregarded, and too high a contact angle would be reported. The obvious remedy for this trouble is to measure the contact angle only with small drops or with drops of different size and to use the value found consistently with the smaller drops. See also the paper by Carroll.[61]

*2.3.1.2. Horizontal Meniscus Technique.* The principle of the tilting plate method can be adapted to large cylinders.[45] The cylinder of diameter $d$ is mounted with its axis parallel to the liquid surface and immersed to a depth at which the meniscus is flat near the three-phase line. The depth $h$ to which the cylinder is immersed is determined mechanically or optically, e.g., by measuring the volume of liquid displaced by the cylinder. The contact angle is then given by

$$\cos \theta = \frac{2h}{d} - 1 \qquad (18)$$

Provision for rotation of the cylinder in both directions makes it possible to establish dynamic advancing and receding contact angles. While this method has not found wide use in general, it has been applied to contact angle measurements of wires and fibers.[62]

*2.3.1.3. Determination from the Equilibrium Meniscus near a Floating Fiber.* A floating cylinder is shown schematically in Figure 15, which also defines the various terms employed. A force balance yields[32]

$$2 \sin(\phi + \theta) + C_{23}R^2[\pi D - \phi + \sin \phi \cos \phi - 2 \sin \phi (Z_0/R)] = 0 \qquad (19)$$

where

$$C_{23} = (\rho_2 - \rho_3)g/\gamma_{LV} \qquad (20)$$

and

$$D = (\rho_1 - \rho_3)/(\rho_2 - \rho_3) \qquad (21)$$

The capillary height (or capillary depression) $Z_0$ may, just as in the case of the vertical plate, be obtained in analytical form:

$$Z_0/R = \pm\{2[1 + \cos(\phi - \theta)]/C_{23}R^2\}^{1/2} \qquad (22)$$

Measuring $Z_0$ by means of a cathetometer and eliminating $\phi$ between equations (19) and (22) by suitable numerical or graphical techniques yields the contact angle.

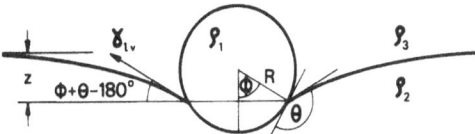

Figure 15.    Cross section through a floating cylinder.

The practical measurement of $Z_0$ may prove to be quite difficult. However, the analysis turns out to be much simpler for small cylinders, i.e., fibers, where we have $C_{23}R^2 \rightarrow 0$. In this case, we have, as may be seen from Figure 15,

$$\phi + \theta = 180° \tag{23}$$

i.e., the liquid surface remains horizontal up to the cylinder, and the contact angle may be obtained from equation (18). The depth of immersion $h$ may be obtained using a cathetometer equipped with a high-power telescope (see above). It is most convenient to measure $h$ using the technique for measuring small vertical distances from the undisturbed level of a liquid, as described for the capillary rise measurements at vertical plates. The magnitude of $h$, however, is small, and so the resolution of the optical system constitutes a serious limitation on the accuracy of this method.

An alternate form of equation (19) has recently be proposed.[63,64] This form of the equation is derived from a free energy analysis of the system. Coupled with equation (22), it may be used just like equation (19) to determine the contact angle.

*2.3.1.4. Wilhelmy Method.* The Wilhelmy technique has only in recent years been applied to the measurements of contact angles of fibers.[65] As noted above, equations (9) and (10) are independent of the exact shape of the solid, so they may be carried over to the case of a fiber. Since we have, for a fiber, a relatively large ratio of perimeter to cross-sectional area, a precise knowledge of the depth of immersion is not necessary. For a fiber of about 10 $\mu$m diameter, the $\Delta M$ due to capillary rise will be of the order of $10^{-4}$ g. Since commercial electrobalances have sensitivities of about $10^{-6}$–$10^{-7}$ g, contact angle measurement by this method is well within the capabilities of such a balance.

The apparatus illustrated in Figure 7 may be used without change. Adsorption of the vapor of the measuring liquid at the suspension and the beam of the balance may become more of a problem than in the case of a plate, because $\Delta M$ is so small. This will be particularly true with liquids of relatively high vapor pressure. To minimize such effects, the suspension of the fiber should present as small a surface as possible and should, if possible, be made of a low-energy solid such as Teflon.

The main problem with the application of the Wilhelmy method to contact angle measurements is the fact that a very precise value for the perimeter of the fiber is needed. This difficulty may be overcome by using the Wilhelmy technique in a separate measurement with a liquid which is known to wet the fiber completely, i.e., to have a contact angle $\theta = 0°$. The total absence of contact angle hysteresis is a convenient test for complete wetting, in most practical cases. Using $\cos\theta = 1$, the perimeter $p$ of the fiber may be calculated from equation (9). The accuracy of the determination of the circumference of the fiber is limited only by the accuracy with which the surface tension of the liquid is known and the absence of swelling of the fiber by the liquid. Assuming $\gamma_{lv}$ to be known to ±0.1%, it is possible to determine the effective circumference of a fiber of a diameter of about 10 $\mu$m to a few hundred angstroms. Since the same considerations apply to the accuracy of the actual contact angle measurement with the Wilhelmy method, it appears that this method should give highly precise contact angles on fibers.

Just as with plates, a continuous immersion of the fiber in the liquid may be used to test the homogeneity of the solid surface (Figure 8). Again, the method may be used to measure continuous changes in contact angles. Thus, if a fiber is dipped into an equilibrated surfactant solution, i.e., a solution in equilibrium with its surface, an observed continuous change in contact angle may be attributed solely to the adsorption of the surface active material on the solid.[65]

The method is relatively easy to employ if the fiber is straight or has straight sections of at least 1- or 2-mm length. If this is not the case, it is possible to suspend a small weight from the free end of the fiber to straighten it out. If the density of the weight is not accurately known, it is necessary to establish the effective volume of the weight. For example, a calibration run with a fiber of known contact angle, e.g., a siliconed glass fiber, may be performed. A determination of the absolute volume of the weight would require a high-accuracy density measurement of the test liquid. It is probably more convenient to determine first the circumference and contact angle of the calibration fiber against the test liquid. Then, after attaching the weight to the free end of the calibration fiber, the portion of the $\Delta M$ due to the buoyancy of the weight is measured. This portion of $\Delta M$ is subtracted in the subsequent measurements with the test fibers. Unfortunately, this portion of the total change in weight $\Delta M$ may be considerably larger than the capillary effect. However, if conditions are carefully controlled, an automatic electrobalance should still produce good results.

If the surface tension of the wetting liquid is not known, it may in principle be determined by measuring the capillary rise simultaneously with the increment of weight (cf. above, the case of flat plate geometry). The numerical integration of the Laplace equation for a meniscus near a vertical fiber of circular cross section is now available.[56] Instead of combining equation (9) with equation (13), as in the case of the wide plate, the former may here be combined with the appropriate parameters of the numerical integration. This adaptation of the method, however, suffers from the disadvantage already noted that the capillary rise may be too small for very accurate measurement, so a separate measurement of $\gamma_{lv}$ may be preferable.

*2.3.1.5. Reflection Method.* The reflection method as developed by Fort and Patterson[49] has also been adapted to measure contact angles of fibers.[58] Observations are made through a horizontal microscope with an internal illuminator. The fiber or filament is mounted horizontally by means of two clamps on a sliding filament holder. The fiber passes through a small eyelet, which serves as the liquid holder (cf. the method of Bascom and Romans[59]). The filament holder and the eyelet are mounted with the eyelet on the axis of a horizontal protractor turntable. The cutoff of the back reflection, and hence the contact angle, are determined by rotating the turntable. Advancing and receding conditions are established with the aid of the sliding motion of the filament holder. The apparatus described by Jones and Porter[66] can be used to measure contact angles up to approximately 75°. Measurements have been made on fibers as thin as 2.5 $\mu$m diameter.

### 2.3.2. Bundles of Fibers, Mats, and Woven Fabrics

*2.3.2.1. General Comment.* The contact angle of a liquid on a fiber is a basic quantity that determines the fluid-repellency of a fabric and its converse, wicking or penetration. These processes and properties of the fabric are, in general, not simple functions of the contact angle; they also depend strongly on the detailed geometry of the fiber array. While it is of interest to investigate wetting phenomena of such arrays in addition to the contact angles on the separate fibers, we will discuss these macroscopic effects only as they yield values of the contact angles on the fibers themselves.

*2.3.2.2. Bundles of Fibers.* It appears from the principles discussed above that only the Wilhelmy method can be usable to measure contact angles. This method might work for a bundle of fibers, because the change in weight observed on immersion is only a function of the liquid surface tension, the contact angle, and the total perimeter of the sample. The number and diameter of the fibers must be known, of course.

A study that is relevant to the use of this method has been made by Princen,[29] who discussed the theory of the capillary rise at certain kinds of complex surfaces. He pointed out that a sand-blasted Wilhelmy plate can be modeled as a flat plate with a multitude of V-shaped grooves in the surface. (Sand-blasting is employed in order that the effective retreating contact angle will be zero for surface tension measurements, even if the advancing angle is not zero.) If the V's are sharp and if adsorption at a flat solid–vapor interface is appreciable for the liquid–solid system, then capillary condensation is likely to occur. This condensation will be the more serious the sharper the wedge angle. The proper reference weight is, then, not the dry weight of the solid (i.e., the fiber bundle), but the weight of the solid equilibrated with the saturated vapor of the wetting liquid. Princen estimated that an error of several percent can arise with a sand-blasted Wilhelmy plate if this reference state is not used.

For a fiber bundle, this effect may lead to an appreciable uncertainty. The region around the line of tangency of two smooth cylinders will be a site of capillary condensation if the contact angle is low. With a fiber bundle, there will be a great many such sites, particularly if the fibers are of small diameter and if the bundle is a tight one. While Princen did analyze the capillary rise of a bundle of vertical cylinders, he did not put his results in the form of an estimated gravimetric error as a function of the bundle parameters. We can recommend this as a problem for experimental and theoretical study. For the present, we must recommend some caution in the use of the Wilhelmy method with fiber bundles.

None of the other methods discussed above for contact angle measurements on *individual* fibers can be applied, since they depend very strongly on *local* shape of the liquid surface.

*2.3.2.3. Mats of Fibers and Woven Fabrics.* The angle exhibited by a drop standing on a woven fabric or a mat of fibers should not be mistaken for the angle the liquid makes with individual fibers. As discussed above, the macroscopic angle is strongly dependent on the local geometry of the fiber array. Consequently, quantitative methods for measuring contact angles on such solids are not available. A useful, qualitative test for the repellency of a fabric consists in putting drops of liquids of different surface tension on a specimen and noting the surface tension of that liquid that just does not penetrate the fabric. This value is a relative index for the liquid repellency of specimens that have the same microgeometry, packing, and pore structure. The lower this limiting value of the surface tension, the better the repellency and the larger the contact angle.[67]

The most convenient method for making this test is to use mixtures of liquids so as to adjust the surface tension $\gamma_{lv}$. The choice of the liquid pair is of some importance, because the surface tension of the solution is not the only variable controlling penetration. The solid–liquid interfacial tension is strongly affected by adsorption of one component or the other. Hence, members of two series of solutions that have the same surface tension may not exhibit the same penetration behavior. Extreme caution in a thermodynamic interpretation of such phenomena is appropriate (see Chapter 1 in this volume).

An alternative experimental technique is to place standardized disks of fabric on the surface of the liquid. The time required for a disk to sink will be the longer the higher the contact angle. This qualitative technique, known as the Draves test, was originally devised to test the wetting power of solutions of surface active agents. This method, when examined in detail, may be seen to be a version of the method based on the rate of capillary penetration, which will be discussed below under the subject of powders. It depends to a very considerable extent on the disks of fabric having the same pore structure, and so it is not particularly useful for measurement of contact angles of *different* kinds of fibers or different textile structures. When unlike fibers are fabricated into textiles, the differences in fibers diameter, stiffness, coefficient of friction, etc. will in general lead to appreciable differences in structure and pore size, even when the intended textiles are the same.

Individual fibers can (in the absence of cross-linking or knotting) be extracted from bundles, mats, and woven structures, so if information about the solid–vapor interface is desired, measurement should be made on the fibers themselves.

## 2.4. Powders

### 2.4.1. Spherical Bodies

Measuring contact angles on solids that are in powder form is more difficult than measurement with fibrous solids. While it is possible to adapt several of the techniques described for flat plates to fibers, this is not generally true for powders.

For sufficiently large, spherical particles, a modification of the tilting plate (or horizontal cylinder) method might be adopted. The rotational symmetry of the sphere about the axis normal to the liquid surface, however, makes observations much more difficult than in the case of the cylinder. A relation similar to equation (18) applies to spheres as well as cylinders.

Measurements utilizing the shape of the meniscus near a floating sphere, in analogy to the floating cylinder, are not simple, since the Laplace equation is not analytically integrable in this case.[32] However, a numerical integration of the Laplace equation is available.[56] Instead of combining the force balance equation for spherical bodies, which is

$$\sin \theta \sin(\theta + \phi) + (C_{23}R^2/6)$$
$$\times [4D - (1 - \cos \phi)^2(2 + \cos \phi) - 3(Z_0/R)\sin^2 \phi] = 0 \quad (24)$$

with the analytically integrated Laplace equation as in the case of a floating cylinder, equation (24) has to be combined with the numerical solution of the Laplace equation. This is done by trial and interpolation, i.e., finding a set of parameters that satisfies the force balance equation and the numerical solution of the Laplace equation simultaneously. The quantity to be determined experimentally is the depth of immersion of the sphere. The quantities that must be known independently are the surface tension of the liquid, the densities of all three phases involved, and the radius of the sphere.

The force balance discussed in connection with the floating cylinder [see equation (19)] shows that for small, floating, spherical particles, we may expect the meniscus to be flat right up to the solid surface. This will also be true for relatively large particles whose density is close to that of the liquid. In this case, we can determine the contact angle from the depth of immersion without use of the Laplace equation. It is, of course, virtually impossible to distinguish between advancing and receding conditions, and for small particles, the depth of immersion is not easy to measure directly.

When the spherical particle is small but not so small that the meniscus is horizontal, there exists no analytical expression for the capillary rise.[64] In this case, the contact angle may be established in a manner identical to the one used when the particle is cylindrical, employing equation (18).

If exceptionally accurate optical techniques are available, the need for expressions for the capillary rise may be cirumvented. The depth of immersion $h$ and the capillary rise need to be measured, the former yielding directly the value of $\phi$. These two values may be inserted in the force balance equation for spheres, and the resulting equation may be solved for the contact angle. This procedure is most useful when the spherical particle is of intermediate size. An equation derived from a free energy analysis may also be used in place of equation (19) to arrive at values of the contact angles.[64]

### 2.4.2. Irregular Particles

*2.4.2.1. Qualitative Methods.* Obviously, the procedures discussed above are not applicable to powders consisting of irregular particles. Since the problem of measuring contact angles on such powders is very difficult, it may be well to consider first the possibilities of simple qualitative tests for wettability.

Working by analogy to the Draves test, one may dust the powder on the surface of the liquid. The time until it sinks will be the longer the larger the contact angle. Unless care is taken, the powder is likely to reach the liquid surface as aggregates, or at least as small, loose piles. If this occurs, the first step of the sinking process may be the capillary penetration of the liquid into the pile or aggregate [cf. below, the method employing the Washburn equation, equation (26)]. If the contact angle is large, the powder may not sink at all. In this case, one can compare the mobility of different powders in a stream of air directed obliquely at the surface of the liquid. The larger the contact angle, the higher the mobility will be, since with large contact angles, the particles will be immersed less deeply, and less work will have to be done against the viscous drag of the liquid.[67]

The degree of wetting of powders may also be related to the sedimentation volume.[68–70] This is the volume a given quantity of the powder assumes a sufficiently long time after it has been dispersed in the liquid and allowed to settle to the bottom of the container or, if less dense than the liquid, to rise to the top. The penetration volume, i.e., the volume of liquid penetrating and saturating a given powder bed[67–69,71–75] gives a related measure of wetting.

### 2.4.2.2. Quantitative Methods

A. Direct Measurement on Compressed Powders. Attempts have repeatedly been made to measure the contact angle directly, by employing a compressed powder cake with a flat upper surface. If the actual contact angle is low, the liquid on the compressed cake normally is not stable but immediately starts to penetrate into the cake. If the contact angle is high, stable drops may be obtained, as in cases of low-energy solids such as organic pigments with liquids whose surface tension is high, or for mercury on most powdered solids. The surfaces of such pressed cakes often are glossy and appear (to the naked eye) to be perfectly smooth. However, microscopic examination shows that the surfaces are porous, and it has been demonstrated thermodynamically[76] that the contact angle on a porous surface will be higher than on a smooth surface that has the same composition. Besides, the plastic distortion of the topmost powder particles by the compression is likely to render the exposed surface quite unlike the surface of the uncompressed powder.

Experimentally, it is found that the contact angles measured on compressed powder cakes do not agree with those obtained on evaporated films of these substances—films which, from scanning electron microscopy, are known to be so smooth that the surface morphology does not influence the contact angles.[77] From this, we must conclude that contact angles measured on pressed cakes, although they may reach a limiting value (i.e., they do not change further if the compacting pressure is increased above a certain value), are to a major extent determined by microscopic roughness (the shapes of the particles) and the porosity. Therefore, they are not the true contact angles of the solids. So measurements of this type can yield, at the best, only qualitative information.

An attempt has been made by Kossen and Heertjes[78] to modify the preparation of compressed cakes to allow contact angle measurements in cases where the liquid penetrates the compressed powder. It was observed that presoaking the cake with the measuring liquid produced solid surfaces such that drops placed on the surface were quite stable. Contact angles on such surfaces can then be determined by the methods described for flat plates (see Section 2.1). Contact angles of the solids were calculated by Kossen and Heertjes from the observed angles using the equivalent of the Cassie method (see Chapter 1), which relates the contact angle observed on a heterogeneous, solid surface to the intrinsic contact angles of the two kinds of domains constituting the surface in the special case when there is no contact angle hysteresis. The implicit assumption was made, that the exposed surfaces of the particles were (a) flat and (b) oriented parallel to the overall "surface" of the compacted saturated mass. This is not a very probable condition. Moreover, with such a heterogeneous surface, there will normally be great hysteresis, so that the conditions for use of the Cassie equation are very far from being met. Therefore, it cannot be expected that the procedure of Kossen and Heertjes will insure the correct angle determinations.

The practical purpose of measuring contact angles on a powder is usually to predict and control the penetration of the liquid into a powder bed or, conversely, the dispersion of the powder in the liquid. For these purposes, it may not be sound to attempt contact angle measurements on the external surface of a highly compressed powder, since the external surfaces may well be uncharacteristic of the surfaces that are not exposed. It seems more promising to investigate the actual process of penetration.

B. Pressure of Displacement. One method that involves actual penetration is to measure the pressure necessary to balance the Laplace pressure, which drives liquid into a capillary bed:

$$\Delta P = \frac{2\gamma_{lv} \cos \theta}{r} \qquad (25)$$

This method was applied to the penetration of powder beds by Bartell *et al.*[79-81] and by Davies and Curtis.[82] The method suffers from the very basic drawback that the effective pore radius $r$ is not a constant from point to point in the bed. If the powder consists of perfectly uniform spheres in a *known* close-packed array, the minimum value of $r$ (corresponding to "entrance radii" for pores) can be computed from the particle diameters. But even with uniform spheres, there are two types of close-packed arrangements—one is body-centered cubic packing (BCC) and the other is face-centered cubic and hexagonal close-packed (FCC and HCP). Therefore, the possibility of estimating $r$ for use in equation (25) is quite remote. In some cases, the packing density changes with exposure to the liquid, so estimates of $r$ from particle size are doubly uncertain. It has been reported that, in actual measurements by this method, the establishment of a stationary meniscus is tedious,[82] and some investigators have been unable to obtain reproducible results.

We may conclude that while, at present, the pressure of displacement method is not a very reliable technique for measuring $\theta$, it is in principle more reliable than the direct method described in Section 2.4.2.2A.

C. Liquid Intrusion. Contact angles on interior pore surfaces, and the surfaces of powder particles, can be measured by a method which draws directly on the technique of liquid-intrusion porosimetry.[83-89] Equation (25) is the basic equation for this type of study, by which the distribution of pore radii [$r$, in equation (25)] may be determined. Mercury is usually used as the penetrating liquid; and the common practice is to assume a contact angle of 130° for mercury, although it has long been recognized that it would be far better to use the actual angle on the solid that is being studied.[87,88] It has also been the common practice to ignore hysteresis, and to use the same angle in both the increasing pressure (intrusion) and decreasing pressure (extrusion) measurements. (See the end of this section, where some relevant results for mercury on silica and on Pyrex glass are summarized.) There is, however, an important topographical aspect of the wetting of porous solids by mercury, which we will consider below, that is at least as important as the uncertainty as to the mercury contact angle on a flat surface.

A method was proposed in 1967, by Rootare and Prenzlow, to determine the specific surface area of a solid by mercury porosimetry.[90-93] Rootare[91,93] has suggested an adaptation of this technique for determining the contact angle; and it seems that this method shows some promise. It was shown in reference 90 that a measure of the area, $A'$, could be obtained by means of the equation

$$A' = -\frac{1}{\gamma_{lv} \cos \theta} \int_0^{V_{max}} P \, dV \tag{26}$$

Here $\gamma_{lv}$ is the surface tension of the intruding liquid, and $P$ is the pressure required to force volume $V$ of mercury into the solid. When a graph of $P$ vs. $V$ is plotted, it is usually found that a limiting volume, $V_{max}$, is reached such that a further increase in pressure causes a negligible increase in the intruded volume. The integral for equation (26) can be determined graphically.

If the maximum pressure that can be generated by the instrument is 50,000 psi (approximately $3.5 \times 10^5$ kPa) the limit of pore size that can be penetrated is a diameter of about 35 Å for a liquid that forms a 130° contact angle.[87]

The specific area of the solid can generally be measured by gas adsorption. If the solid does not have micropores, the BET surface area,[94] $A_{BET}$, should be approximately equal to $A'$. Rootare and Prenzlow substituted $A_{BET}$ into equation (26), and on rearranging, they obtained

$$\cos \theta = \frac{-1}{\gamma_{lv} A_{BET}} \int_0^{V_{max}} P \, dV \tag{27}$$

For pressure in psi and area in m$^2$/gm, and if $\gamma_{lv} = 485$ dynes/cm,

$$\cos \theta = \frac{-0.0141}{A_{BET}} \int_0^{V_{max}} P \, dV$$

For pressure in kilopascals, the expression is

$$\cos \theta = \frac{-0.00205}{A_{BET}} \int_0^{V_{max}} P \, dV$$

The data of Rootare and Prenzlow lead to estimates of contact angles ranging from 115° to 160°[91] for mercury on various porous solids. Some anomalously high results found in reference 91, such as 154° for mercury on porous Vycor glass, call for explanation. Recent work by Good and Mikhail[95,96] has shown that anomalies such as this probably are consequences of the pore topology; for example, the pores in a solid such as porous Vycor resemble interconnected wormholes. As mercury penetrates down a larger pore, at a particular pressure, it does not penetrate smaller pores that branch off it, but forms new liquid–vapor area at the pore entrances. This behavior resembles that discussed by Shuttleworth and Bailey[76] and by Cassie and Baxter.[97,98] Cassie's equation may be used to describe the situation:

$$\cos \theta' = \sigma_1 \cos \theta_1 + \sigma_2 \cos \theta_2 \qquad (28)$$

For a solid having two different types of areas, such that the contact angle on type 1 is $\theta_1$ and that on type 2 is $\theta_2$, and if the fractional areas of the two types are $\sigma_1$ and $\sigma_2$, respectively, equation (28) gives the contact angle that should be observed. Cassie[98] also concluded that if the solid was porous, with $\sigma_1$ being the fractional area of flat surface and $\sigma_2$ the fractional area of void, then $\theta_2$ would be 180°; for mercury on such a solid,

$$\cos \theta' = \sigma_1 \cos \theta_1 + \sigma_2 \qquad (29)$$

We may refine Cassie's model, by pointing out that the mercury-vapor surface at the entrance to a small pore will in general be convex, and the ratio of mercury-vapor surface to mercury-solid surface will be larger than $\sigma_2/\sigma_1$. Correction for this configuration can be made by multiplying $\sigma_2$ by a factor between 1, which would correspond to a flat liquid surface at the pore entrance, to 2, which would correspond to a hemispherical liquid surface. Then we must rewrite equation (29) in the form

$$\cos \theta' = \sigma_1 \cos \theta_1 + \alpha \sigma_2, \quad 1 < \alpha < 2 \qquad (30)$$

Since relative to the larger pores, the smaller pores can be expected to have a lesser fraction of their internal surfaces consisting of the mouths of yet smaller pores, the ratio $\sigma_1/\alpha\sigma_2$ will, in general, decrease as the mercury intrusion volume increases. So, according to the discussion just given, the effective contact angle $\theta'$ for the first increments of intruding mercury (i.e., when $V \ll V_{max}$) will be only slightly dependent on $\theta_1$, and will be close to 180°. In any case, $\theta'$ will be greater than $\theta_1$; and $\theta'$ should be a decreasing function of $V$.

Rootare's derivation assumed constancy of $\theta_1$ throughout the intrusion process. It is reasonable to say that this approximation will be much less valid for some porous solids than for others; and for porous Vycor, it will not be correct, particularly for the walls of macropores. If $\theta$ is not a constant, equation (26) must be replaced by

$$A' = -\frac{1}{\gamma_{lv}} \int_0^{V_{max}} \frac{P \, dV}{\cos \theta} \qquad (31)$$

In this case, equation (27) will not be valid. If equation (27) were used, it would yield a weighted mean value of cos $\theta$, with the weighting function being very difficult to determine except in certain cases where the pore geometry was already known. (For mercury porosimetry, the mean value of a varying contact angle would be of little direct use for determining pore size distribution.)

It appears that a fruitful direction for research on contact angles in porous media will be the examination of the distribution of $\theta$ as a function of fractional void volume, for various geometries of the pore space. The result of such a study will be an improvement in the interpretation of the data of mercury intrusion porosimetry.

It is also worth mentioning that Good and Paschek[51] found, using the reflection method described above in Section 2.3.3, that the contact angle of mercury on glass is dependent on the partial pressure of water in the gas phase, for fused silica and Pyrex glass. The advancing angle increased from 135° to 139° with an increase of relative humidity from 20 to 100%, and the retreating contact angle increased from 120 to 124° for the same humidity range. See reference 51 for details of these results. This work indicates that there is need for further study of the contact angle of mercury as a function of humidity, on various porous solids.

D. Rate of Capillary Penetration: The Washburn Equation. A relative method employing the unopposed penetration of a liquid into a powder bed shows considerable promise. Although, as has been pointed out,[99] there exists at present no completely satisfactory description of the process, a simple equation generally credited to Washburn[54] is useful. Washburn combined the capillary driving force for a cylindrical tube of radius $r$, equation (25), with the Poisseulle equation for viscous drag in conditions of steady flow. For the case when the tube is vertical and gravity opposes the flow, Washburn obtained

$$\frac{dh}{dt} = \frac{r^2}{8\eta h}\left(\frac{2\gamma_{lv}\cos\theta}{r} - \Delta\rho\, gh\right) \tag{32}$$

where $\eta$ is viscosity, $\Delta\rho$ the difference in density between the liquid and the surrounding medium, and $h$ the height of penetration at time $t$.

If the radius of the capillary or, in the case of a powder bed, an effective capillary radius is known, the contact angle $\theta$ may be obtained from the rate of penetration $dh/dt$. The effective radius is obtained by making the same measurement with a liquid that is known to have zero contact angle, and assuming the pore structure and the penetration process to be the same in the runs with the different liquids.

If the bed or capillary is horizontal, the term $\Delta\rho\, gh$ in equation (32) is absent. We can neglect this term with a vertical bed if the pore radius is small enough. In either case, we can write

$$\frac{dh}{dt} = \frac{r\gamma_{lv}\cos\theta}{4h\eta} \tag{33}$$

Since $r$ varies from point to point in a powder bed, the observed rate $dh/dt$ must correspond to an average value of $r$. Rather than dignify that quantity with a symbol such as $\bar{r}$ or calling it an effective radius, Ely and Pepper.[100] suggested replacing it by a "tortuosity constant" $K$ and writing

$$\frac{dh}{dt} = \frac{K\gamma_{lv}\cos\theta}{h\eta} \tag{34}$$

$K$ is, as already noted, determined from the rate measured with a liquid for which $\theta = 0$. Rearranging equations (33) and (34),

$$\cos\theta = \frac{4h\eta}{r\gamma_{lv}}\frac{dh}{dt} = \frac{h\eta}{K\gamma_{lv}}\frac{dh}{dt} \tag{35}$$

It may be noted that the dependence on $r$ is different in equation (32) from that in equation (33). This difference suggests that, if the two terms in the parentheses in equation (32) are of about the same size, it might be possible to determine contact angles by an appropriate combination of measurements with horizontal and vertical powder beds.

In principle, the Washburn method can be adapted to the study of liquid–liquid–solid contact angles. There does not appear to have been any demand, as yet, for this adaptation. It is to be expected that the flow process in the experiment would be slow and that problems of mutual saturation and of adsorption would arise that might interfere with the measurement.

Rate data in studies of this kind generally exhibit a serious statistical scatter[101,102] A major reason for this scatter is the change in structure of the packed bed with wetting, as mentioned above. One solution to this trouble is to work in the very early stages of the penetration process. The objection to this variant of the method has been raised, that the Washburn equation breaks down when $t$ approaches zero; in the limit $h \to 0$, $dh/dt \to \infty$ in equation (26). This breakdown is a mathematical consequence of omitting inertia and acceleration terms in deriving the equation. A more exact analysis was given[103] that includes such terms; it was found that in all cases of interest, the inertia forces decay in time intervals much smaller than 1 sec. This means that we may apply the Washburn equation even to earliest stages of the penetration process that are accessible in ordinary laboratory measurements.

Liquid penetration of powder beds may be studied gravimetrically. In the work of Weber and Neumann,[104] the bed, consisting of uniform, fine glass beads in a glass tube, was suspended from an electrobalance, and the liquid was raised until it made contact (cf. the Wilhelmy method described in

Section 2.2.1). Close control of the depth of immersion was found to be essential; this was accomplished by an automatic switch that turned off the motor just as soon as the contact was established and the voltage input to the recorder started to change. Values of the contact angles for the liquids were determined for glass (in plate form) of the same composition as the beads. Initial rates of penetration multiplied by viscosities were plotted against $\gamma_{lv} \cos \theta$. It was found that for measurements with various liquids and the same batch of glass beads (i.e., with constant $r$ or $K$), these plots resulted in a smooth curve for each batch, i.e., with constant $r$, as expected from the Washburn equation. The use of smaller or larger beads, i.e., changing the effective capillary radius, was found to change the rate of penetration appreciably only at relatively large values of $\gamma_{lv} \cos \theta$. This indicates that, at low values of $\gamma_{lv} \cos \theta$, the determination of the effective pore radius is not critical. The smoothness of the curves obtained with glass beads also indicated that there was no appreciable rate dependence of the contact angle in the range of rates that were observed. This is not always the case with penetration phenomena.[105] The conclusion from this study was a confirmation of the Washburn equation as a method for measuring contact angles.

Recently, Good and Lin[101,102,106] have shown that, with a powder that is initially devoid of an adsorbed film of the wetting liquid, the rate of penetration can be greater than that predicted by the Washburn equation. The theory of this effect is related to the "initial spreading coefficient" argument of Harkins[107]: In the absence of an adsorbed film, the solid contributes free energy $\gamma_s$, not $\gamma_{sv}$, to the driving force for penetration. If dissipative processes, such as diffusion ahead of the advancing liquid, are not as rapid as the advance of the liquid itself, then there is a contribution to the driving force for penetration, which is given numerically by the spreading pressure $\pi_e$. Experimental confirmation of this effect so far has been limited to systems in which $\theta = 0$. It will be interesting to establish whether an appreciable effect is also present with systems for which $\theta > 0$.

*2.4.2.3. Thermodynamic Methods of Studying the Wetting of Powders.* The heat of immersion $\Delta H^i$, which is the negative of the heat evolved per square centimeter of powder immersed in a liquid, is a measure of the wettability of the powder by the liquid.[108] It has been shown[109] that the heat of immersion is related to the contact angle and its temperature derivative:

$$\cos \theta = \frac{1}{\varepsilon_{lv}} \left( T \gamma_{lv} \frac{d \cos \theta}{dT} - \Delta H^i \right) \qquad (36)$$

where $\varepsilon_{lv}$ is the total surface energy of the liquid,

$$\varepsilon_{lv} = \gamma_{lv} + \frac{T d \gamma_{lv}}{dT} \qquad (37)$$

$\Delta H^i$ is negative for almost all liquid–solid system combinations. If the right side of equation (36) is greater than 1, that means that $\theta = 0$. For all known high-energy solids in all liquids that have been studied, $\Delta H^i$ is large and negative, and so it would be preducted that $\theta$ will be zero—which is in agreement with experiment.

In general, the temperature dependence of contact angles will be available only if the contact angle itself is already known, and while $d\theta/dt$ is generally small, it cannot be assumed to be identically zero except when $\theta = 0$. Therefore, the accurate determination of contact angles from heats of immersion is not feasible at this time. However, heat of immersion remains a valuable tool for comparing wettability and total surface energy of those surfaces on which the contact angle is zero.

### 2.5. Consolidated Porous Solids

A porous body may have considerable mechanical integrity. Sandstone and porous limestone are two natural examples; brick and other nonglazed ceramic bodies and porous Vycor are synthetic examples. For such solids the same conclusions, given above with respect to powders, apply. Contact angles measured directly on the external surface are of little use for studying the properties of the solid–vapor interface within the pores. The pore structure of a sintered powder or other consolidated porous solid does not change with time, so the drawback of changes in structure of a packed bed during wetting or penetration is absent. That is to say, for example, there is no change with exposure to the liquid during a run, so measurements in which the pore radius $r$ or the tortuosity constant $K$ is determined with a reference liquid are more likely to be valid than are measurements on powders.

## 3. Preparation of Liquids

We have already mentioned the fact that the purity of the liquids whose contact angle is being measured affects the measurements. A major, early achievement of the laboratory of W. A. Zisman, that helped make possible the advancement of contact angle studies to their present state, was to establish the fact that, for reproducible and scientifically significant work, ordinary laboratory-grade liquids simply are not pure enough. We have already noted above, the need for continuous purification by an electric method during vertical plate measurements.

Modern laboratory solvents are available, in many cases, that need little further purification. Spectrograde solvents are usually satisfactory for routine studies. However, for the most reliable results, it is necessary to employ purification steps such as passing an organic solvent through a bed of activated alumina or silica or charcoal. Zone refining and gas chromatography may be required in extreme cases. Protection from light in storage is necessary for most or all organic liquids. Suitable low-actinic bottles may be procured from the Corning Glass Works by special order. Purity should be checked from time to time. A relevant, easy test of purity is to measure the surface tension of the liquid. Probably the most valid test is to measure the interfacial tension versus water, because the impurities that can easily be generated or picked up, and that can do the most harm to the accuracy of wetting studies, are usually much more surface active at the oil–water interface than at the interface with air. See Chapter 3 in this volume for the interfacial tension method that is recommended the most highly.

## 4. Preparation of Solid Surfaces

Two distinct major uses of contact angle measurements must be identified at this point. One is to characterize *real* surfaces, and by this is meant, for example, surfaces that may or may not be homogeneous and smooth. Thus, for example, the experimenter might wish to study the effect of roughness or heterogeneity, or contamination on adhesion using the contact angle as a tool.

The other is to study the energetics of *ideal* solid surfaces, i.e., surfaces in which any lack of homogeneity or smoothness either does not affect the properties or else can be taken into account in some unequivocal manner. For the latter use, there is a standing need to develop methods of preparing surfaces that are free from contaminants and are also of controlled (and minimal) roughness and heterogeneity.

The question of sensitivity of contact angles to impurities that are present in trace amounts as fractional *surface* concentration can be given only a qualitative answer. The general answer must depend on the properties of the principal component and of the impurities, as well as those of the wetting liquid. For example, in a hydrocarbon surface (say, crystalline polyethylene), a patch of hydroxyl groups will have a far greater effect on wetting by water than will a patch of chlorine atoms. Moreover, a particular concentration of an impurity in a bulk phase may lead to different concentrations of that impurity at the different interfaces that bound that phase.

White[110] has recently given an extensive discussion of the effect of organic contaminants on the contact angle of water on high-energy solids such as mica, glass, and oxidized metals. He showed that surfaces such as alumina or oxidized Nichrome were, qualitatively, much more sensitive to atmosphere-borne contaminants than were mica or glass. He also discussed the effect of oxidation (as well as dehydration) on contact angles of water on solids such as metallic silicon, silica, and gold.

It should by this point be evident that the proper handling of real surfaces is a matter of the highest importance, and that this is particularly so for high-energy solids. Most obviously, surfaces should never be touched with bare fingers, since transferred organic material can affect the contact angle of any solid. Surgical rubber gloves that have been cleaned in chromic acid cleaning solution may be acceptable for handling low-energy solids, but even these are not recommended. Laboratory air must be well filtered, and sources of oil vapors (e.g., mechanical vacuum pumps) must be excluded. It may be desirable in critical cases to carry out sample preparation in a glove box. In extreme cases, elaborate precautions may be necessary, such as depositing or cleaning in a well baked-out vacuum system that is well protected from any sources of oil, stopcock grease, etc., and measuring the contact angle *in situ*.

For *routine* measurements, however, it is probably satisfactory to exercise normal precautions against touching the face of the sample and to be sure that sources of airborne oil in the laboratory are absent. It is always desirable to protect the sample from the laboratory atmosphere during measurement; for example, with the sessile drop method, the stage on which the sample is mounted should be in an enclosed chamber.

The following discussion illustrates selected aspects of surface preparation. A comprehensive discussion of surface preparation, here, is out of the question. Indeed, as already noted, one of the main purposes of contact angle measurements is to study this very question (e.g., the effect of surface preparation on composition and structure and on the properties that are related to these variables. Thus, it is possible to prepare Teflon surfaces (by etching with sodium aryl compounds) to which other substances can be effectively glued; without such treatment, the adhesion of any practical adhesive to Teflon is weak. The contact angle of water on this treated surface is low. On the other hand, a Teflon surface can be given a wide variety of treatments, ranging from gentle washing to soaking in chromic acid cleaning solution, with negligible effect on contact angles on the solid and without improvement of its poor characteristics as a substrate for adhesion.

Preparation of a surface by cleavage does not necessarily produce an ideal surface; indeed, probably the only case in which a cleavage surface resembles an ideal surface is mica[111] cleaved in vacuum. Energy of cleavage experiments show that when the experiment is carried out in air, an adsorbed film (nitrogen, oxygen, water, or other gas) forms very rapidly.[111] Also, there is evidence for locally nonuniform distribution of charge over a freshly cleaved mica surface[112] Finally, there is direct evidence[113] that irregular steps in the surface can be produced in the cleavage of mica, the step height corresponding exactly to the *c*-spacing of the unit cell. The atomically planar areas between steps are generally of the order of a few hundred angstroms to a few microns wide. There is sufficient variability in this surface structure that it is recommended for workers to expend some effort in characterizing the surface before claiming fundamental significance for their results on mica.

Other brittle solids besides mica may appear to cleave nicely, but even the best cleavage surfaces contain irregular steps that are much more closely spaced than those in mica. Moreover, considerable plastic deformation occurs in the vicinity of a propagating crack front. It is for this reason that the cleavage energy of a crystal is generally several orders of magnitude larger than the energy required for the simple formation of the new surface by the breaking of bonds. A fracture surface generally has a high density of dislocations. Even worse, cleavage surfaces are likely to be very active in chemisorption. It is well known that the fracture of diamond breaks covalent bonds, and that the resulting surface reacts very rapidly with air, water vapor, etc. It is less well known that surfaces such as cleaved MgO or Nacl are also very reactive.

Mechanical polishing produces a very heavily deformed surface whose properties depend on the history of the sample. The surface may also contain traces of the polishing agent. Annealing a polished sample causes the surface to recrystallize, and often to become faceted. Chemical etching commonly roughens a surface by preferential attack at grain boundaries and dislocations. Electropolishing is a technique that, for metals, can produce a smooth surface that is free from the deformed layer left by mechanical polishing. The field-enhanced dissolution can combat the tendency of the chemical agents present to attack dislocations and grain boundaries, and in addition, it will preferentially remove peaks and ridges. This method can also be used to remove the deformed layer from a cleavage surface.

An important limitation with electropolishing is, of course, the fact that it must be done in contact with water or some other conducting solvent. Since the metal is the anode, there is the possibility of leaving at least a monolayer of oxide. In some cases, the oxide can subsequently be reduced, e.g., by heating with hydrogen, but then there is danger of producing a surface metal hydride or of depositing discrete crystals on the flat surface. Etching with a reagent that dissolves the oxide but not the metal is a better procedure. There seems to be no general solution to this dilemma.

Casting a solid from the molten state using a polished surface in the mold is an obvious method of preparing smooth surfaces. But it has serious drawbacks. If the polished surface is free from adsorbed material, the object that is being molded is very likely to adhere strongly to it. If it separates from the mold at all, it does so by a fracture mechanism if the solid is brittle, or by ductile failure if the solid is not brittle. In either case, the surface that results is chemically active—a high-energy surface that reacts with (chemisorbs) components from the air, including water vapor. Such reactions produce a surface that is likely to be heterogeneous and of a composition that depends on the atmosphere to which it has been exposed; and in topography, it will be quite rough.

The standard method of producing clean release of the solid from the mold surface is to use a mold-release or parting agent. Such an agent usually functions by being weak as regards its own cohesion, so when the solid comes out of or off the mold, it has on its surface a layer of foreign matter, which must be elimated if the true surface of the solid is to be studied.

Moreover, the solid surface is likely to have been affected by the surface against which it was formed, whether bare solid or release agent. One way to study this effect is to mold the solid against a thin sheet of the solid with the polished surface (e.g., to use a metal foil) in the absence of any release agent, and then to dissolve the foil on other mold material away chemically. Schonhorn[114,115] has used this technique with certain polymers; he found that the wetting behavior depended upon the nature of the mold surface he employed.

Various cleaning methods have been employed to remove impurities so that the true surface nature of a solid can be studied. One would want, for example, to remove films of the mold-release agents mentioned above, or lubricants that had been added in order to facilitate rolling of a film or foil. A few words of caution, however, are in order. First, a study of contact angles on cleaned surfaces may develop into a piece of serious research on cleaning methods, before a way of preparing a reproducible surface is found; it would be naive to assume that the optimum method has been chosen without comparing at least two methods, even when there is precedent in the literature for a particular choice. Some methods that have been employed, as reported in the literature, include washing with mild or strong detergents, soaking in chromic acid cleaning solution, extracting with solvent, and heating. The experimenter must, of course, be wary of unexpected changes in the solid. For example, Good *et al.*[55] concluded, on the basis of scanning electron microscopic examination, that acetone (which had been used to extract Teflon FEP film prior to contact angle measurements) was causing the roughness of the surface to increase in a peculiar manner.

For solids that can be evaporated without decomposition, it may be possible to prepare nicely reproducible surfaces by condensation. The roughness of the resulting surface varies with the experimental conditions, e.g., temperature and rate, and some effort may be required in order to find how to produce smooth surfaces. The scanning electron microscope is an invaluable adjunct for this purpose. The following description is based on experiments by one of the authors.[77]

A setup to produce smooth solid surfaces of organic materials is shown in Figure 16. In the vacuum chamber consisting of the base plate (1) and the

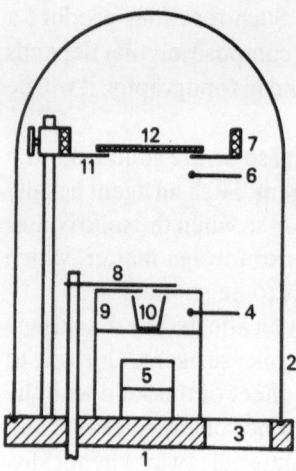

Figure 16. Vacuum coater for depositing smooth and homogeneous solid surfaces; (1) base plate, (2) bell jar, (3) vacuum connection, (4 and 6) thermocouples, (5) heating element, (7) heater for substrate holder, (8) removable lid, (9) Knudsen cell, (10) crucible, (11) substrate holder, (12) substrate.

bell jar (2), a Knudsen furnace is mounted above the heating wires (5). The organic material is contained in a small ceramic beaker, which is placed underneath the aperture of the Knudsen furnace; the aperture has a diameter of about 0.5 cm. Since smooth surfaces are usually only obtained if the rate of evaporation is kept constant, the aperture is kept closed by means of the movable cover (8) until the thermocouple (4) indicates constant temperature. When the Knudsen furnace is opened, the vapor stream impinges on the surface of the substrate (12). (A glass plate is often used as the substrate, because natural glass surfaces are the most conveniently available solid surfaces that are sufficiently smooth for these purposes.) The substrate rests on the specimen holder (11). Since the smoothness of the evaporated films also depends on the substrate temperature, the temperature of the glass plate is adjusted by means of the heating device (7) and measured using the thermocouple (6). Optimum temperatures of the substrate in the Knudsen furnace must be found empirically for each substance. Substrate specimens of about 2 × 3 cm are quite convenient.

In general, evaporated films of satisfactory quality can be obtained only on very clean glass slides; therefore, a glass plate must be cleaned very thoroughly prior to deposition of the film. This may be done, for example, by first cleaning the plate mechanically in order to rid it of particulate dirt, then treating it in hot chromic acid to remove organic contaminants, then in distilled water, and finally in a high-purity organic liquid of high vapor pressure such as methanol or acetone. In some cases, it is advisable to keep the cleaned glass plate in this last liquid until immediately before introducing it into the vacuum chamber in order to minimize contamination in the atmosphere. The liquid clinging to a glass plate after it is removed from the last bath must be removed quickly, since dust particles impinging on the liquid surface will be deposited on the solid when the liquid evaporates. Such deposits sometimes prevent the formation of satisfactory evaporated layers. (Treating glass and other substrates in a radio-frequency glow discharge is an alternative means of removing organic contaminants from solid surfaces.) For nonpolar substances such as hexatriacontane[35] or organic compounds consisting of comparatively large molecules, e.g., cholesteryl acetate,[34] or organic pigments,[70,77] the vapor deposition procedure has been found to yield very smooth surfaces.

In cases where there is strong adsorption of the molecules of the subtrate that is to be investigated, other and simpler techniques are often satisfactory. Certain silicone compounds may be adsorbed onto glass slides in such a way that the contact angle hysteresis is negligible[36,117] and that the surface may, as far as wetting is concerned, be considered to be homogeneous. Long-chain organic acids, alcohols, and amines may be adsorbed from solution onto glass or metals. Smooth metal surfaces for use as substrates for this deposition may be obtained by evaporation of films in vacuum onto glass plates. When mechanical polishing of a substrate is employed, care must be taken that no abrasive is left on the solid surface.

The deposition of long-chain polar organic materials may be done in two different ways. The high energy solid may be immersed either (a) into the melt of the organic substance or (b) into a solution of the organic material in a solvent that has a surface tension above the $\gamma_c$ of the organic film[2] and retracted slowly. In either case, if the liquid phase does not wet the deposited film, a monolayer is left on the solid. The mechanism of this deposition, for forming a monomolecular film, depends on the adsorption of the polar group on the high-energy surface of the solid substrate and the dense packing of the nonpolar ends of the molecules. The final, low-energy solid surface is made up of the nonpolar ends (e.g., terminal $CH_3$ or $CF_3$ groups) of the adsorbed molecules. This simple and (for these substances) efficient method of preparing solid surfaces was devised by Zisman and his co-workers.[118–125]

Such films are not always extremely stable and sometimes are partly removed by a test liquid. This partial removal may be recognized by the existence of very large hysteresis, and by the fact that, after the surface has been exposed to a liquid, the advancing contact angle in a second measurement is different from that in the first test.

The methods for preparing solid surfaces described so far are applicable only to metals or low molecular weight organic solids; polymer surfaces cannot be prepared in these ways. Thermoplastic polymers can be pressed at elevated temperatures, between clean glass plates, in order to obtain smooth surfaces for contact angle measurements. For polytetrafluoroethylene and other fluorocarbon polymers, temperatures of 150°C and pressures of about 1000 psi have proved to be satisfactory.[1] However, it is sometimes found that the polymer adheres to the glass strongly enough that, when the polymer is separated from it, the glass is covered with a layer of the polymer.[126] As already noted, the polymer surface that results must be a fracture surface, and so cannot be considered characteristic of the polymer *per se*. Pressing the polymer between siliconed glass plates may reduce this difficulty. Polishing of polymers is possible in some cases if they are fairly hard. Care must be taken that no traces of abrasive be left on the surface of the polymer. Satisfactory results have been obtained by polishing poly-tetrafluoroethylene on a roughened glass plate.[37]

The surface of a polymer formed in contact with air may be smooth and homogeneous enough to yield well-defined contact angle values. This is often the case with glossy paint or varnish films.[33] Solidification in contact with air, however, must not be taken as automatically leading to a smooth surface, even for a glossy polymer, without independent verification. Cross-linked polymers present a special difficulty, since deformation, such as by polishing, must break covalent bonds. Measurements have recently been made on polystyrene cross-linked with divinylbenzene polymerized in contact with mercury, with glass, and with air.[127] It was found that there was an appreciable difference between contact angle results for the three types of contacting surface.

A centrifuge spinning technique has been used extensively in the laboratory of one of the authors (A.W.N.) in the preparation of smooth surfaces of polymers, proteins, and occasionally low molecular weight solids. In principle, the technique can be used whenever the solid of which a smooth surface is to be prepared can be dissolved in a volatile solvent. Solute concentrations of the order of 1% usually prove satisfactory. A suitable substrate, e.g., a clean glass microscope slide, is mounted horizontally on the rotor of a small tabletop centrifuge; the centrifuge is set at a moderate speed; and a drop of the solution is deposited in the center of the glass slide. The centrifugal force spreads the solution in a thin layer over the glass slide, the solvent evaporates, and a thin film of the solute remains. In some instances, polymer films were obtained that were so smooth that no surface features could be recognized under the scanning electron microscope. The choice of the substrate is arbitrary as long as it has a smooth surface and is wettable by the solution. On very hydrophobic substrates, such as siliconed glass, thin films cannot be obtained by this technique.

## 5. Standards

Completely satisfactory standards of equilibrium contact angles are not yet available. The main difficulties arise from our as yet incomplete understanding of contact angle hysteresis. However, some contact angles have

*Table 2. Contact Angles of* n-*Alkanes on Teflon at Room Temperature*

| Alkane | $\theta_a$, degrees[116] | $\theta$, degrees[11]a |
|---|---|---|
| Hexadecane | 46.0 | 46 |
| Tetradecane | 44.8 | 44 |
| Dodecane | 42.7 | 42 |

[a] Contact angle measured by sessile drop method, with deposition of drop by Zisman's method; see text.

been repeatedly observed and therefore may be taken as a useful guide when setting up an apparatus. The systems that have been investigated the most often are polytetrafluoroethylene–higher $n$-alkanes. Typical results from two different laboratories for the advancing contact angle at room temperature are given in Table 2. These were obtained with the methods of capillary rise at a vertical plate,[116] and of the sessile drop.[1] Contact angles with short-chain $n$-alkanes depend to some degree on the experimental conditions (see above, and also Neumann *et al.*[116]). They are not suitable as standards. Contact angle measurements on Teflon using liquids of relatively high surface tension tend to be less reproducible than those with higher alkanes.

The observed lack of hysteresis leads to the conclusion that the contact angle of water on the solids $n$-hexatriacontane[35] and cholesteryl acetate[34] prepared by vacuum evaporation and siliconed glass[36] should qualify as standards. See Table 3 and also the discussion in Chapter 1, of the absence of adsorption of water on ideal, low-energy surfaces.

## 6. Note on Liquid–Liquid–Solid Systems

The material above has been written from the viewpoint of measurements in gas–liquid–solid systems. With one exception, everything can be carried over to liquid–liquid–solid systems. The exception is connected with hysteresis—the contact angle hysteresis is, in general, very much greater.

*Table 3. Equilibrium Contact Angles of Water on Different Solid Surfaces at 25°C*

| Surface | $\theta$, degrees |
|---|---|
| $n$-Hexatriacontane[35] | 105.3 |
| Cholesteryl acetate[34] | 103.1 |
| Siliconed glass[36] | 107.4 |

The reason this is so is not known. One possibility is that fluid-mechanical effects associated with the discontinuities of properties (density, viscosity, etc.) at the two-phase interfaces and at the three-phase line may be much more serious when the two fluid phases are of comparable density and viscosity than when one phase is a gas. Another is that adsorption–desorption kinetic effects should be much more important (cf. Chapter 1).

The displacement of one liquid from a solid, by another liquid, is of obvious importance in detergency, in oil recovery (by the waterflooding method), and in many other applications. Hence, it is to be expected that the direct measurement of liquid–liquid–solid contact angles may receive considerably more attention in the future.

## References

1. H. W. Fox and W. A. Zisman, *J. Colloid Sci.* **5**, 520 (1950).
2. W. A. Zisman, *Advan. Chem. Ser.* **43**, (1964).
3. R. H. Ottewill, quoted by A. W. Adamson, *The Physical Chemistry of Surfaces* (3rd ed.), p. 343, Wiley Interscience, New York (1976).
4. R. J. Good, *J. Colloid Interface Sci.* **52**, 308 (1975).
5. J. J. Bikerman, *Surface Chemistry* (2nd ed.), p. 343, Academic Press, New York (1958); *Ind. Eng. Chem. Anal. Ed.* **13**, 443 (1941).
6. J. F. Padday, in *Surface and Colloid Science* (E. Matijevic, ed.), Vol. 1, p. 151, Wiley Interscience, New York (1969).
7. R. E. Baier, personal communication.
8. F. Neumann, *Vorlesungen über die Theorie der Capillarität*, B. G. Teubner, Leipzig (1893).
9a. A. W. Neumann and R. J. Good, *J. Colloid Interface Sci.* **38**, 341 (1972).
9b. L. Boruvka and A. W. Neumann, *J. Colloid Interface Sci.* **65**, 315 (1978).
10. G. McDougall and C. Ockrent, *Proc. Roy. Soc.* London **A180**, 151 (1942).
11. V. R. Gray, *For. Prod. J.* **12**, 425 (1962); *Chem. Ind.* 969 (1965).
12. G. W. Longman and R. P. Palmer, *J. Colloid Interface Sci.* **29**, 185 (1967).
13. G. J. Jameson and M. C. G. Del Cerro *J. Chem. Soc. Faraday Trans. I* **72**, 833 (1976).
14. P. S. de Laplace, *Mechanique Celeste, Supplement to Book 10*, J. B. M. Duprat, Paris (1806).
15. F. Bashforth and J. C. Adams, *An Attempt to Test the Theory of Capillary Action*, Cambridge University Press, Cambridge, England (1892).
16. C. A. Smolders, Ph.D. Thesis, Rijksuniversiteit, Utrecht (1961).
17. C. A. Smolders and J. Th. G. Overbeek, *Rec. Trav. Chim.* **80**, 635 (1961).
18. R. Ehrlich, *J. Colloid Interface Sci.* **28**, 5 (1968).
19. D. N. Staicopolus, *J. Colloid Sci.* **17**, 439 (1962); **18**, 793 (1963); **23**, 453 (1967).
20. K. G. Parvatikar, *J. Colloid Interface Sci.* **23**, 274 (1967).
21. E. Lefebre du Prey, *Rev. Inst. Français Petrole*, 701 (1969).
22. C. Maze and G. Burnet, *Surface Sci.* **13**, 451 (1969).
23. J. F. Padday, *Proc. Roy. Soc. London* **A330**, 561 (1972).
24. J. F. Padday, *J. Colloid Interface Sci.* **48**, 170 (1974).
25. R. S. Burdon, *Surface Tension and the Spreading of Liquids*, p. 15. Cambridge University Press, Cambridge, England (1949).

26. L. Wilhelmy, *Ann. Physik* **119**, 177 (1863).
27. A. W. Neumann, Ph.D. Thesis, University of Mainz (1962); *Z. Physik. Chem. N. F.* **41**, 339 (1964); **43**, 71 (1964).
28. D. D. Jordan and J. E. Lane, *Austral. J. Chem.* **17**, 7 (1964).
29. H. M. Princen, *Austral. J. Chem.* **23**, 1789 (1970); *J. Colloid Interface Sci.* **30**, 69 (1969); **30**, 359 (1969).
30. J. Guastalla, *J. Chim. Phys.* **51**, 583 (1954); *II Intern. Congr. Surface Activity* (*London*), Vol. **3**, p. 143 (1957).
31. R. E. Johnson, Jr., and R. H. Dettre, in *Surface and Colloid Science* (E. Matijevic and F. R. Eirich, eds.), Vol. 2, p. 85, Academic Press, New York (1969).
32. H. M. Princen, in *Surface and Colloid Science* (E. Matijevic and F. R. Eirich, eds.), Vol. 2, p. 1, Academic Press, New York (1969).
33. W. Funke, G. E. H. Hellwig, and A. W. Neumann, *Angew. Macromolec. Chem.* **8**, 185 (1969).
34. G. E. H. Hellwig and A. W. Neumann, *Kolloid-Z. Z. Polym.* **229**, 40 (1969).
35. G. E. H. Hellwig and A. W. Neumann, *V Intern. Congr. Surface Activity, Barcelona, 1968*, Vol. 2, p. 687, Ediciones Unidas (1969).
36. A. W. Neumann and D. Renzow, *Z. Physik. Chem. N. F.* **68**, 11 (1969).
37. A. W. Neumann and W. Tanner, *J. Colloid Interface Sci.* **34**, 1 (1970).
38. A. W. Neumann, *Advan. Colloid Interface Sci.* **4**, 105 (1974).
39a. A. C. Hardy and F. H. Perrin, *The Principles of Optics*, p. 130, McGraw-Hill, New York (1932).
39b. G. L. Kehl, *Principles of Metallographic Laboratory Practice*, (3rd ed.), McGraw-Hill, New York (1952).
40. A. J. T. Scarr, *Metrology and Precision Engineering*, Chapter 5, McGraw-Hill, New York (1967).
41a. A. A. Michelson, *J. Opt. Soc. Amer. Rev. Sci. Instr.* **8**(1), 321 (1924).
41b. A. A. Michelson, *Studies in Optics*, University of Chicago Press, Chicago (1927).
42. A. W. Neumann and W. Tanner, *Tenside* **4**, 220 (1967).
43. J. Kloubek and A. W. Neumann, *Tenside* **6**, 4 (1969).
44. N. K. Adam and G. Jessop. *J. Chem. Soc.* **1925**, 1863 (1925).
45. N. K. Adam, *The Physics and Chemistry of Surfaces* (3rd ed.), Oxford University Press, Oxford, England (1941).
46. F. M. Fowkes and W. D. Harkins, *J. Amer. Chem. Soc.* **62**, 337 (1940).
47. A. L. Spreece, C. P. Rutkowski, and G. L. Gaines, Jr., *Rev. Sci. Instr.* **28**, 636 (1957).
48. I. Langmuir and V. J. Schaeffer, *J. Am. Chem. Soc.* **59**, 2400 (1937).
49. T. Fort, Jr., and H. T. Patterson, *J. Colloid Sci.* **18**, 217 (1963).
50. R. J. Good and G. V. Ferry, *Advan. Cryogen. Eng.* **8**, 306 (1963).
51. R. J. Good and J. K. Paschek, in *Wetting, Spreading and Adhesion* (J. F. Padday, ed.), p. 147, Academic Press, New York (1978).
52. W. D. Harkins and A. E. Alexander, in *Physical Methods of Organic Chemistry* (A. Weissberger ed.) (3rd ed.), Vol I, part I, p. 757, Interscience, New York (1959).
53. F. E. Bartell and H. Y. Jennings, *J. Phys. Chem.* **38**, 495 (1934).
54. E. W. Washburn, *Phys. Rev.* **17**, 374 (1921).
55. R. J. Good, J. A. Kirkstad, and W. O. Bailey, *J. Colloid Interface Sci.* **35**, 314 (1971).
56. C. Huh and L. E. Scriven, *J. Colloid Interface Sci.* **30**, 323 (1969).
57. A. M. Schwartz and F. W. Minor, *J. Colloid Sci.* **14**, 572 (1959).
58. A. M. Schwartz and C. A. Rader, *IVth Intern. Congr. Surface Activity, Brussels, 1964*, Vol. 2, p. 383, Gordon and Breach, New York (1964).
59. W. D. Bascom and J. B. Romans, *Ind. Eng. Chem. Prod. Res. Develop.* **7**, 172 (1968).
60. R. J. Roe, *J. Colloid Interface Sci.* **50**, 70 (1975).

61. B. J. Carroll, *J. Colloid Interface Sci.* **57**, 488 (1976).
62. N. K. Adam and H. L. Schute, *Symposium on Wetting and Detergency, 1937*, p. 53, Chemical Publishing Company, New York (1937).
63. A. V. Rapacchietta, A. W. Neumann, and S. N. Omenyi, *J. Colloid Interface Sci.* **59**, 541 (1977).
64. A. V. Rapacchietta and A. W. Neumann, *J. Colloid Interface Sci.* **59**, 555 (1977).
65. A. W. Neumann and W. Tanner, *Vth Internat. Conference on Surface Activity, Barcelona 1968*, Vol. 2, p. 727, Ediciones Unidas, Barcelona (1969); A. W. Neumann, *Chemie. Ing.-Tech.* **42**, 969 (1970).
66. W. C. Jones and M. C. Porter, *J. Colloid Interface Sci.* **24**, 1 (1967).
67. E. Weber, *Habilitationsschrift*, University of Stuttgart (1968).
68. K. L. Wolf, *Physik und Chemie der Grenzflächen*, Vol. 2, p. 322, Springer, Heidelberg (1959).
69. T. Steudel, *IVth International Congress Surface Activity, Brussels, 1969*, Vol. 2, p. 1251, Gordon and Breach, New York (1964).
70. G. E. H. Hellwig and A. W. Neumann, *Farbe Lack* **73**, 823 (1967).
71. H. Freundlich, O. Schmidt, and G. Lindau, *Kolloid-Beiheft* **36**, 43 (1932).
72. H. Freundlich, O. Enslin, and G. Lindau, *Kolloid-Beiheft* **37**, 242 (1933).
73. E. Sauer and W. Gussmann, *Kolloid Z.* **82**, 253 (1938).
74. M. von Stackelberg and K. H. Frangen, *Forschungsbericht des Landes Nordrhein-Westfalen*, no. 166 (1955).
75. F. Kindervater, *Farbe Lack* **69**, 21 (1965).
76. R. Shuttleworth and G. L. J. Bailey, *Discuss Faraday Soc.* **1948**(3), 16, (1948).
77. A. W. Neumann, D. Renzow, H. Reumuth, and I.-E. Richter, *Fortschr. Kolloide Polym.* **55**, 49 (1971).
78. N. W. F. Kossen and P. M. Heertjes, *Chem. Eng. Sci.* **20**, 593 (1965).
79. F. E. Bartell and H. J. Osterhof, *Ind. Eng. Chem.* **19**, 1277 (1927).
80. F. E. Bartell and H. J. Osterhof, *J. Phys. Chem.* **34**, 544 (1930); *J. Phys. Chem.* **37**, 543 (1933).
81. F. E. Bartell and C. E. Whitney, *J. Phys. Chem.* **36**, 3115 (1933).
82. N. S. Davies and H. A. Curtis, *Ind. Eng. Chem.* **24**, 1137 (1932).
83. H. L. Ritter and L. C. Drake, *Ind. Eng. Chem., Anal. Ed.* **17**, 782 (1945).
84. L. C. Drake and H. L. Ritter, *Ind. Eng. Chem., Anal. Ed.* **17**, 787 (1945).
85. L. C. Drake, *Ind. Eng. Chem.* **41**, 780 (1949).
86. L. G. Joyner, E. P. Barrett, and R. Skold, *J. Amer. Chem. Soc.* **73**, 3155 (1951).
87. C. Orr, *Powder Technol.* **3**, 117 (1969–70).
88. C. Orr and J. M. Dalla Valle, *Fine Particle Measurement*, Macmillan, New York (1959).
89. S. Diamond, *Cement Concrete Res.* **1**, 531 (1971).
90. H. M. Rootare and C. F. Prenzlow, *J. Phys. Chem.* **71**, 2733 (1967).
91. H. M. Rootare, *Aminco Laboratory News*, Aminco Reprint, no. 439 (1968).
92. H. M. Rootare and J. Suencer, *Powder Tech.* **6**, 17 (1972).
93. H. M. Rootare, personal communication.
94. S. Brunauer, *The Adsorption of Gases and Vapors*, Princeton University Press (1945).
95. R. J. Good and R. Sh. Mikhail, in preparation.
96. R. Sh. Mikhail and R. J. Good, in preparation.
97. A. B. D. Cassie and S. Baxter, *Trans. Faraday Soc.* **40**, 546 (1944).
98. A. B. D. Cassie, *Disc. Faraday Soc.* **3**, 11 (1948).
99. V. Ludviksson and E. N. Lightfoot, *AIChE J.* **14**, 674 (1968).
100. D. D. Ely and D. C. Pepper, *Trans. Faraday Soc.* **42**, 697.
101. R. J. Good and N. J. Lin, *J. Colloid Interface Sci.* **54**, 52 (1976).

102. R. J. Good and N. J. Lin, in *Proceedings of the 50th Colloid and Interface Science Symposium, Colloid and Interface Sci.* (M. Kerker, ed.), Vol. 3, p. 277, Academic Press, New York (1976).
103. J. Szekely, A. W. Neumann, and Y. K. Chuang, *J. Colloid Interface Sci.* **35**, 273 (1971).
104. E. Weber and A. W. Neumann, *Giesserei* **56**, 628 (1969).
105. W. Rose and R. W. Heins, *J. Colloid Sci.* **17**, 39 (1962).
106. R. J. Good, *J. Colloid Interface Sci.* **42**, 473 (1973).
107. W. D. Harkins, *The Physical Chemistry of Surface Films*, Reinhold, New York (1952).
108. W. D. Harkins and G. Jura, *J. Amer. Chem. Soc.* **66**, 1362 (1944).
109. R. J. Good and L. A. Girifalco, *J. Phys. Chem.* **62**, 1418 (1958).
110. M. L. White, in *Clean Surfaces* (G. Goldfinger ed.), p. 361, Marcel Dekker, New York (1969).
111. A. I. Bailey, *DECHEMA Monogr.* **51**, 210 (1964).
112. A. I. Bailey, private communication.
113. F. P. Bowden and D. Tabor, *The Friction and Lubrication of Solids*, Vol. 1, Clarendon Press, Oxford (1958).
114. H. Schonhorn, *J. Polymer Sci.* **B5**, 919 (1967).
115. H. Schonhorn, *Macromolecules* **1**, 145 (1968).
116. A. W. Neumann, G. Haage, and D. Renzow, *J. Colloid Interface Sci.* **35**, 379 (1971).
117. W. J. Herzberg, J. E. Marian, and T. Vermeulen, *J. Colloid Interface Sci.* **33**, 164 (1970).
118. W. C. Bigelow, D. L. Pickett, and W. A. Zisman, *J. Colloid Sci.* **1**, 513 (1946).
119. W. C. Bigelow, E. Glass, and W. A. Zisman, *J. Colloid Sci.* **2**, 563 (1947).
120. E. G. Shafrin and W. A. Zisman, *J. Colloid Sci.* **4**, 571 (1949).
121. H. R. Baker, E. G. Shafrin, and W. A. Zisman, *J. Phys. Chem.* **56**, 405 (1952).
122. E. G. Shafrin and W. A. Zisman, Hydrophobic monolayers and their adsorption from aqueous solution, in *Monomolecular Layers*, H. Sobotka (ed.), AAAS Symposium Publication, Washington, D.C. (1954).
123. O. Levine and W. A. Zisman, *J. Phys. Chem.* **61**, 1068 (1957).
124. E. G. Shafrin and W. A. Zisman, *J. Phys. Chem.* **64**, 519 (1960).
125. C. O. Timmons and W. A. Zisman, *J. Phys. Chem.* **69**, 984 (1965).
126. R. J. Good and N. W. Han, unpublished work.
127. R. J. Good and E. D. Kotsidas, *J. Adhesion* (1978), in press; E. D. Kotsidas, M.S. Thesis, State University of New York at Buffalo (1977).

# 3

# Pendant Drop Technique for Measuring Liquid Boundary Tensions

*Durga S. Ambwani and Tomlinson Fort, Jr.*

## 1. Introduction

### 1.1. Fundamental Principles and Outstanding Advantages

Boundary tension tends to make a liquid drop hanging (pendant) from the tip of a capillary tube take the form of a sphere. In a gravitational field, the sphere is distorted into an elongated "tear drop" shape. This distortion increases with effective drop density and decreases as liquid boundary tension becomes greater. Consequently, it is possible to calculate boundary tension if gravity and liquid densities are known and drop shape can be accurately measured. These measurements and calculations are the subject of this report.

The pendant drop technique is neither the oldest nor most used method for determining boundary tension. However, it has outstanding advantages,[1] some of which are unique:

1. Very accurate (±0.15% or less) boundary tension measurements can be made.

_Durga S. Ambwani_ • Shell Development Company, Emeryville, California 94608. _Present address_: Asia Development Corporation, Kaiser Center, Oakland, California 94612. _Tomlinson Fort, Jr._ • Department of Chemical Engineering, Carnegie-Mellon University, Pittsburgh, Pennsylvania 15213.

2. Complete mathematical analysis of the relation of boundary tension and drop shape is available. Boundary tension may be calculated without recourse to empirical correction factors.
3. Measurements may be made rapidly.
4. The locus of the measurement is removed from any point of contact with materials other than the fluids.
5. Successive measurements of a given interface may be made without disturbing that surface.
6. Boundary tensions of nearly any magnitude may be observed.
7. Photographs of the drop from which measurements are made may serve as a permanent record.
8. Only small samples are required.
9. The shape of the drop is independent of the contact angle made by the hanging drop with the capillary tip from which it hangs.

The only apparent disadvantages are:

1. The lowest surface age that can be investigated is approximately 5 sec.
2. Some setup effort and expense is necessary to make accurate measurements.

In addition, no commercial pendant drop equipment is available and there has been no practical and current guide to use of the method.

## 1.2. Applications of Pendant Drop Technique

The pendant drop method of measuring boundary tension has found increasing acceptance in recent years because of its outstanding advantages. It has been used to study both liquid–vapor and liquid–liquid boundary tensions. It has been applied to materials ranging from organic liquids to molten metals and salts and from pure solvents to concentrated solutions. While most of the work has been carried out at normal temperatures and pressures, the method has been applied successfully in vacuum and at high pressures and temperatures. Almost everyone who has worked with the technique seems to agree that it is an exact and accurate method for the determination of boundary tension. Padday,[2] while comparing various methods of measuring boundary tension, concluded that the pendant drop is a true equilibrium method and is particularly suitable for measuring the boundary tension between two liquids. Douglas[3] has cited the advantages of the method, as have Hauser and Michaels.[4] The present authors' experience with the technique has convinced them of its utility and they hope this report will lead to its wider use as a convenient, accurate, and primary method for the evaluation of liquid boundary tensions.

## 2. Previous Experimental Studies

### 2.1. Surface Tension Studies

Andreas *et al.*[1] used the pendant drop technique in 1938 to measure tensions of 10-sec-old surfaces of benzene, ethanol, methanol, and toluene. Interfacial tensions of mercury–benzene, mercury–water, water–carbon tetrachloride, and water–toluene systems were also determined. The agreement with the literature was fairly good. While they claimed an accuracy of 1.0% for surface tensions, they believed that the method was capable of an accuracy of 0.1%. Since that time, pendant drop technique has been modified and improved for special studies and better accuracy.

Smith and Sorg[5] and Smith[6] also used the pendant drop method to determine the surface tensions of organic liquids. Smith and Sorg measured surface tensions of normal aliphatic alcohols from $C_1$ to $C_{12}$ and the readily obtainable isomers below $C_6$. They attributed values that were higher than literature values to ultrapure liquids. The surfaces they photographed were 10 sec old. Smith determined surface tensions of fifteen highly purified hydrocarbons in the range $C_5$ to $C_8$. He felt that his precision was that of the empirical $H$–$S$ (see Section 3) tables then available.

### 2.2. Interfacial Tension Studies

Several workers have applied the method to interfacial tensions. Donahue and Bartell[7] studied water–organic liquid interfaces. Interfacial tensions in these systems were found to be a linear function of the logarithm of the degree of miscibility of water with the organic liquids. Bartell and Davis[8] studied the interfacial tensions of water–$n$-heptane, water–benzene, and water–$n$-amyl ketone systems. Bartell and Niederhauser[9] used the pendant drop technique to measure interfacial tensions between crude petroleum oils and water. Recent applications of the method[10,11] include studies of oil–water systems whose interfacial tension may be as low as $10^{-4}$ dyn/cm.

While studying the hysteresis of contact angles in mercury–benzene–water systems, Bartell and Bjorklund[12] found a time effect in the systems mercury–benzene saturated with water and mercury–water saturated with benzene. Bartell and Bard[13] determined the effects of atmospheric gases on the interfacial tension between mercury and several different liquids. Mercury–heptane interfacial tension was constant when heptane contained either pure nitrogen, pure oxygen, or pure carbon dioxide. The interfacial tension decreased if heptane was saturated with water or with atmospheric air.

Butler,[14] in a study of contact angles and interfacial tensions in the mercury–water–benzene system, found results that were quite contrary to those obtained by Bartell and Bjorklund.[12] Interfacial tension between mercury and water saturated with benzene changed with temperature as did the interfacial tension between mercury and benzene saturated with water and as did the interfacial tension between water and benzene. None of the interfacial tensions changed with time. The interfacial tensions between mercury and benzene and between mercury and water did not change with time or with temperature over the range 10–50°C.

### 2.3. Studies of Highly Viscous Materials

Roe[15,16] has successfully applied the pendant drop technique to the determination of surface tensions of homopolymers above their melting points and the interfacial tension between polymer liquids. The polymers studied were, in order of decreasing surface tension: polyethylene oxide, linear polyethylene, branched polyethylene, polyisobutylene, polypropylene, and polydimethylsiloxane. The pendant drop method is inherently suited for polymer liquids, because the liquids involved are very viscous and therefore the attainment of equilibrium constitutes the main limiting factor in obtaining reliable data.

Davis and Bartell[17] applied the pendant drop technique to solidified melts. They assumed that the shape of a molten drop did not change upon cooling, although the volume and density did change. This assumption appeared to be justified in the case of isotropic substances. The procedure was to heat a small amount of solid or viscous material until it melted or flowed sufficiently to form a pendant drop. The drop was allowed to solidify, then drop dimensions were measured at room temperature. This method was used for glasses, waxes, polystyrene, electrolytic iron, lead oxide, antimony trioxide, and lead chloride. The glasses and waxes gave values that agreed with literature values.

### 2.4. Studies in Vacuum and at High Temperatures and Pressures

In 1951, Michaels and Hauser[18] reported investigations of interfacial tensions measured at high pressures and temperatures. They studied the benzene–water and *n*-decane–water systems over a pressure range of 700 atm and at temperatures ranging from 20°C to 130°C. Later results for the same systems obtained by Jennings[19] showed that the effect of temperature was much greater than that of the pressure. A general equation was presented for the interfacial tension as a function of temperature and pressure over the interval 25–176°C and 1–817 atm. Jennings'[20] apparatus, of new design, permitted measurement of interfacial tension values as low as 0.001 dyn/cm. Interfacial tensions at higher pressures and temperatures for the water–methane system were determined by Hough *et al.*[21] Stegemeier[22] determined surface and interfacial tensions as low as 0.007 dyn/cm for several light paraffin hydrocarbons in the neighborhood of the critical region. Good and Wieser[23] have recently applied the method to study of the pressure coefficient of interfacial tension at pressures from 1 to 60 atm.

Stage[24] used the pendant drop method to obtain surface tension of mercury in high vacuum. The value she obtained was several dynes per centimeter lower than the accepted surface tension for pure mercury. Stage also obtained adsorption isotherms for *n*-heptane and ethanol, both singly and from mixtures of the two vapors, by observing the dependence of the surface tension lowering of mercury on the partial pressures of these vapors and subsequent application of the Gibbs adsorption equation to these data.

## 2.5. Studies of Adsorption from Solutions

The pendant drop technique has been used for solutions where the solute was either positively or negatively adsorbed at the surface. Ward and Tordai[25] measured boundary tensions for interfaces between water and solutions of long-chain fatty acids in hexane over a wide range of concentrations. These results were then used with the Gibbs adsorption equation to obtain information about the kinetics and the equilibrium of adsorption at liquid–liquid interfaces. Studies of adsorption of *n*-alkanols at the air–aqueous solution interface were reported by Clint *et al.*[26] The data were used to calculate standard thermodynamic quantities for transfer of *n*-alkanols to the interface. They correctly reported that the accuracy of the pendant drop method was limited only by the precision with which the dimensions of the drop could be measured. Ambwani *et al.*[27] and Bardasz[28] successfully adopted this method to study the adsorption of long-chain polar compounds from cyclohexane solution onto mercury surfaces.

To summarize, the pendant drop method has been used to study both liquid–vapor and liquid–liquid boundary tensions. It has been applied to materials ranging from organic liquids to molten metals and from pure solvents to concentrated solutions. Apart from normal temperatures and pressures, it has been successfully applied in vacuum and at elevated temperatures and pressures.

## 3. Theory

### 3.1. Historical Development

The possibility of calculating liquid boundary tensions from pendant drop shapes was considered as early as 1881 by Worthington.[29] However, early students of the method employed poor experimental techniques and unsatisfactory mathematical procedures, which yielded generally unreliable results. The first practical studies were reported in 1937 by Andreas *et al.*[1] These authors[30,31] made a careful comparison of five procedures that had been suggested for determining liquid boundary tensions from measurements of pendant drops and, after considering the advantages and limitations of each, developed the method of the selected plane. Some variation of the selected plane technique is used for nearly all pendant drop boundary tension measurements made today.

## 3.2. Mathematical Derivations

The mathematical treatment of pendant drop shape is based on the fundamental equation of capillarity, which relates the boundary tension at a point on a surface to the pressure difference across the surface and to the two principal radii of curvature of the surface at that point. The equation for a point $P$ (Figure 1) in terms of the pressure difference across the interface at the reference point $A$ can be written as

$$(p_A^1 - g\rho^1 z) - (p_A^2 - g\rho^2 z) = \gamma\left(\frac{1}{R_{1P}} + \frac{1}{R_{2P}}\right) \tag{1}$$

where $p_A^1$ and $p_A^2$ are the pressure on the concave and convex sides of the surface at point $A$, respectively; $\gamma$ is the boundary tension; $R_{1P}$ and $R_{2P}$ are the two principal radii of curvature of the surface at point $P$; $\rho^1$ and $\rho^2$ are the density of the fluid on the concave and convex side of the surface,

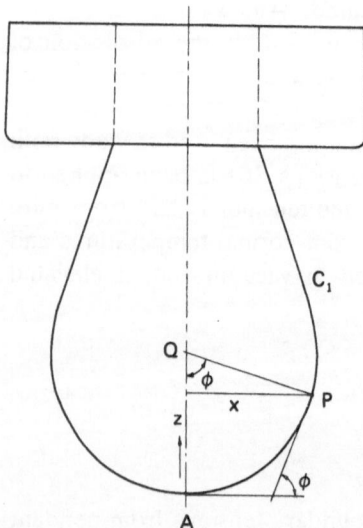

Figure 1. Geometric treatment of the pendant drop.

respectively; $g$ is the acceleration due to gravity; and $z$ is the ordinate of point $P$ in the coordinate system where point $A$ is chosen as origin. Since the reference point $A$ lies on the axis of revolution, the two principal radii of curvature must be equal at this point.

Let

$$R_A^1 = R_A^2 = b$$

From equation (1), the quantity $(p_A^1 - p_A^2)$ is then equal to $2\gamma/b$. Equation (1) can be rewritten

$$\gamma\left(\frac{1}{R_{1P}} + \frac{1}{R_{2P}}\right) = \frac{2\gamma}{b} + gz\,\Delta\rho \tag{2}$$

where $\Delta\rho = \rho^2 - \rho^1$.

In principle, a knowledge of the principal radii of curvature at two points on the surface would enable calculations of $\gamma$ and $b$. However, their measurement from the photographic image of a pendant drop is difficult, and therefore, it is necessary to substitute into equation (2) quantities that can be measured easily.

Let $R_{1P}$ be the radius of curvature at point $P$ of the curve $C_1$ (Figure 1) and $C_{P1}$ be the curve that passes through $P$ and is perpendicular to $C_1$ at point $P$. Since $QP$ is normal to both curves at $P$, and since $Q$ is on the axis of revolution, $P$ remains on cuve $C_{P1}$ when $QP$ rotates about the axis $AQ$. Therefore,

$$QP = x/\sin\phi$$

is the other radius of curvature of the surface at point $P = R_{2P}$. Equation (2) can then be written as

$$\gamma\left(\frac{1}{R_{1A}} + \frac{\sin\phi}{x}\right) = \frac{2\gamma}{b} + gz\,\Delta\rho \tag{3}$$

or

$$\frac{1}{R_{1A}/b} + \frac{\sin\phi}{x/b} = 2 - \frac{\beta z}{b} \tag{4}$$

where

$$\beta = -\frac{\Delta\rho\,gb^2}{\gamma} \tag{5}$$

Using the standard formula for the radius of curvature of a curve in the $x$–$z$ plane,

$$R_{1A} = \frac{[1 + (dz/dx)^2]^{3/2}}{d^2x/dx^2} \tag{6}$$

Since $\tan \phi = dz/dx$,

$$\sin \phi = \frac{\tan \phi}{(1 + \tan^2 \phi)^{1/2}} = \frac{dz/dx}{[1 + (dz/dx)^2]^{1/2}} \tag{7}$$

Substitution of equations (6) and (7) into equation (4) yields

$$\frac{d^2z}{dx^2} + \frac{dz/dx}{x}\left[1 + \left(\frac{dz}{dx}\right)^2\right] = \left(\frac{2}{b} - \beta\frac{z}{b^2}\right)\left[1 + \left(\frac{dz}{dx}\right)^2\right]^{3/2} \tag{8}$$

It is apparent that a solution of equation (8) would give $z$ as a function of $x$ containing $b$ and $\beta$ as parameters. The measurement of the coordinates of two points on the periphery of a photographic image of a pendant drop would enable calculation of $b$ and $\beta$ for that drop. Boundary tension could then be calculated from equation (5).

Bashforth and Adams,[32] whose derivation is followed above, evaluated $x/b$, $z/b$ and $\phi$ for a number of values of $\beta$. However, early attempts to use the pendant drop method were unsatisfactory due to lack of proper optical equipment and to the laborious calculations involved.

### 3.3. Method of the Selected Plane

Andreas et al.[1] circumvented the mathematical difficulties by modification of equation (5) and empirical evaluation of the resulting function. They defined a function of drop shape,

$$S = \frac{d_s}{d_e} \tag{9}$$

where $d_e$ is the maximum (equatorial) diameter of the pendant drop and $d_s$ is the diameter of the pendant drop in a selected plane at a distance $d_e$ from the apex of the drop.

The dimension $d_e$ and $d_s$ (Figure 2) can be readily determined from the photographed profile of a drop. To eliminate the quantity $b$ from equation (5), they defined another function of drop shape,

$$H = -\beta\left(\frac{d_e}{b}\right)^2 \tag{10}$$

Substitution of equation (10) into equation (5) yields

$$\gamma = \frac{\Delta\rho \, g d_e^2}{H} \tag{11}$$

from which the boundary tension can be calculated if the value of $1/H$ corresponding to the value of $S$ for the pendant drop in question is known. Andreas et al.[1] evaluated $1/H$ corresponding to a range of values of $S$ from 0.700 to 1.00 by determining $S$ for drops of conductivity water of various

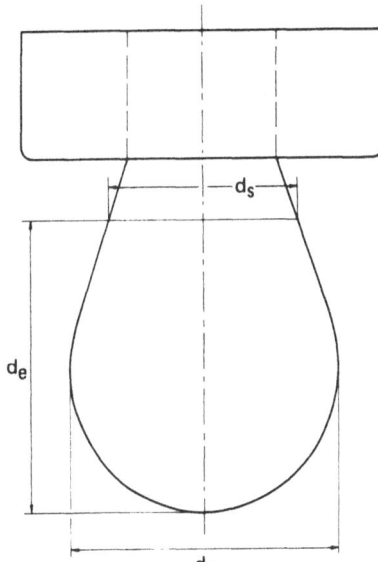

Figure 2. Method of selected plane for pendant drop boundary tension calculations.

shapes and sizes and by calculating $1/H$ from equation (11). Accuracy of their values for $1/H$ is limited by the assumption as to the value of the surface tension of water (taken as 72.0 dyn/cm) and by the accuracy with which the measurements were made.

## 3.4. Tables and Equations for the Method of the Selected Plane

Niederhauser,[9,33] using the method of Bashforth and Adams, calculated the values of $\phi$, $x/b$, and $z/b$ for a number of values of $\beta$ not covered by former authors and completed a table of values for the evaluation of $1/H$ as a function of $S$ at intervals of 0.001 units from 0.670 to 1.002. An identical table was published by Fordham.[34] The two independently published tables agreed in the fourth decimal place over the entire range of values and thus established their accuracy. The empirical tables of Andreas *et al.* were not accurate over the entire range. The results of Fordham[34] are shown in Table 1. Stauffer,[35] using a theoretical method of calculation, published a table containing $S$ values ranging from 0.30 to 0.66 (Table 2). Other tables, published by Mills,[36] contain $S$ values from 0.46 to 0.66.

Recently, Padday[37–39] solved equation (4) by a different procedure than Fordham, Mills, and Stauffer and provided tables for many more values of $\beta$ within the experimentally determined range of values of $S$. These tables give $x/b$ and $z/b$ for equal intervals of $\phi$, the interval $\Delta\phi$ being 5°.

Misak[40] divided the combined tables of Fordham, Stauffer, and Mills into five groups and derived an equation that would best fit the data in each group. The equations were developed by assuming that $\log(1/H)$ versus

Table 1.  Fordham's Table of 1/H in Terms of S [a]

| S | 0 | 1 | 2 | 3 | 4 | 5 | 6 | 7 | 8 | 9 |
|---|---|---|---|---|---|---|---|---|---|---|
| 0.66 | 0.93828 | 0.93454 | 0.93082 | 0.92712 | 0.92345 | 0.91979 | 0.91616 | 0.91255 | 0.90895 | 0.90538 |
| 0.67 | 0.90183 | 0.89830 | 0.89478 | 0.89129 | 0.88782 | 0.88436 | 0.88092 | 0.87751 | 0.87411 | 0.87073 |
| 0.68 | 0.86737 | 0.86403 | 0.86070 | 0.85739 | 0.85410 | 0.85083 | 0.84758 | 0.84434 | 0.84112 | 0.83792 |
| 0.69 | 0.83473 | 0.83156 | 0.82841 | 0.82527 | 0.82215 | 0.81905 | 0.81596 | 0.81289 | 0.80983 | 0.80679 |
| 0.70 | 0.80376 | 0.80075 | 0.79776 | 0.79478 | 0.79182 | 0.7887 | 0.78594 | 0.78302 | 0.78011 | 0.77722 |
| 0.71 | 0.77435 | 0.77149 | 0.76864 | 0.76581 | 0.76300 | 0.76019 | 0.75741 | 0.75463 | 0.75187 | 0.74912 |
| 0.72 | 0.74639 | 0.74367 | 0.74097 | 0.73828 | 0.73560 | 0.73293 | 0.73028 | 0.72764 | 0.72502 | 0.72240 |
| 0.73 | 0.71980 | 0.71722 | 0.71464 | 0.71208 | 0.70953 | 0.70700 | 0.70447 | 0.70196 | 0.69946 | 0.69697 |
| 0.74 | 0.69449 | 0.69202 | 0.68957 | 0.68713 | 0.68470 | 0.68228 | 0.67988 | 0.67748 | 0.67510 | 0.67273 |
| 0.75 | 0.67037 | 0.66803 | 0.66569 | 0.66337 | 0.66105 | 0.65875 | 0.65646 | 0.65418 | 0.65191 | 0.64965 |
| 0.76 | 0.64740 | 0.64516 | 0.64294 | 0.64072 | 0.63851 | 0.63632 | 0.63413 | 0.63195 | 0.62979 | 0.62763 |
| 0.77 | 0.62549 | 0.62335 | 0.62122 | 0.61911 | 0.61700 | 0.61490 | 0.61281 | 0.61074 | 0.60867 | 0.60661 |
| 0.78 | 0.60457 | 0.60253 | 0.60050 | 0.59848 | 0.59647 | 0.59447 | 0.59248 | 0.59049 | 0.58852 | 0.58656 |
| 0.79 | 0.58460 | 0.58265 | 0.58072 | 0.57879 | 0.57687 | 0.57496 | 0.57305 | 0.57116 | 0.56927 | 0.56739 |
| 0.80 | 0.56553 | 0.56366 | 0.56181 | 0.55997 | 0.55813 | 0.55630 | 0.55448 | 0.55266 | 0.55086 | 0.54906 |
| 0.81 | 0.54727 | 0.54549 | 0.54371 | 0.54195 | 0.54019 | 0.53844 | 0.53669 | 0.53496 | 0.53323 | 0.53151 |
| 0.82 | 0.52979 | 0.52808 | 0.52638 | 0.52469 | 0.52300 | 0.52132 | 0.51965 | 0.51799 | 0.51634 | 0.51469 |
| 0.83 | 0.51305 | 0.51142 | 0.50970 | 0.50817 | 0.50656 | 0.50496 | 0.50336 | 0.50176 | 0.50018 | 0.49860 |
| 0.84 | 0.49703 | 0.49546 | 0.49390 | 0.49234 | 0.49090 | 0.48926 | 0.48772 | 0.48619 | 0.48467 | 0.48316 |
| 0.85 | 0.48165 | 0.48015 | 0.47865 | 0.47716 | 0.47567 | 0.47420 | 0.47272 | 0.47126 | 0.46980 | 0.46834 |
| 0.86 | 0.46690 | 0.46545 | 0.46402 | 0.46259 | 0.46116 | 0.45974 | 0.45833 | 0.45692 | 0.45552 | 0.45412 |
| 0.87 | 0.45273 | 0.45134 | 0.44996 | 0.44858 | 0.44721 | 0.44584 | 0.44448 | 0.44313 | 0.44178 | 0.44044 |
| 0.88 | 0.43910 | 0.43777 | 0.43644 | 0.43512 | 0.43380 | 0.43249 | 0.43118 | 0.42988 | 0.42858 | 0.42729 |
| 0.89 | 0.42600 | 0.42471 | 0.42344 | 0.42216 | 0.42089 | 0.41963 | 0.41837 | 0.41712 | 0.41587 | 0.41462 |
| 0.90 | 0.41338 | 0.41214 | 0.41091 | 0.40968 | 0.40846 | 0.40724 | 0.40602 | 0.40481 | 0.40360 | 0.40240 |
| 0.91 | 0.40121 | 0.40001 | 0.39882 | 0.39764 | 0.39646 | 0.39528 | 0.39411 | 0.39294 | 0.39177 | 0.39061 |
| 0.92 | 0.38946 | 0.38831 | 0.38716 | 0.38601 | 0.38487 | 0.38374 | 0.38260 | 0.38147 | 0.38035 | 0.37922 |
| 0.93 | 0.37810 | 0.37699 | 0.37588 | 0.37477 | 0.37366 | 0.37256 | 0.37146 | 0.37037 | 0.36928 | 0.36819 |
| 0.94 | 0.36711 | 0.36602 | 0.36494 | 0.36387 | 0.36280 | 0.36173 | 0.36066 | 0.35960 | 0.35854 | 0.35748 |
| 0.95 | 0.35643 | 0.35538 | 0.35433 | 0.35328 | 0.35224 | 0.35120 | 0.35016 | 0.34913 | 0.34809 | 0.34706 |
| 0.96 | 0.34604 | 0.34501 | 0.34399 | 0.34297 | 0.34195 | 0.34093 | 0.33992 | 0.33890 | 0.33789 | 0.33688 |
| 0.97 | 0.33588 | 0.33487 | 0.33387 | 0.33287 | 0.33186 | 0.33086 | 0.32987 | 0.32887 | 0.32787 | 0.32688 |
| 0.98 | 0.32588 | 0.32489 | 0.32389 | 0.32290 | 0.32191 | 0.32092 | 0.31992 | 0.31893 | 0.31794 | 0.31695 |
| 0.99 | 0.31595 | 0.31496 | 0.31396 | 0.31296 | 0.31196 | 0.31095 | 0.30994 | 0.30893 | 0.30792 | 0.30690 |
| 1.00 | 0.30588 | 0.30484 | 0.30381 | 0.30276 | — | — | — | — | — | — |

[a] Table reprinted from S. Fordham, *Proc. Roy. Soc.(London)* **194A**, 1 (1948).

log $S$ would plot a straight line. In this manner, the five following equations were derived:

For $S = 0.401$ to $S = 0.46$

$$1/H = (0.32720/S^{2.56651}) - 0.97553S^2 + 0.84059S - 0.18069 \quad (12)$$

For S > 0.46 to $S = 0.59$

$$1/H = (0.31968/S^{2.59725}) - 0.46898S^2 + 0.50059S - 0.13261 \quad (13)$$

For $S > 0.59$ to $S = 0.68$

$$1/H = (0.31522/S^{2.62435}) - 0.11714S^2 + 0.15756S - 0.05285 \quad (14)$$

For $S > 0.68$ to $S = 0.90$

$$1/H = (0.31345/S^{2.64267}) - 0.09155S^2 + 0.14701S - 0.05877 \quad (15)$$

For $S > 0.90$ to $S = 1.00$

$$1/H = (0.30715/S^{2.84636}) - 0.69116S^3 + 1.08315S^2 - 0.18341S$$
$$- 0.20970 \quad (16)$$

The $1/H$ data were computed from the five equations for $S = 0.401$ to $S = 1.000$ in increments of $0.001$ and compared to the values of Fordham, Mills, and Stauffer. The greatest deviation is $\pm 0.00033$ between $S = 0.601$

*Table 2. Stauffer's Table of 1/H in Terms of S* [a]

| S | 0 | 1 | 2 | 3 | 4 | 5 | 6 | 7 | 8 | 9 |
|------|---------|---------|---------|---------|---------|---------|---------|---------|---------|---------|
| 0.30 | 7.09837 | 7.03966 | 6.98161 | 6.92421 | 6.86746 | 6.81135 | 6.75586 | 6.70099 | 6.46672 | 6.59306 |
| 0.31 | 6.53998 | 6.48748 | 6.43556 | 6.38421 | 6.33341 | 6.28317 | 6.23347 | 6.18431 | 6.13567 | 6.08756 |
| 0.32 | 6.03997 | 5.99288 | 5.94629 | 5.90019 | 5.85459 | 5.80946 | 5.76481 | 5.72063 | 5.67690 | 5.63364 |
| 0.33 | 5.59082 | 5.54845 | 5.50651 | 5.46501 | 5.42393 | 5.38327 | 5.34303 | 5.30320 | 5.26377 | 5.22474 |
| 0.34 | 5.18611 | 5.14786 | 5.11000 | 5.07252 | 5.03542 | 4.99868 | 4.96231 | 4.92629 | 4.89061 | 4.85527 |
| 0.35 | 4.82029 | 4.78564 | 4.75134 | 4.71737 | 4.68374 | 4.65043 | 4.61745 | 4.58479 | 4.55245 | 4.52042 |
| 0.36 | 4.48870 | 4.45729 | 4.42617 | 4.39536 | 4.36484 | 4.33461 | 4.30467 | 4.27501 | 4.24564 | 4.21654 |
| 0.37 | 4.18771 | 4.15916 | 4.13087 | 4.10285 | 4.07509 | 4.04759 | 4.02034 | 3.99334 | 3.96660 | 3.94010 |
| 0.38 | 3.91384 | 3.88786 | 3.86212 | 3.83661 | 3.81133 | 3.78627 | 3.76143 | 3.73682 | 3.71242 | 3.68824 |
| 0.39 | 3.66427 | 3.64051 | 3.61696 | 3.59362 | 3.57047 | 3.54752 | 3.52478 | 3.50223 | 3.47987 | 3.45770 |
| 0.40 | 3.43572 | 3.41393 | 3.39232 | 3.37089 | 3.34965 | 3.32858 | 3.30769 | 3.28698 | 3.26643 | 3.24606 |
| 0.41 | 3.22582 | 3.20576 | 3.18587 | 3.16614 | 3.14657 | 3.12717 | 3.10794 | 3.08886 | 3.06994 | 3.05118 |
| 0.42 | 3.03258 | 3.01413 | 2.99583 | 2.97769 | 2.95969 | 2.94184 | 2.92415 | 2.90659 | 2.88918 | 2.87192 |
| 0.43 | 2.85479 | 2.83781 | 2.82097 | 2.80426 | 2.78769 | 2.77125 | 2.75496 | 2.73880 | 2.72277 | 2.70687 |
| 0.44 | 2.69110 | 2.67545 | 2.65992 | 2.64452 | 2.62924 | 2.61408 | 2.59904 | 2.58412 | 2.56932 | 2.55463 |
| 0.45 | 2.54005 | 2.52559 | 2.51124 | 2.49700 | 2.48287 | 2.46885 | 2.45494 | 2.44114 | 2.42743 | 2.41384 |
| 0.46 | 2.40034 | 2.38695 | 2.37366 | 2.36047 | 2.34738 | 2.33439 | 2.32150 | 2.30870 | 2.29600 | 2.28339 |
| 0.47 | 2.27088 | 2.25846 | 2.24613 | 2.23390 | 2.22176 | 2.20970 | 2.19773 | 2.18586 | 2.17407 | 2.16236 |
| 0.48 | 2.15074 | 2.13921 | 2.12276 | 2.11640 | 2.10511 | 2.09391 | 2.08279 | 2.07175 | 2.06079 | 2.04991 |
| 0.49 | 2.03910 | 2.02838 | 2.01773 | 2.00715 | 1.99666 | 1 98623 | 1.97588 | 1.96561 | 1.95540 | 1.94527 |
| 0.50 | 1.93521 | 1.92522 | 1.91530 | 1.90545 | 1.89567 | 1.88596 | 1.87632 | 1.86674 | 1.85723 | 1.84778 |
| 0.51 | 1.83840 | 1.82909 | 1.81984 | 1.81065 | 1.80153 | 1.79247 | 1.78347 | 1.77453 | 1.76565 | 1.75683 |
| 0.52 | 1.74808 | 1.73938 | 1.73074 | 1.72216 | 1.71364 | 1.70517 | 1.69676 | 1.68841 | 1.68012 | 1.67188 |
| 0.53 | 1.66369 | 1.65556 | 1.64748 | 1.63946 | 1.63149 | 1.62357 | 1.61571 | 1.60790 | 1.60014 | 1.59242 |
| 0.54 | 1.58477 | 1.57716 | 1.56960 | 1.56209 | 1.55462 | 1.54721 | 1.53985 | 1.53253 | 1.52526 | 1.51804 |
| 0.55 | 1.51086 | 1.50373 | 1.49665 | 1.48961 | 1.48262 | 1.47567 | 1.46876 | 1.46190 | 1.45509 | 1.44831 |
| 0.56 | 1.44158 | 1.43489 | 1.42825 | 1.42164 | 1.41508 | 1.40856 | 1.40208 | 1.39564 | 1.38924 | 1.38288 |
| 0.57 | 1.37656 | 1.37028 | 1.36404 | 1.35784 | 1.35168 | 1.34555 | 1.33946 | 1.33341 | 1.32740 | 1.32142 |
| 0.58 | 1.31549 | 1.30958 | 1.30372 | 1.29788 | 1.29209 | 1.28633 | 1.28060 | 1.27491 | 1.26926 | 1.26364 |
| 0.59 | 1.25805 | 1.25250 | 1.24698 | 1.24149 | 1.23603 | 1.23061 | 1.22522 | 1.21987 | 1.21454 | 1.20925 |
| 0.60 | 1.20399 | 1.19875 | 1.19356 | 1.18839 | 1.18325 | 1.17814 | 1.17306 | 1.16801 | 1.16300 | 1.15801 |
| 0.61 | 1.15305 | 1.14812 | 1.14322 | 1.13834 | 1.13350 | 1.12868 | 1.23389 | 1.11913 | 1.11440 | 1.10969 |
| 0.62 | 1.10501 | 1.10036 | 1.09574 | 1.09114 | 1.08656 | 1.08202 | 1.07750 | 1.07300 | 1.06853 | 1.06409 |
| 0.63 | 1.05967 | 1.05528 | 1.05091 | 1.04657 | 1.04225 | 1.03796 | 1.03368 | 1.02944 | 1.02522 | 1.02102 |
| 0.64 | 1.01684 | 1.01269 | 1.00856 | 1.00446 | 1.0037  | 0.99631 | 0.99227 | 0.98826 | 0.98427 | 0.98029 |
| 0.65 | 0.97635 | 0.97242 | 0.96851 | 0.96463 | 0.96077 | 0.95692 | 0.95310 | 0.94930 | 0.94552 | 0.94176 |
| 0.66 | 0.93803 | 0.93431 | 0.93061 | 0.92693 | 0.92327 | 0.91964 | 0.91602 | 0.91242 | 0.90884 | 0.90528 |

[a] Table reprinted from C. E. Stauffer, *J. Phys. Chem* **69**, 1933 (1965).

to $S = 0.893$, the $S$ values most commonly encountered. These equations are sufficiently accurate to be used in the most exacting boundary tension calculations.

Stegemeier[22] derived an approximate equation for slide-rule calculations by assuming that a plot of log $(1/H)$ versus log $S$ is a single straight line over the entire range of $S$ values:

$$1/H = 0.31270S^{-2.6444} \qquad (17)$$

Although this equation is convenient to use, its results deviate considerably from the more precisely determined values, particularly for very high and very low values of $S$.

### 3.5. More Recent Methods

Winkel[41] considered refinement of the pendant drop method by considering the ratio $d\min/d_e$ instead of $d_s/d_e$, where $d\min$ is the minimum diameter of the drop. Although conceptually sound, the method of Winkel is impractical because of difficulties in forming a stable drop having a minimum diameter. Even when such a drop is formed, the plane of the minimum diameter is often so near the capillary tip that its measurement is affected by irregularities in the tip. Roe *et al.*[42] have published tables of $1/H$ as a function of $0.8S$, $0.9S$, $1.0S$, $1.1S$, and $1.2S$. The special value of these tables is that identical values of boundary tension calculated from different values of $S$ establish the fact of hydrostatic equilibrium in the pendant drop. Very recently, Arundel and Bagnall[43] have described a method to estimate the surface area of pendant drops from drop photographs. A relationship between drop surface area and drop volume is also demonstrated for a fixed fluid system and syringe tip diameter. This enables drops of given surface area to be created by control of drop volume alone.

## 4. Experimental Apparatus

### 4.1. General Survey

One of the most complete treatments of the experimental aspects of pendant drop technique is the work of Andreas *et al.*[1] Other descriptions of experimental setups for work at normal temperatures and pressures may be found in the papers of Bartell and Bard,[13] Smith and Sorg,[5] Douglas,[3] Hough and Heathington,[44] Ambwani and Fort,[45] and Jennings.[20] For work at elevated temperatures and pressures, the work of Hough *et al.*,[21] Michaels and Hauser,[18] Stegemeier,[22] and Good and Wieser[23] may be consulted. Roe[15] has given a complete account of his determination of

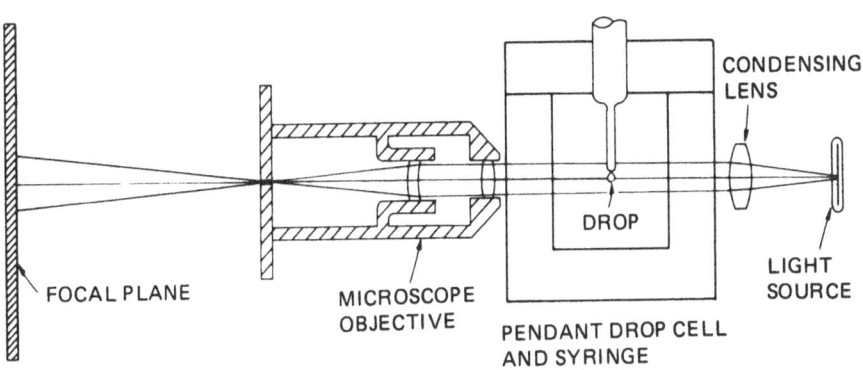

Figure 3. Schematic diagram of pendant drop apparatus.

surface tension of homopolymers above their melting points and the inter-facial tension between polymer liquids. Surface and interfacial tension measurements for polymer melts at temperatures up to 200°C are also described by Wu.[46] In Section 2, references are given for special situations, e.g., high vacuum and adsorption from solution, for which the pendant drop method has been found useful.

## 4.2. Basic Elements

The apparatus consists essentially of a light source, a pendant drop cell and syringe assembly contained in a constant temperature chamber, and a precise photomicrographic arrangement. The essential elements are shown in Figure 3. Among the auxiliary equipment that determines to a large extent the precision and accuracy of the results are the vibration-free support on which the pendant drop cell and syringe assembly is mounted and the arrangement for the production of pendant drop profiles from which the dimensions $d_e$ and $d_s$ can be measured. A typical pendant drop setup is shown in Figure 4.

## 4.3. Light Source and Collimating Lens

It is essential to project a clear image of the pendant drop with rays of light that are parallel to the optical axis. Early experimenters[1,5] used high-pressure mercury vapor lamps in combination with Wratten filters to obtain a monochromatic light. Equally good results were obtained using a concentrated arc lamp.[13,14] For a parallel beam of light, a projection lens system is used, e.g., 400-mm Leitz compound lens.[5] Work with a pulsed laser beam[47] has also given very good results. In investigations where the

Figure 4. Typical pendant drop assembly.

heat from the light source could affect the sample, a glass or water heat filter is normally used.

### 4.4. Pendant Drop Cell

For normal temperatures and pressures, a Pyrex glass cell can be used.[1,27] For higher temperatures and pressures, glass (except for optical windows) is replaced by stainless steel.[18,21] Figures 5 and 6 show completely assembled and sealed systems used by Ambwani et al.[27] to study the adsorption of long-chain polar compounds at the mercury–hydrocarbon interface. In the flow model (Figure 5), the hydrocarbon can be purged with dry nitrogen prior to the formation of mercury seals in the sidearms. Thus, boundary tension can be studied under controlled conditions. Spectrophotometer cells can be used in a simpler apparatus when contamination from the atmosphere is not a problem.

### 4.5. Pendant Drop Syringe

Usually, the drops are formed at the end of a precision-bore capillary tube or stainless steel hypodermic needle. Liquid is forced into the tube or needle from a syringe controlled by a micrometer screw arrangement. The end of the tube or needle is carefully ground square. The correct bore is usually in the range of 0.0125–0.100 cm[3] depending on the relative densities and boundary tension of the system. The drops hang from the inner radius (bore) of the tip if the tip is not wetted by the pendant drop and from the outer radius of the tip if the tip is wetted by the pendant drop.

From equation (11), it is noted that the diameter $d_e$ of a drop is determined by $S$ and $\gamma/\Delta\rho$. This relationship is presented graphically in Figure 7. The capillary tip radius can be correlated with $d_e$ using an equation for maximum drop volume

$$V_{\max} = \gamma r/g \, \Delta\rho \, F \qquad (18)$$

where $\gamma$ is boundary tension, $r$ is capillary tip radius (bore radius if the pendant drop does not wet the tip, outer radius if the drop wets the tip), $g$ is gravitational acceleration, $\Delta\rho$ is the density difference between the pendant drop and the surrounding medium, and $F$ is the correction factor.[48] Assuming that the volume $V$ of the drop is 20% greater than that of a sphere having diameter $d_e$, the relationship of minimum diameter of the tip to diameter of the drop is shown in Figure 8. With the aid of these data, the correct tip size can be selected.

For measurements of ultralow interfacial tensions, the capillary tubes from which the drops are suspended must be of very small diameter. At Carnegie-Mellon University, good success has been achieved in measuring oil–water interfacial tensions as low as $10^{-4}$ dyn/cm by suspending the drops from 10–20 $\mu$m diameter capillaries. These capillaries are made by drawing down Pyrex melting point tubes to the required diameter. In these special,

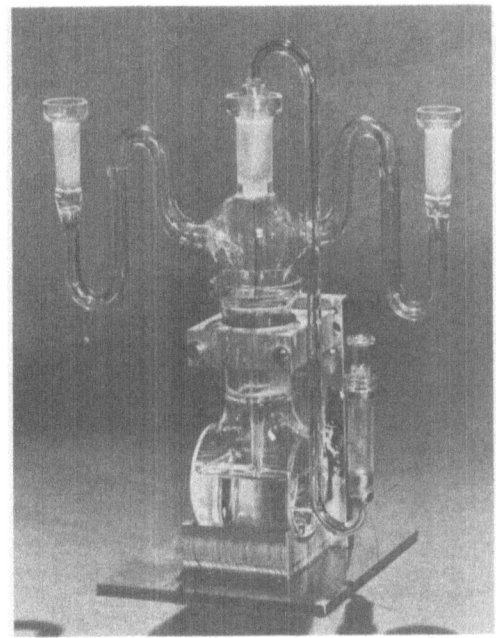

Figure 5. Pendant drop cell and syringe (flow model).

low-tension systems, the drops hang from the outside of the capillary and not from the tube end, which makes the requirement of a square ground tube end noncritical.

For best alignment of the cell and syringe assembly with the light source and the photomicrographic equipment, the entire assembly should be mounted on an optical bench.

## 4.6. Photomicrographic Arrangement

One of the essential parts of pendant drop equipment is a photomicrographic setup capable of producing an enlarged, undistorted photographic image of the profile of a pendant drop. In order to avoid distortion resulting from false perspective, it is necessary that the drop be photographed in parallel light. Some experimenters[1,5] have used a telecentric stop at the back focus of the objective lens to give the desired effect of passing the rays that are parallel to the optical axis and excluding all others.

The focal length of the objective and the lens to film distance are selected from consideration of the size of the pendant drop and the size of the photographic image desired for the chosen method of measuring $d_e$ and $d_s$.

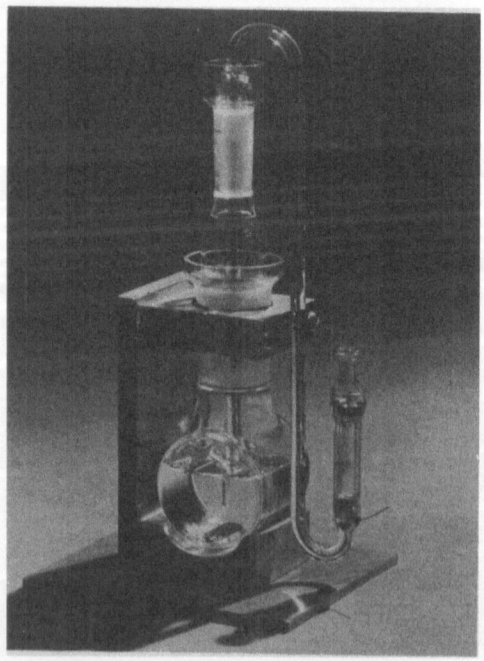

Figure 6. Pendant drop cell and syringe (nonflow model).

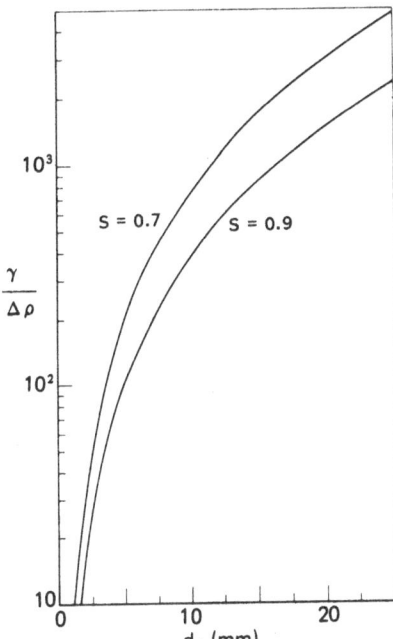

Figure 7. Relationship between drop diameter, boundary tension, and effective density.

The dimensions $d_e$ and $d_s$ are measured either directly by projecting the drop onto a screen or perhaps more conveniently by photographing on a film, film plate, Polaroid transparency, or Polaroid paper. From films, the dimensions of the drop profile are measured by an optical comparator[49] or microprojector.[45] Alternately, images of pendant drops on film or glass plates or Polaroid paper can be measured by using a coordinate overlay as shown in Figure 9. The overlay is made by photographic reduction of an appropriately sized graph paper grid onto a 5 × 7 in. glass plate to yield approximately 0.5-mm divisions.[47] A magnifying glass is used to increase the accuracy of the measurements. The precision with which $d_e$ and $d_s$ are measured largely determines the accuracy of the boundary tension measurements.

### 4.7. Temperature and Vibration Control

Control of temperature and vibration is essential for determination of a meaningful boundary tension. Normally, a two-unit water[1,14] or air[27,49] thermostat is used to control the temperature. For vibration control, some investigators have carried out their experiments in a basement laboratory,[14] and others have used special vibration-damping techniques.[13,45] A successful vibration-damping mount at Carnegie-Mellon University consists of a 3 ft × 3 ft × 1 ft block of granite (weighing approximately 1800

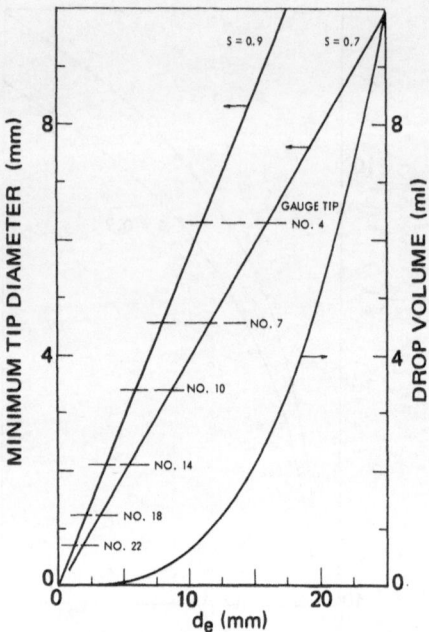

Figure 8. Relationship between minimum diameter of syringe tip and diameter of drop.

lb) supported by four automobile inner tubes (Figure 10). To avoid disturbing such a support during the experiments, the camera and microscope assembly is mounted on a completely separate platform.

### 4.8. Tests of the Experimental Methods

A new pendant drop apparatus should be tested and calibrated before beginning measurements of boundary tensions. Possible sources of error include distortions caused by the projection lens, any optical imperfections in the cell and thermostat windows, nonalignment of the various components of the system, and uncertainties caused by the nature of the light and of the photographic process. Also, it is necessary to know if the light is imaging the place of maximum diameter. To test a new apparatus at Carnegie-Mellon University, pictures of the following three setups are taken on the same photographic film[27,45]:

1. A micrometer disk (American Optical Company, Buffalo, New York) with a 5-mm scale divided into 50 divisions mounted on a steel base and placed at the same position at which the pendant drop is to be formed.

2. A round steel ball (Industrial Tectonics, Inc., Ann Arbor, Michigan) of precisely known diameter glued to the tip of a glass tube placed in the same position at which the pendant drop is to be formed. The diameter of the

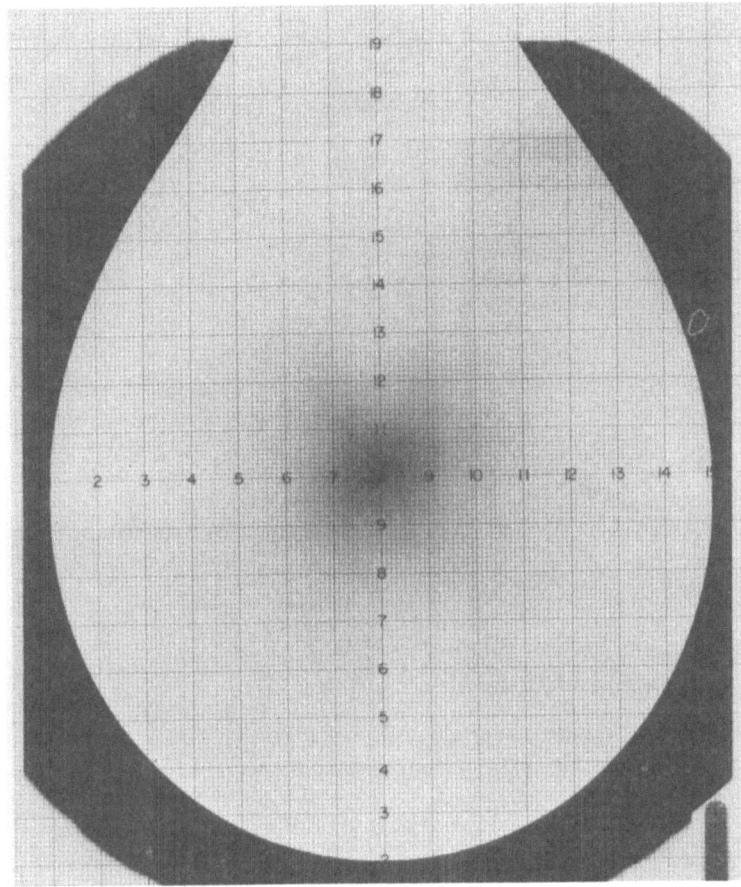

Figure 9. Pendant drop negative with coordinate overlay for measuring $d_e$ and $d_s$.

ball is chosen to be approximately equal to the maximum diameter of the pendant drop formed during the experiments. This choice is made so as to minimize errors in magnification factor caused by the fact that the eye sees as the phase boundary the point of maximum gradient in the optical density curve.

3. A pendant drop of a liquid of known surface tension.

The diameter of the ball calculated from calibration with the plane micrometer disk should be very close to the actual diameter. Calculated surface tension of the test liquid should, of course, check the known surface tension of the liquid.

Figure 10. Overview of pendant drop assembly with vibration control mount.

## 5. Calculation of Boundary Tension

The boundary tension is calculated from the measured profile dimensions $d_e$ and $d_s$ and using the following equations:

$$S = \frac{d_s}{d_e} \tag{9}$$

$$\gamma = \Delta\rho \, g d_e^2 \frac{1}{H} \tag{11}$$

Besides the measured $d_e$ and $d_s$, calculation of boundary tension requires the density of the pendant drop and the surrounding fluid, acceleration due to gravity, and a conversion factor to obtain actual maximum diameter of the drop from the measured $d_e$. Knowing $S$, $H$ is obtained from Table 2 or 3. A digital computer program can be written to solve the equations (9) and (11).[50]

### Sample Calculations

System:   Carbon tetrachloride in air at 25°C[1]

Density of $CCl_4$ . . . . . . . . . . . . . 1.585 g/cm$^3$
Density of air . . . . . . . . . . . . . . 0.001 g/cm$^3$

Effective density $(\Delta\rho)$ . . . . . . . . . 1.584 g/cm$^3$
Diameter of drop at equator $d_e$ . . . . . 0.228 cm†
Diameter of drop at selected plane $d_s$ . . 0.224 cm†

$$S = \frac{d_s}{d_e} = \frac{0.224}{0.228} = 0.9824$$

from Table 2, $1/H = 0.3235$

$$\gamma = \Delta\rho\, gd_e^2\, (1/H)$$

$$= (1.584)\,(980.4)\,(0.228)^2\,(0.3235)$$

$$= 26.12 \text{ dyn/cm}$$

## 6. Error Analysis

### 6.1. Method of Analysis

The error in the value of the boundary tension due to the errors in the measured quantities can be calculated using the calculus of variations.[22,50] Equations for the calculation of boundary tension are

$$S = \frac{d_s}{d_e} \tag{9}$$

$$\gamma = \Delta\rho\, gd_e^2\, \frac{1}{H} \tag{11}$$

$$\frac{1}{H} = f(S) \tag{19}$$

The general equation for error in boundary tension is

$$\Delta_q(\gamma) = \frac{\partial\gamma}{\partial q}\, \delta_q \tag{20}$$

where $\Delta_q(\gamma)$ is the error in $\gamma$ due to error in $q$, $\delta_q$ the error in variable $q$, and $q$ is any variable on which $\gamma$ depends.

† Actual values to which magnification factor has been applied.

## 6.2. Error Due to Uncertainties in Known Quantities

The error in boundary tension caused by error in the gravitational constant may be calculated as follows:

$$\Delta_g(\gamma) = \frac{\partial \gamma}{\partial g} \delta_g$$

$$\frac{\partial \gamma}{\partial g} = \Delta\rho \frac{1}{H} d_e^2$$

$$\Delta_g(\gamma) = \left( \Delta\rho \frac{1}{H} d_e^2 \right) \delta_g$$

$$= \frac{\gamma}{g} \delta_g \tag{21}$$

The error in boundary tension caused by error in density difference is

$$\frac{\partial \gamma}{\partial (\Delta\rho)} = g \frac{1}{H} d_e^2$$

$$\Delta_{\Delta\rho}(\gamma) = \left( g \frac{1}{H} d_e^2 \right) \delta_{\Delta\rho}$$

$$= \frac{\gamma}{\Delta\rho} \delta_{\Delta\rho} \tag{22}$$

## 6.3. Error Due to Measurements of Pendant Drop Dimensions

The error in boundary tension caused by error in the measurement of maximum diameter $d_e$ is

$$\frac{\partial \gamma}{\partial d_e} = \Delta\rho \, g \frac{1}{H} 2d_e + \Delta\rho \, g d_e^2 \frac{\partial(1/H)}{\partial d_e} \tag{23}$$

Differentiation of equation (19) with respect to $S$ gives

$$\frac{d(1/H)}{dS} = f'(S) \tag{24}$$

The partial derivative of equation (9) with respect to $d_e$ gives

$$\frac{\partial S}{\partial d_e} = -\frac{d_s}{d_e^2}$$

$$\partial S = -\frac{d_s}{d_e^2} \partial d_e \tag{25}$$

From equations (24) and (25) is obtained

$$\frac{\partial(1/H)}{\partial d_e} = -\frac{d_s}{d_e^2}f'(S) \tag{26}$$

From equations (25) and (26) is obtained

$$\frac{\partial \gamma}{\partial d_e} = \Delta\rho\, g \frac{1}{H} 2d_e - \Delta\rho\, gd_e^2 \frac{d_s}{d_e^2} f'(S)$$

$$\frac{\partial \gamma}{\partial d_e} = \Delta\rho\, g\left(2d_e \frac{1}{H} - d_s f'(S)\right) \tag{27}$$

An equation for $1/H$ derived by Misak[40] and discussed in Section 3.4 is†

$$\frac{1}{H} = \frac{0.31345}{S^{2.64267}} - 0.09155S^2 + 0.14701S - 0.05877 \tag{15}$$

On differentiation of this equation with respect to $S$, there is obtained

$$\frac{d(1/H)}{dS} = 0.31345(-2.64267)S^{-3.64267} - 0.18310S + 0.14701 \tag{28}$$

From equations (26), (27), and (28) is obtained

$$\frac{\partial \gamma}{\partial d_e} = \Delta\rho\, g\left[2d_e \frac{1}{H} + d_s(0.828345S^{-3.64267} + 0.18310S - 0.14701)\right]$$

Therefore,

$$\Delta d_e(\gamma) = \left[\frac{2\gamma}{d_e} + \Delta\rho\, gd_s(0.828345S^{-3.64267} + 0.18310S - 0.14701)\right]\delta d_e \tag{29}$$

The error in boundary tension caused by error in the measurement of the selected plane diameter $d_s$ is

$$\frac{\partial \gamma}{\partial d_s} = \Delta\rho\, gd_e^2 \frac{\partial(1/H)}{\partial d_s} \tag{30}$$

Also,

$$\frac{\partial S}{\partial d_s} = \frac{1}{d_e} \tag{31}$$

---

† This equation was used in analyzing error in the author's experiments[27] with mercury–hydrocarbon interfacial tensions since the $S$ values for mercury drops were within the range of 0.68 to 0.90. For other systems, appropriate equations [equations (12)–(16)] should be used, depending upon the $S$ value of the pendant drop.

From equations (31) and (24),

$$\frac{\partial(1/H)}{\partial d_s} = \frac{1}{d_e} f'(S) \tag{32}$$

On combination of equations (24), (28), and (30), there is obtained

$$\frac{\partial \gamma}{\partial d_s} = \Delta\rho \, gd_e (0.828345S^{-3.64267} + 0.18310S - 0.14701)$$

or

$$\Delta d_s(\gamma) = -[\Delta\rho \, gd_e (0.828345S^{-3.64267} + 0.18310S - 0.14701)] \, \delta d_s \tag{33}$$

### 6.4. Expression for Overall Probable Error

Probable error in the value of the boundary tension caused by errors in the different variables can be obtained from the following expression:

$$\Delta\gamma = \{[\Delta_g(\gamma)]^2 + [\Delta_{\Delta\rho}(\gamma)]^2 + [\Delta_{d_e}(\gamma)]^2 + [\Delta_{d_s}(\gamma)]^2\}^{1/2}$$

$$= \left[\left(\frac{\gamma}{g}\delta_g\right)^2 + \left(\frac{\gamma}{\Delta\rho}\delta_{\Delta\rho}\right)^2 + (D_E \, \delta_{d_e})^2 + (D_S \, \delta_{d_s})^2\right]^{1/2} \tag{34}$$

Here,

$$D_E = \frac{2\gamma}{d_e} + \Delta\rho \, gd_s H_S \tag{35}$$

$$D_S = -\Delta\rho \, gd_e H_S \tag{36}$$

and

$$H_S = 0.828345S^{-3.64267} + 0.18310S - 0.14701 \tag{37}$$

### 6.5. An Example of Error Calculation

An example of error calculation will be given using data for a pendant mercury drop in a hydrocarbon medium as an example. Similar calculations can be made for other systems to determine the influence of various factors on boundary tension and to determine and/or improve the performance of a particular experimental setup.

The uncertainty in calculated boundary tension can be expressed as $\pm\Delta\gamma$ dyn/cm. Then, if

$$g = 980.4 \text{ cm/sec}^2$$

$$\Delta\rho = 12.7695 \text{ gm/cm}^3$$

$$d_e = 0.3903 \text{ in.}$$

$$d_s = 0.2846 \text{ in.}$$

$$\gamma = 351.7425 \text{ dyn/cm}$$

$$\text{conversion factor} = 0.505561 \text{ cm/in.}$$

and assuming

$$\delta_g = +0.05 \text{ cm/sec}^2$$

$$\delta_{\Delta\rho} = +0.005 \text{ gm/cm}^3$$

$$\delta_{d_e} = +0.0001 \text{ in.}$$

$$\delta_{d_s} = +0.0001 \text{ in.}$$

the error calculation is made as follows: Calculated quantities are $H_S = 2.603801348$, $D_E = 8253.343536$, and $D_S = -6429.327$. Then

$$\Delta\gamma = \left[ \left( \frac{\gamma}{g} \delta_g \right)^2 + \left( \frac{\gamma}{\Delta\rho} \delta_{\Delta\rho} \right)^2 + (D_E \delta_{d_e})^2 + (D_S \delta_{d_s})^2 \right]^{1/2}$$

$$= [(0.00032) + (0.0189794) + (0.17410) + (0.10565)]^{1/2}$$

$$= 0.547$$

$$\gamma = 351.74 \pm 0.55 \text{ dyn/cm}$$

These calculations, based on the precision of measurements currently made at Carnegie-Mellon University, show that the error in the value of boundary tension caused by uncertainties in the various experimental quantities is approximately ±0.15%. If the precision of measurement of $d_e$ and $d_s$ were improved (the theoretical limit is the grain size of the photographic film), accuracy of the calculated boundary tension could be further increased.

ACKNOWLEDGMENTS

The authors thank Mr. G. S. Ronay of Shell Development Company, Emeryville, California, for a critical review of the paper and for additions to Section 4 and Dr. W. C. Simpson, Shell Development Company, and Mr. R. A. Jao, Case Western Reserve University, for reading the manuscript and their helpful comments. One of us (Tomlinson Fort, Jr.) acknowledges National Science Foundation Grant GK-17312, which supported his part of this work.

## Symbols

| | | | |
|---|---|---|---|
| $b$ | Radius of curvature at apex of drop | $g$ | Acceleration due to gravity |
| $d_e$ | Maximum diameter of pendant drop | $H$ | Defined by equation (10) |
| $d_{min}$ | Minimum diameter of pendant drop | $H_S$ | Defined by equation (37) |
| $d_s$ | Diameter of pendant drop at $d_e$ from apex | $S$ | Defined by equation (9) |
| | | $\beta$ | Defined by equation (5) |
| $D_E$ | Defined by equation (35) | $\gamma$ | Boundary (surface or interfacial) tension |
| $D_S$ | Defined by equation (36) | | |
| $F$ | Correction factor | $\rho$ | Density |

## Subscript

$q$     Variables on which $\gamma$ depends

## References

1. J. M. Andreas, E. A. Hauser, and W. B. Tucker, *J. Phys. Chem.* **42**, 1001 (1938).
2. J. F. Padday, in *Surface and Colloid Science*, Vol. 1 (E. Matijević, ed.), Wiley Interscience, New York (1969).
3. H. W. Douglas, *J. Sci. Instr.* **27**, 67 (1950).
4. E. A. Hauser and A. S., Michaels, *J. Phys. Chem.* **53**, 590 (1949).
5. G. W. Smith and L. V. Sorg, *J. Phys. Chem.* **45**, 671 (1941).
6. G. W. Smith, *J. Phys. Chem.* **48**, 168 (1944).
7. D. J. Donahue and F. E. Bartell, *J. Phys. Chem.* **56**, 480 (1952).
8. F. E. Bartell and J. M. Davis, *J. Phys. Chem.* **45**, 1321 (1941).
9. F. E. Bartell and D. O. Niederhauser, *Fundamental Research on Occurrence and Recovery of Petroleum, 1946–47*, American Petroleum Institute, New York (1949).
10. C. D. Manning, *Measurement of Ultralow Interfacial Tensions in Surfactant–Oil–Brine Systems*, M.S. thesis, Department of Chemical Engineering, University of Minnesota, Minneapolis, Minnesota (1976).
11. C. A. Miller, R. N. Hwan, W. J. Benton, and T. Fort, Jr., *J. Colloid Interface Sci.* **61**, 554 (1977).
12. F. E. Bartell and C. W. Bjorklund, *J. Phys. Chem.* **56**, 453 (1952).
13. F. E. Bartell and R. J. Bard, *J. Phys. Chem.* **56**, 532 (1952).
14. E. B. Butler, *J. Phys. Chem.* **67**, 1419 (1963).
15. R. J. Roe, *J. Phys. Chem.* **72**, 2013 (1968).
16. R. J. Roe, *J. Colloid Interface Sci.* **31**, 228 (1969).
17. J. K. Davis and F. E. Bartell, *Anal. Chem.* **20**, 1182 (1948).
18. A. S. Michaels and E. A. Hauser, *J. Phys. Chem.* **55**, 408 (1951).
19. H. Y. Jennings, *J. Colloid Interface Sci.* **24**, 323 (1967).
20. H. Y. Jennings, *Rev. Sci. Instr.* **28**, 774 (1957).
21. E. W. Hough, M. J. Rzasa, and B. B. Wood, *Petroleum Trans., AME* **192**, 57 (1951).
22. G. L. Stegemeier, Ph.D. thesis, University of Texas, Austin, Texas (1959).
23. R. J. Good and J. D. Wieser, *The Thickness of a Surfactant Film as Measured by the Pressure Coefficient of Interfacial Tension*, Abstracts, Paper 14, Division of Colloid and Surface Chemistry, 174th ACS National Meeting, Chicago, Illinois, August 29–September 1, 1977.

24. D. V. Stage, Ph.D. thesis, Iowa State College, Ames, Iowa (1955).
25. A. F. H. Ward and L. Tordai, *Rec. Trav. Chim.* **71**, 396 (1952).
26. J. H. Clint, J. M. Corkill, G. F. Goodman, and J. R. Tate, The adsorption of *n*-alkanols at the air/aqueous solution interface. Presented at the 155th ACS Meeting, San Francisco, California (1968).
27. D. S. Ambwani, R. A. Jao and T. Fort, Jr., *J. Colloid Interface Sci.* **42**, 8 (1973).
28. E. A. Bardasz, Ph.D. thesis, Case Western Reserve University, Cleveland, Ohio (1974).
29. A. M. Worthington, *Proc. Roy. Soc. London* **32**, 362 (1881).
30. J. M. Andreas, Sc.D. thesis, Massachusetts Institute of Technology, Cambridge, Massachusetts (1938).
31. W. B. Tucker, Sc.D. thesis, Massachusetts Institute of Technology, Cambridge, Massachusetts (1938).
32. F. Bashforth and J. C. Adams, *An Attempt to Test the Theories of Capillary Action*, University Press, Cambridge, England (1883).
33. D. O. Neiderhauser, Ph.D. thesis, Univ. Michigan, Ann Arbor, Michigan (1947).
34. S. Fordham, *Proc. Roy. Soc. (London)* **A194**, 1 (1948).
35. C. E. Stauffer, *J. Phys. Chem.* **69**, 1933 (1965).
36. O. S. Mills, *Brit. J. Appl. Phys.* **3**, 358 (1952).
37. J. F. Padday, *Phil. Trans. Roy. Soc. (London)* **A269**, 265 (1971).
38. J. F. Padday, *J. Electroanal. Chem.* **37**, 313 (1972).
39. J. F. Padday and A. R. Pitt, *Phil. Trans. Roy. Soc. (London)* **A275**, 489 (1973).
40. M. D. Misak, *J. Colloid Interface Sci.* **27**, 141 (1968).
41. D. Winkel, *J. Phys. Chem.* **69**, 348 (1965).
42. R. J. Roe, V. L. Bacchetta and P. M. G. Wong, *J. Phys. Chem.* **71**, 4190 (1967).
43. P. A. Arundel and R. D. Bagnall, *J. Phys. Chem.* **81**, 2079 (1977).
44. E. W. Hough and K. Heathington, *Producers Monthly*, p. 21, May 1961.
45. D. S. Ambwani and T. Fort, Jr., *J. Colloid Interface Sci.* **42**, 1 (1973).
46. S. Wu, *J. Colloid Interface Sci.* **31**, 153 (1969).
47. G. S. Ronay, Shell Development Company, Emeryville, California, private communication.
48. W. D. Harkins, in *Physical Methods of Organic Chemistry* (3rd ed.) (A. Weissberger, ed.), Chap. 14, Interscience, New York (1959).
49. G. L. Mack, J. M. Davis and F. E. Bartell, *J. Phys. Chem.* **45**, 846 (1941).
50. D. S. Ambwani, Ph.D. thesis, Case Western Reserve University, Cleveland, Ohio (1969).

# 4

# Electrophoresis of Particles in Suspension

## Arthur M. James

## 1. Introduction

Electrophoresis is the movement of charged particles suspended in a liquid under the influence of an applied electric field. The usual aims of electrophoresis experiments are the obtaining of information on the electrical double layer surrounding the particles, the analysis of a mixture, or the identification of surface components on the particles (e.g., biological cells).

The electrophoretic mobility $\bar{v}$, defined as the (electrophoretic) velocity $v_p$ per unit field strength $X$, is the quantity that is obtained directly from experimental observations.† It is necessary to consider briefly the theory relating the mobility to the structure of the double layer, the surface potential, and the $\zeta$-potential.

## 2. Concept of the Electrical Double Layer

"Electrokinesis" is the general term applied to a group of phenomena that have a common origin in the asymmetrical distribution of ions at an interface, the electrical double layer. At an interface between two phases (e.g., electrolyte and solid), on the time average, ions of one sign will be

---

† SI units are used throughout this article. The use of these units and their relation to cgs units is discussed in Section 11, Nomenclature and Units, with particular reference to the physical quantities under study.

---

*Arthur M. James* • Department of Physical Chemistry, Bedford College, University of London, Inner Circle, Regent's Park, London NW1 4NS, England.

predominantly associated with one phase and those of the other sign with the other phase. Thus, on the application of an electric field, there will be a tangential movement of one phase relative to the other at a velocity depending on the boundary potential across the plane of shear—the $\zeta$-potential. Conversely, if one phase is moved mechanically relative to the other, then a potential, known as the sedimentation potential, is established.

Substances acquire a surface electric charge either by the ionization of surface charged groups (e.g., biological cells, ion exchange particles) or by adsorption of ions (e.g., colloidal carbon, air bubbles) when in contact with an aqueous medium. A system containing particles in suspension is no different from a simple electrolyte solution in that overall it must be electrically neutral. However, one of the species has dimensions larger than simple ions and a net charge many times greater than that of a simple univalent ion. The ions of opposite charge to that on the particle, which must be present to maintain electrical neutrality, are known as "counter" or "gegen" ions. Ions of similar sign to that on the particle (co-ions) are also present in solution. There must be a neutralizing excess of counterions over co-ions in the locality of a particle in suspension.

The structure of the double layer, now widely accepted, is that developed by Stern,[1] which embodies the principles put forward by Helmholtz,[2] Perrin,[3] Gouy,[4] and Chapman.[5] Helmholtz predicted that the double layer constituted a parallel plate condenser; this was criticized by Gouy and Chapman on the basis that thermal energy would prevent the formation of such a compact double layer and would tend to distribute the ions throughout the atmosphere. In place, they postulated a diffuse ionic atmosphere in which the potential falls to zero over the distance $1/\kappa$, the statistical thickness of the double layer (Figure 1).

Excess gegen ions near the charged surface screen the electrostatic attraction for the gegen ions lying further away from the particle surface. This results in the concentration of excess counter ions and hence the electrical potential falling off rapidly at first and then more slowly with increasing distance until the region of uniform charge distribution is reached. This simple picture predicts a capacity of the double layer that is higher than that measured experimentally, leading to absurdly high ionic concentrations in the environment of the surface. Stern attributed this discrepancy to the treatment of the ions as point charges. He showed that, alone, neither the sharp not the diffuse double layer theory was adequate, and he developed a theory embodying the general characteristics of both. He considered the possibility of specific ion adsorption, which produced a layer of gegen ions attached to the surface, partly by electrostatic and partly by van der Waals forces, strongly enough to overcome thermal forces. The double layer is thus divided into two parts. One, approximately a single ion in thickness (the Stern layer), remains almost in contact with the surface; in

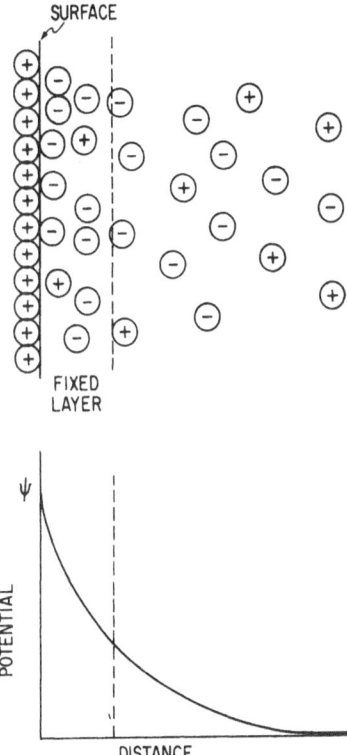

Figure 1. Schematic representation of a diffuse electrical double layer.

this layer there is a sharp fall in potential from $\psi_0$ to $\psi_\delta$ (Figure 2). The second part, which extends into the liquid phase, is diffuse, and here the potential falls from $\psi_\delta$ to zero. In this diffuse atmosphere, to which the Gouy–Chapman theories are applicable, thermal agitation permits free movement of the ions. The distribution of the positive and negative ions is not uniform, however, since the electrostatic field at the surface results in attraction of ions of the opposite sign. The potential change in the Stern layer increases with the concentration and valence type of the electrolyte. With polyvalent counterions it is possible for reversal of charge to occur within the Stern layer, i.e., for $\psi_0$ and $\psi_\delta$ to have opposite signs. Figure 3 summarizes the asymmetrical distribution of ions around a negatively charged particle in suspension.

The charge on the surface is equal in magnitude but opposite in sign to the sum of the charges in the fixed and diffuse parts of the double layer. It is thus apparent that when an external electric field is applied to a particle suspended in an electrolyte solution, the migration velocity is related to that part of the potential gradient across the shearing plane where the potential is $\zeta$. Thus, the numerical value of the $\zeta$-potential is dependent on the position of the shearing plane with respect to the demarcation plane between the

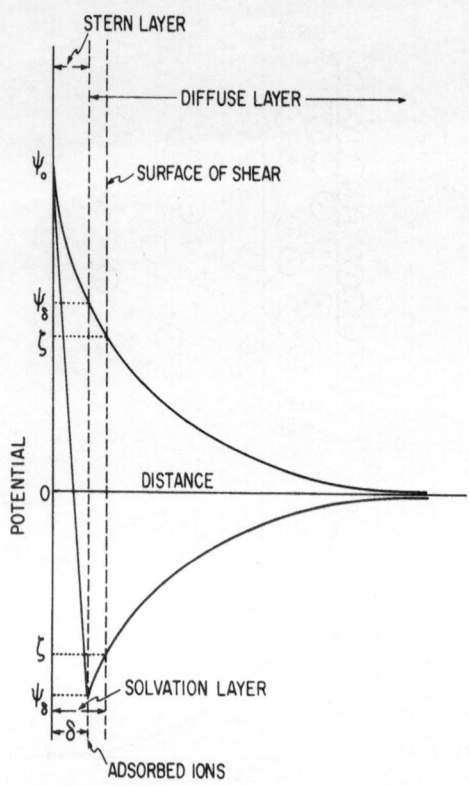

Figure 2. Potential decay curves for an electrical double layer associated with a particle of surface potential $\psi_0$. The lower curve results from strong adsorption within the Stern layer.

fixed and diffuse layers. A layer of water about one molecule thick is probably bound to the surface by charge–dipole interaction. This means that the shearing plane is outside the Stern layer, and thus $\zeta$ is smaller than $\psi_s$. An increase in the electrolyte concentration produces a lowering of the $\zeta$-potential, since more of the potential drop occurs in the immobile part of the double layer.

The ratio of $\zeta/\psi_0$ found experimentally in several systems[6–8] in low electrolyte concentrations is about 0.5, suggesting that specific adsorption effects are low at concentrations less than 0.1 mol dm$^{-3}$. Above this concentration, for these particular systems, however, specific effects become important and reversal of charge occurs.

## 3. Mathematical Treatment of Migration in an Electric Field

The early theories relating the migration velocity to the $\zeta$-potential have been reviewed by Abramson *et al.*[9] Smoluchowski, regarding electrophoresis as the reverse of electroosmosis, derived the equation

$$\bar{v} = \frac{v}{x} = \frac{\varepsilon\zeta}{4\pi\eta} = \frac{\varepsilon_r\varepsilon_0\zeta}{\eta} \tag{1}$$

relating the electrophoretic mobility to the $\zeta$-potential, where $\varepsilon$ is the permittivity, $\varepsilon_r$ the relative permittivity, and $\varepsilon_0$ the permittivity of a vacuum and $\eta$ the dynamic viscosity. Equation (1) is applicable to a particle of any shape or orientation, as well as to an oriented cylinder, provided that the particle is of "easy" shape and that the radius of curvature at all points on the surface is much greater than the thickness of the electrical double layer. The derivation of equation (1) involves the following assumptions:

1. the product of the reciprocal thickness of the double layer [defined in equation (4)] and the radius of curvature of the surface $a$ is large,
2. the particle is nonconducting,
3. the conductivity, permittivity, and viscosity have the same values within the double layer as in the bulk liquid,
4. the applied electric field can be simply added to the field of the electric double layer,
5. the velocity gradient begins at the particle surface.

Debye and Hückel,[10] from a consideration of the forces acting on a charged particle, derived the equation

$$\bar{v} = \frac{C\varepsilon\zeta}{\eta} \tag{2}$$

which, apart from the constant $C$, is identical in form to that of Smoluchowski. Equation (2) is valid independent of the shape and size of the particles, but $C$ depends on both these factors. Analysis of the hydrodynamic equations for a sphere and a cylinder gave values of $C$ as $\frac{1}{6}\pi$ and $\frac{1}{4}\pi$, respectively. Later, Henry[11] found that $C$ also depended on the

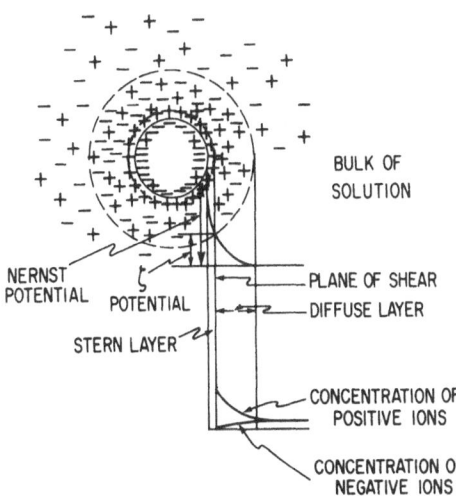

NERNST POTENTIAL

$\zeta$ POTENTIAL

STERN LAYER

BULK OF SOLUTION

PLANE OF SHEAR

DIFFUSE LAYER

CONCENTRATION OF POSITIVE IONS

CONCENTRATION OF NEGATIVE IONS

Figure 3. Schematic representation of charge distribution around a particle with a net negative charge.

orientation, and for a cylinder oriented broadside to the field, obtained the value of $\frac{1}{8}\pi$.

The limited experimental evidence is in agreement with the Smoluchowski equation (1); Abramson[12] and Abramson and Michaelis[13] have, for example, shown that, for white blood cells and also for inert particles coated with protein, the mobility is independent of shape, size, and orientation. Mooney[14] and Henry[11] independently showed that the two theories applied to different limiting cases. While Smoluchowski had assumed that the radius of curvature of a particle was great compared with the thickness of the double layer, it was implicit in the treatment of Debye and Hückel that the radius of curvature be small compared with the thickness of the double layer. These latter authors further assumed that the applied potential was unaffected by the presence of the migrating particle; this is only strictly valid when the conductivity of the particle is identical with that of the medium or when the particle is so small that no appreciable distortion of the field occurs within the double layer. Since, in general, it is assumed that the conductivity of the migrating particle is zero, this theory only applies to very small spheres. Clearly, the mobility of particles under conditions intermediate between the limiting cases will depend on the relative values of the radius of the particles and the thickness of the double layer.

Henry[11] considered the effect of distortion of the applied electric field and derived the equation:

$$\bar{v} = \frac{\varepsilon \zeta}{6\pi\eta} f(\kappa a) \tag{3}$$

where $a$ is the radius of curvature of the particle and $\kappa$ the reciprocal thickness of the double layer, given by

$$\kappa = \left(\frac{8\pi e^2 N_A}{1000\varepsilon kT}\right)^{1/2} I^{1/2} \tag{4}$$

where $I$ is the ionic strength of the electrolyte, $N_A$ the Avogadro constant, $k$ the Boltzmann constant, $e$ the electronic charge. At 298.16 K for water, equation (4) becomes†

$$\kappa/m^{-1} = 1.034 \times 10^8 (I/\text{mol m}^{-3})^{1/2} \tag{5}$$

The term $f(\kappa a)$, which represents a power series of $a$ obtained on integra-

---

† For new workers in this field, considerable time and confusion will be saved if SI units are used from the outset (see Section 11); this applies particularly to the potential measured in esu and absolute units. Concentrations and volumes are expressed in the text in the practical working units of mol dm$^{-3}$ and either cm$^3$ or dm$^3$, respectively.

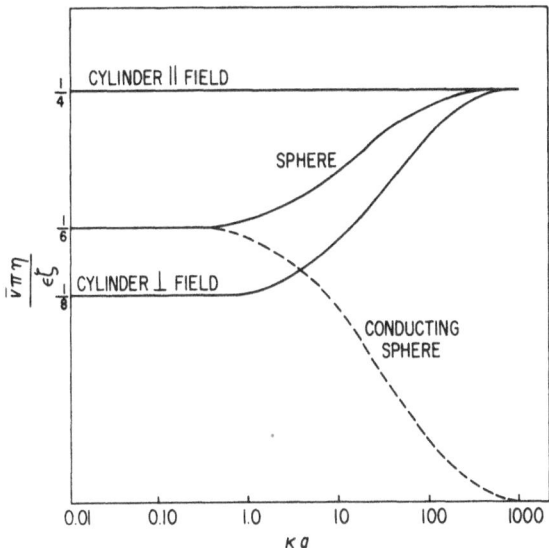

Figure 4. Variation of mobility with $\kappa a$. Reprinted from reference 11, p. 125.

tion, depends upon the shape, size, and orientation of the particles. Comparing equations (2) and (3), it is apparent that $C = f(\kappa a)/6\pi$. Figure 4 shows the variation of $\bar{v}\pi\eta/\varepsilon\zeta$ with $\kappa a$ for cylindrical and spherical particles. From this, it is apparent that the Smoluchowski equation is valid (i) for a cylinder moving parallel to the field for all values of $\kappa a$ and (ii) a spherical particle or a cylindrical particle moving at right angles to the field, provided that $\kappa a$ is greater than 100.

For a particle of radius 1 $\mu$m, the limiting equation of Smoluchowski may be applied, provided observations are made in an electrolyte solution of ionic strength exceeding 1 mol m$^{-3}$. In the range $0.1 < \kappa a < 100$, Henry's correction becomes appreciable. In Figure 5, the radius of the particle for $\kappa a = 100$ and $\kappa a = 0.1$ is plotted as a function of $\kappa$ and the concentration of a 1:1 electrolyte.

The Smoluchowski equation (1) is valid for bacteria and cells in suspension in salt solutions at the concentrations usually used, provided that $a$ is taken as the radius of the particle. However, if $a$ is taken as the radius of curvature of the fine structure on a biological surface,[15] then Henry's equation must be used, even for bacteria and biological cells that have been shown to have fine structure under the electron microscope.

Associated with the high concentration of ions near a charged surface will be an additional conductance, known as the surface conductance $\kappa_s$. This may significantly reduce the value of the electrophoretic mobility by a

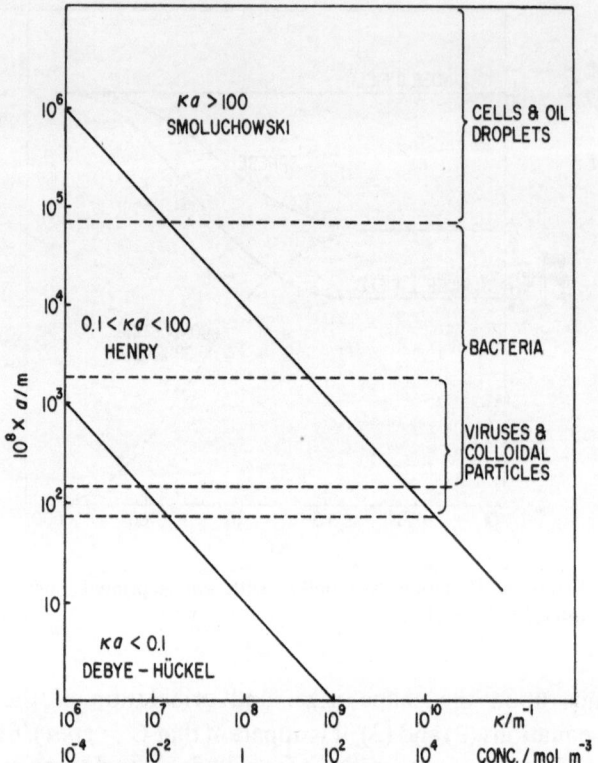

Figure 5. The product $\kappa a$ plotted as a function of $\kappa$ and $a$, showing regions where the equations of Smoluchowski, Henry and Debye–Hückel are applicable. Reprinted from Brinton and Lauffer,[15] courtesy of Academic Press, Inc.

parallel short circuiting mechanism. Henry[16] showed that for an insulating particle,

$$\bar{v} = \frac{\varepsilon\zeta}{4\pi\eta}\frac{\kappa_0}{\kappa_0 + \kappa_s/a} \tag{6}$$

where $\kappa_0$ is the conductivity of the suspension medium.

The surface conductance for each part of the double layer can be calculated from an equation of the type

$$\kappa_s = \frac{\sigma l_+}{\mathscr{F}} \tag{7}$$

where $\sigma$ is the charge density—either $\sigma_D$ [equation (9)] or $\sigma_s$ [equation (11)], $l_+$ the molar conductivity of the counterion, and $\mathscr{F}$ the Faraday constant.

If the surface conductance is only the contribution from the diffuse part of the double layer, then the correction [equation (6)] will be insignificant. Only for large particles when $\kappa a$ is greater than 100 is the surface conductance correction important; $\zeta$ calculated without allowing for $\kappa_s$ may be too low.[17] It has been observed[18] that the $\zeta$-potential calculated from equation (3) passes through a maximum as the ionic strength is lowered. This maximum has been attributed[17] to the neglect of $\kappa_s$ in the evaluation of the $\zeta$-potentials. Recent measurements of the surface conductance of model particles[19–21] show that the experimentally determined values are 100 to 1000 times greater than the contribution from the diffuse part of the double layer [equation (7)]. This suggests that the greater proportion of the surface conductance can be assigned to the fixed part of the double layer. These measurements[22] have been based on the theory and method suggested by Fricke and Curtis.[23]

The surface charge density ($\sigma_0$) is the sum of the charge densities in the Stern ($\sigma_S$) and the diffuse ($\sigma_D$) layers:

$$\sigma_0 = \sigma_S + \sigma_D \tag{8}$$

The charge density in the diffuse layer is given,[9] for a surface of infinite radius, by the equation

$$\sigma_D = \left(\frac{N_A \varepsilon kT}{2000\pi}\right)^{1/2}\left\{\sum c_i\left[\exp\left(\frac{-z_i e\psi_\delta}{kT}\right) - 1\right] + \sum c_j\left[\exp\left(\frac{z_j e\psi_\delta}{kT} - 1\right)\right]\right\}^{1/2} \tag{9}$$

where $c_i$ and $c_j$ are the concentrations and $z_i$ and $z_j$ the valences of the anions and cations, respectively. For electrolytes consisting of ions of single valence $z$ and concentration $c$, equation (9) reduces to the simple form

$$\sigma_D = \left(\frac{N_A \varepsilon kT}{500\pi}\right)^{1/2} c^{1/2} \sinh\frac{z e\psi_\delta}{kT} \tag{10}$$

The charge density in the Stern layer has been calculated[24] using the Langmuir adsorption isotherm, in which it is assumed that the adsorption is monomolecular, as

$$\sigma_S = \frac{n_s z \mathscr{F}}{1 + (1/c)\{\exp[(z\mathscr{F}\psi_\delta + \phi_+)/RT]\}}$$
$$- \frac{n_s z \mathscr{F}}{1 + (1/c)\{\exp[(-z\mathscr{F}\psi_\delta - \phi_-)/RT]\}} \tag{11}$$

where $n_s$ is the number of sites available for adsorption per square centimeter, and $\phi_+$ and $\phi_-$ are the adsorption potentials of cations and anions. The contribution from ions of the same sign as the surface is relatively unimportant, and so one of the terms in equation (11) can be neglected. This equation cannot be applied to real systems, since the values of $n_s$ and $\phi_+$ and $\phi_-$ are generally unknown.

All the equations so far considered are only applicable to a surface that is impenetrable to ions. Haydon[25] has considered modifications to equation (9) to allow for this factor. He showed that the diffuse double layers of various planes of charge will overlap and affect each other's potentials to an extent that depends on the ionic strength. No matter what model he chose, the surface charge was always higher than that calculated for an impenetrable surface, although the general shape of the $\sigma$-concentration curves remained unaltered except when charges of opposite sign to the surface charge were located below the surface. He concluded that the surface charge density, given by equation (9), underestimates the true charge density by a factor $[1 + (1 - \alpha)^{1/2}]$, where $\alpha$ is the fraction of the total space within the surface that is not available to counterions. For a situation when $\alpha = 0$, the surface is completely penetrable to counterions, and thus the charge density has twice the value calculated from equation (9). Attempts to determine $\alpha$ by titration methods have not been successful, and without a knowledge of $\alpha$, this correction cannot be applied. The assumption of impenetrability is more valid at low ionic strengths.

## 4. Calculation of $\zeta$-Potentials and Surface Charge Densities

The numerical calculations of the $\zeta$-potential from the mobility and of the surface charge density from $\psi_\delta$ are complicated first, by the correct choice of equation, second, by the relation of $\zeta$ to $\psi_0$ or $\psi_\delta$, and third, by the fact that the permittivity and the coefficient of viscosity of the medium may have abnormal values within the double layer.

In the double layer, the local electrical field strength may be very high, since the potential may fall as much as 100 mV in a distance of 2 nm from the surface. In such circumstances, the viscosity may well be greater than and the permittivity less than the corresponding value in the bulk phase. If both $\varepsilon$ and $\eta$ vary with the field strength, then equation (1) becomes

$$\bar{v} = \frac{1}{4\pi}\int_0^{\psi_0} \frac{\varepsilon'}{\eta'}\,\partial\psi \qquad (12)$$

where the integration is carried to $\psi_0$, since there is no sharply defined plane of shear. But $\zeta$ from equation (1) may be redefined as

$$\zeta = \frac{4\pi\bar{v}\eta}{\varepsilon} = \frac{\eta}{\varepsilon}\int_0^{\psi_0}\frac{\varepsilon'}{\eta'}\,\partial\psi \qquad (13)$$

where $\eta$ and $\varepsilon$ refer to the values for the bulk solvent and $\eta'$ and $\varepsilon'$ the corresponding values in the double layer.

In the first place, the viscosity of water rises steeply at a field strength of $10^6$ V cm$^{-1}$.[26] Such a field is reached at 0.3 nm from the plane of fixed charges if $\psi_0 = 200$ mV when the concentration is 10 mol m$^{-3}$. Nearer the surface, there is effectively a monolayer of soft ice. In this region, however, it is likely that the many counterions will disorient this soft ice, making it less viscous, although no correction is available for this factor.

The permittivity falls off as the potential increases; several estimates of this effect are available.[27,28] The net result of these variations is that the ratio $\varepsilon/\eta$ falls off sharply with rise in potential gradient.

Equation (13) also permits a theoretical evaluation of the ratio $\zeta/\psi_0$, showing that the tendency is for the ratio to increase as either $I$ or $\psi_0$ is lowered.[6-8] Haydon[8,29] concluded that, for surfaces with charge densities less than 1 C m$^{-2}$, a good approximation to the true surface potential may be obtained, assuming the bulk value of the $\varepsilon/\eta$ ratio. At charge densities in excess of this, the $\zeta$-potential will be less than $\psi_0$.

Despite these difficulties, most workers assume that the values of $\varepsilon$ and $\eta$ in the double layer are the bulk values and that $\psi_\delta$ and $\zeta$ can be equated. Most experimental work is designed to fulfill the requirements of the Smoluchowski equation (1). Thus, using the following values of the physical constants at a temperature of 298 K: $\eta = 0.8937 \times 10^{-3}$ N s m$^{-2}$; $\varepsilon = \varepsilon_r \varepsilon_0$, $\varepsilon_r = 78.54$, $\varepsilon_0 = 8.854 \times 10^{-12}$ F m$^{-1}$; $e = 1.602 \times 10^{-19}$ C; $k = 1.380 \times 10^{-23}$ J K$^{-1}$; $N_A = 6.022 \times 10^{23}$ mol$^{-1}$, the $\zeta$-potential (in millivolts) can be calculated from the measured mobility ($x \times 10^{-8}$ m$^2$ s$^{-1}$ V$^{-1}$) using equation (1), which becomes

$$\zeta/\text{mV} = 12.85x \tag{14}$$

and the charge density (in coulomb per square meter) in the diffuse part of the double layer can be calculated by equation (10), which becomes

$$\sigma_D/\text{C m}^{-2} = (3.713 \times 10^{-3})c^{1/2} \sinh[(\zeta/\text{mV})/51.3] \tag{15}$$

where $c$ is the concentration of the 1:1 electrolyte measured in the units of moles per cubic meter. In the cgs system, the charge density was quoted in electrostatic units per square centimeter (1 C m$^{-2}$ = $3.335 \times 10^{-6}$ esu) (see Section 11).

On account of these uncertainties, most experimenters now prefer to record and report mobility values and to discuss their results in terms of these experimentally measured quantities.

## 5. Principles of Experimental Methods

The electrophoresis of colloidal particles is very similar to ionic migration, not only with respect to the nature and mechanism of the phenomenon, but also as regards the values of the mobility, which are of the order of micrometers per second under an applied field of $1 \, \mathrm{V \, m^{-1}}$. In spite of the similarity between ionic migration and electrophoresis, the methods adopted for investigating these phenomena are rather different. In any determination of a mobility, the velocity of the ion or particle must be measured under a known applied field.

There are two variants of electrophoresis, depending on whether or not the particles are visible under a microscope, namely, the moving boundary method and the ultramicroscopic or microscopic method. The moving boundary method[30] is normally used for the characterization and isolation of proteins from biological systems, but it has also been used to measure the mobility values of both bacteria and red blood cells.[31] This method is not very suitable when it is necessary to use suspension media of high ionic strength or when large particles are to be studied; for such systems, the microscopic method, known as particulate electrophoresis, is preferred. In this method, an individual particle is observed and its rate of migration is measured under a known applied electric field strength. With a suitable apparatus, the technique is rapid and can be used with suspension media over a wide range of pH and ionic strength. High-power magnification is used, and thus, the shape, size, and orientation of the particle and the presence of mixed populations can be observed directly in an environment that is constant with respect to pH and ionic strength over the small time (less than 10 s) required for a measurement.

Basically, the particulate electrophoresis apparatus consists of a closed transparent chamber (usually glass), mounted between two reversible, nongassing electrodes, in which a suspension of particles can be observed microscopically under controlled conditions. The chamber or cell is either rectangular or cylindrical in cross section.

The particulate electrophoresis technique is applicable to suspensions of particles of all types, provided that the particles are visible under the normal light microscope, e.g., inorganic particles, latex particles, oil droplets, air bubbles, and biological cells. Most of the work in the author's laboratory has been concerned with the study of biological cells and bacteria, and it is generally true that the largest use of the technique has been in the study of biological cell surfaces. There are obviously points of special importance and also special precautions that are very relevant to such biological surfaces; these are mentioned where appropriate.

## 5.1. Theory of the Closed Cell

Since the walls of the cell or cuvette assume a charge relative to the contained liquid, there will be, on the application of an electric field, an electroosmotic transfer of liquid along the walls and, in a closed system, a

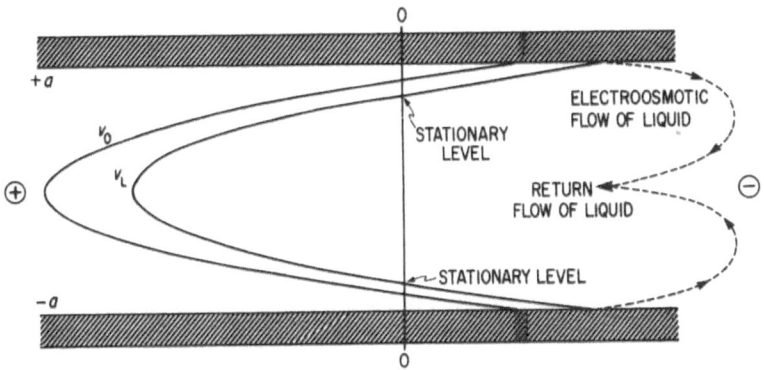

Figure 6. Cross section of flat electrophoresis cell, showing magnitude and direction of liquid flow ($v_L$) and the observed velocity ($v_0$) of the particle as a function of cell depth. The difference between the curves is the true electrophoretic velocity.

return flow through the center of the cell. The observed particle velocity $v_0$ is the algebraic sum of two superimposed velocities, namely, the absolute electrophoretic velocity of the particle relative to a stationary liquid $v_p$ and the velocity of the liquid relative to the stationary surface $v_L$:

$$v_0 = v_p + v_L \tag{16}$$

$v_L$ varies throughout the depth of the cell, while $v_p$ is constant at all depths (Figure 6). Thus, it is impossible to measure the true electrophoretic velocity by observation at any random depth in the cell. There will be a certain location between the walls and the axis of the cell where no net flow of liquid occurs; this is a cylindrical null plane in the cylindrical cell and two flat planes (one on either side of the axis) in a rectangular cell. This null position is termed the "stationary level," and it is the only level in the cell that the measured (observed) velocity is the true electrophoretic velocity. The stationary levels are at different relative positions in cells of rectangular and cylindrical cross section.

The velocity of a particle relative to a liquid was first studied by Ellis,[32] who pointed out that the average observed migration of particles at all depths relative to the cell must be the true migration velocity of the particle relative to the liquid, since the total liquid flow in a closed cell is zero, i.e.,

$$\int_{-a}^{+a} v_L \, dx = 0 \tag{17}$$

Thus,

$$v_p = \frac{1}{x} \int_{-a}^{+a} v_0 \, dx$$

where $2a$ is the total depth of the cell (i.e., $x$ varies from $+a$ to $-a$), and $v_0$ is the observed velocity at a depth $x$. By measuring $v_0$ at different depths in the cell, a velocity–depth curve may be constructed; $v_p$ can be calculated by integration of the equation to the curve [see Section 6.5]. The position of the stationary levels can be obtained by substitution of the value of $v_p$ from equation (18) into the quadratic equation for the velocity–depth curve and solving for $x$, provided that the ratio of width to depth of the cell $(k)$ is large.[33]

The variation of the liquid velocity with cell depth has been studied by Smoluchowski,[34] who considered a flat rectangular cell in which the width is large compared to its depth. Thus, only electroosmotic flow across the two plane surfaces is considered; any such flow around the walls is, under these conditions, negligible. From the definition of viscosity in differential form,

$$F = \eta A \frac{dv_L}{dx} \tag{19}$$

where $F$ is the force, $A$ the area of two liquid planes a distance $dx$ apart, and $v_L$ is the velocity at a depth $x$. In the closed cell, the tendency for the liquid to move due to electroosmosis results in a pressure difference $P$ between the two ends of the cell; this, in turn, causes the liquid to flow in the reverse direction. Considering a layer of liquid of length $l$, width $w$ $(A = lw)$, and thickness $dx$, it follows that the difference between the forces on the top and bottom of this layer, or viscous drag, is given by $lw\eta(\partial/\partial x)(dv_L/dx)\,dx$. This drag must balance the electroosmotic pressure multiplied by the area of the end of the layer, i.e., $Pw\,dx$. Thus,

$$Pw\,dx = lw\eta \frac{\partial}{dx}\left(\frac{dv_L}{dx}\right) dx \tag{20}$$

or

$$\frac{P}{l} = \eta \frac{\partial^2 v_L}{\partial x^2} \tag{21}$$

Integration of equation (21) gives $v_L$ as a function of $x$:

$$v_L = \frac{P}{2\eta l}x^2 + Ax + B \tag{22}$$

where $A$ and $B$ are integration constants. When $x$ is measured from the center of the cell, $A = 0$, since the flow is symmetrical about the center. At the boundaries where $x = +a$, the liquid velocity is the electroosmotic velocity $U$. Thus, equation (22) becomes

$$v_L = \frac{P}{2\eta l}(x^2 - a^2) + U \tag{23}$$

Substituting this value for $v_L$ into equation (17) and integrating gives

$$P = \frac{3U\eta l}{a^2} \qquad (24)$$

when this value of $P$ is substituted into equation (23), the value of $v_L$,

$$v_L = \frac{U}{2}\left(\frac{3x^2}{a^2} - 1\right) \qquad (25)$$

is obtained as a function of the distance $x$ from the center of the cell for a given electroosmotic velocity.

Equation (25) is the equation of a parabola symmetric about the central axis (Figure 6). The levels at which $v_L = 0$ are obtained by equating the expression in brackets [equation (25)] to zero, since $U \neq 0$; thus, $3x^2/a^2 = 1$ or $x = a/3^{1/2} = \pm 0.578a$ from the center. The stationary levels thus occur at depths of 0.21 and 0.79 of the whole cell depth from an inside surface, positions that are independent of the cell depth and the electroosmotic velocity. This result is valid for cells for which the ratio of width to depth is large. Komagata[35] considered a more general case and predicted that the positions of the stationary levels in a flat, closed cell are

$$y\,(v_L = 0) = \pm a\left[\frac{1}{3}\left(1 + \frac{384}{\pi^5 k}\right)\right]^{1/2}$$

where $y$ is the level measured from the central axis and $k$ the ratio of width to thickness. In terms of the fractional depth measured from the top inside face, the depths are given by

$$\frac{1}{2} \pm \left(\frac{1}{12} + \frac{32}{\pi^5 k}\right)^{1/2} \qquad (26)$$

This equation takes into account the side wall effects and gives the following values for the position of the stationary levels:

| $k$ | 10 | 15 | 20 | 50 | 100 | ∞ |
|---|---|---|---|---|---|---|
| Stationary levels from top inside face | 0.194 0.806 | 0.199 0.801 | 0.202 0.798 | 0.208 0.792 | 0.210 0.790 | 0.211 0.789 |

For cells in which $k$ is greater than 40, the Smoluchowski equation (25) is valid.

For a closed cell of cylindrical cross section of radius $R$, the liquid velocity is given[36] by

$$v_L = \frac{P}{4\eta}\left(r^2 - \frac{R^2}{2}\right) \qquad (27)$$

The velocity of the liquid at the surface $U$, where $r = R$, is given by $PR^2/8\eta$; hence, equation (27) becomes

$$v_L = U(2r^2/R^2 - 1) \tag{28}$$

Again, this is the equation of a parabola symmetrical about the central axis. As before, $v_L = 0$ when $2r^2 = R^2$ or $r = R/2^{1/2} = 0.707R$ from the axis or $0.293R$ from the wall of the tube.

The parabolic variation of $v_L$ with depth in either type of cell leads in practice to a large change in the observed velocity of the particle ($v_0$) with the position in the chamber, particularly in the region of the stationary level. The thinner the cell, the steeper will be the velocity–depth curve at the position of the stationary level; thus, errors involved in locating this position accurately can be serious. The use of thicker cells or tubes overcomes this but introduces the practical difficulty of the requirement of an objective with a large working distance. Further, in deep rectangular cells, there is the additional problem of electroosmosis around the edges of the cell.

## 5.2. Types of Closed Cell

Subsequent to the pioneer work of Ellis,[32] many different types of cell have been described.[37,38] Hydrodynamically, the cylindrical cell is preferred on account of its symmetry, but the complications in its use offset this advantage. Unless special optical precautions are taken, there is considerable distortion of the image of the particle in suspension, and there is difficulty in focusing on the stationary level as the distance from the roof of the cell increases.[39] Further, as has been discussed, the stationary level is a cylindrical plane within the tube and is therefore not flat in the field of view of the microscope.

In an attempt to overcome errors involved in focusing at the stationary levels that occur on the steep portion of the $v_0$–depth curve, cells have been designed in which the return flow of liquid occurs not along the center of the observation tube but through an auxiliary tube.[40–43] The true particle velocity can be observed at the median plane in the cell, errors through inaccurate focusing being thereby reduced. The operation of such a cell assumes that, during the course of a series of measurements, the surface conditions in both tubes are identical.

When the internal surface of a 1.2-mm-inner-diameter capillary tube is coated with a 0.2-mm-thick layer of agarose, the electroosmotic flow is reduced to a negligible value, and timings can be made at any depth in the cell[43a]; this reduces errors due to the exact location of the stationary levels. Since agarose has a very low mobility, the return electroosmotic flow along the walls of the capillary is negligibly small. During prolonged use, loss of agarose from the glass surface eventually giving rise to bare glass areas and adsorption of ions and other material from the suspension could be potential sources of error.†

Many types of particulate electrophoresis apparatus based upon the rectangular cell pattern have been described. The earliest were made either from glass plates in which a groove was cut and covered with another glass plate or from microscope slides cemented together.[37] Such cells are not capable of any high degree of accuracy, since the use of cement makes adequate cleaning difficult and the unevenness of the surface interferes with the hydrodynamic behavior of the liquid. The first glassblown cell was that of Northrop and Kunitz[44] who fused microscope slides together with a space between and joined the cell so formed to the electrode compartments by rubber tubing. The rubber connections gave rise to difficulties in cleaning, in vibration, and in support.

The first type of apparatus capable of giving accurate results was designed by Abramson[45]; a complete description of this apparatus and experimental procedure has been given by Moyer.[46] Most rectangular cells have been based on this design, including that used in the author's laboratory for many years,[47] which will be described in detail later.

The cylindrical chamber, first used by Mooney[48] and later modified by Alexander and Saggers,[49] has been fully described by Seaman and his collaborators.[38,50] This type of observation chamber consists essentially of uniform-bore capillary tube in which a flat face has been ground in the thick wall to enable observation with a microscope.

The horizontal mounting of the rectangular cell, which rests directly on the microscope stage,[46,47] is suitable if the particles under observation are small so that little settling occurs during their excursion time across the field of vision in the microscope. If the cell is mounted laterally,[51] i.e., short axis in the vertical plane, then any settling and most convective disturbances are perpendicular to the direction of the velocity gradient in the cell. Measurement of the horizontal component of the particle velocity at the stationary level gives the electrophoretic velocity directly of a particle that is settling in the gravitational field. In this position, good thermostatting is essential to minimize the effect of convection currents.

---

† *Note by the Editor.* In the method recommended by van Oss *et al.*,[43a] the 2.5% agarose gel is prepared using the same buffer solution that is used to suspend the particles that are being studied. It is believed[43a] that the electroosmotic flow is not eliminated, but is confined within the stationary gel layer, and so the electroosmotic flow is decoupled from the flow of liquid in the lumen. No disturbance has been observed, which could be attributed to the irregularities of the agarose layer; this is as would be expected on account of the suppresion of flow in the lumen. As employed in the laboratory of van Oss, there is no need to use the same agarose layer for long periods of time, so possible loss of agarose need present no problem.

Whatever type of cell is used, it must be firmly mounted, often on the stage or modified stage of the microscope, so that it cannot move relative to the microscope and is not subject to outside vibrations.

## 5.3. Electrode Systems

Considerable difficulty has been experienced with the electrode and electrode assemblies that have been employed. It is essential that any electrode system must be reversible and nongassing. Platinum electrodes have been commonly used with the cylindrical type cells.[38,50] The area of the electrodes must be such that the current density is sufficiently low to obviate serious risk of polarization, gassing, etc.

Agar plugs saturated with KCl have been used to prevent the diffusion of the electrode liquid $ZnSO_4$ in $Zn/ZnSO_4$ or $CuSO_4$ in $Cu/CuSO_4$ electrodes[45,52] into the observation chamber. Plaster of Paris plugs saturated with $Na_2SO_4$ to render them conducting have been used to separate the $Cu/CuSO_4$ electrode system from the suspension medium.[46] The difficulty of such systems is that eventually the electrode solution containing divalent copper ions (which are toxic to biological systems) will diffuse through the plug and contaminate the suspension (thereby also altering the ionic strength) or the plug will drop out (Figure 7a). In either event, frequent maintenance is required. Replacement of these plugs with sintered glass disks and the use of a silver–silver chloride electrode system[53] has resulted in a permanent assembly (Figure 7b) that does not

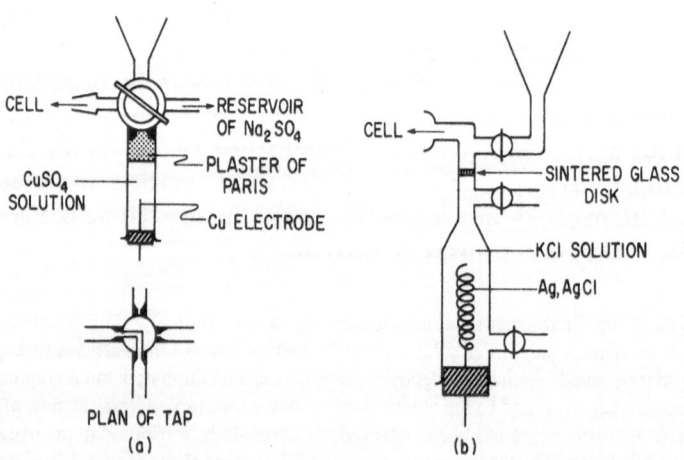

Figure 7. Typical electrode systems: (a) using Cu, $CuSO_4$ electrodes with plaster of Paris plugs[46]; (b) using Ag, AgCl, KCl electrodes with sintered glass disks.[53]

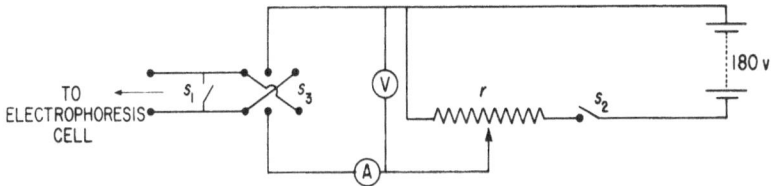

Figure 8. Electrical circuit for particulate electrophoresis. $S_1$ and $S_2$ are toggle switches, $S_3$ a reversing switch, $r$ is a variable 10-k$\Omega$, potentiometer, $A$ a milliammeter, $V$ a high-resistance voltmeter (only included in circuit for use with cylindrical cells that incorporate platinum electrodes).

require frequent renewal of the mechanical boundary.[47] This electrode system in which potassium chloride is the electrolyte has the additional advantage that slight leakage of the potassium chloride solution will not be so serious as with copper sulfate solution.

The use of palladium in the form of plates precharged with hydrogen to a $H_2/Pd$ ratio of 0.5, has been described.[54] Electrodes thus charged are nongassing and can be used as anodes for more than 12 h at 10 mA before the hydrogen is used up.

### 5.4. Electrical Circuit

Suitable circuits including milliammeter, reversing switch, and controls have been described.[46,47] The simple circuit currently used in the author's laboratory (Figure 8) has proved entirely satisfactory. The circuit is supplied from high-tension batteries in series to give about 180 V; the current flowing is controlled by the wire-wound resistance ($r$) of 10 k$\Omega$ dimensioned for 10 W. The constant value of the current flowing is recorded on the milliammeter.

$S_1$ is a shorting switch for the electrodes, $S_2$ the battery switch, and $S_3$ the reversing switch, which does not interfere with the direction of the current through the milliammeter ($A$). The milliammeter recommended is a works-calibrated Sangamo-Weston type, with ranges 0–1, 0–2, 0–5, 0–10, 0–20, and 0–50. The high-resistance voltmeter ($V$) is usually only included in the circuit for use with cylindrical cells, which have a high internal resistance.

The field strength, $X/V\,m^{-1}$ is best calculated[46] from the application of Ohm's law to the cell in the form

$$X = \frac{I}{q\kappa} \tag{29}$$

where $I/A$ is the current flowing, $\kappa/\Omega^{-1}\,m^{-1}$ is the conductivity of the suspension and $q/m^2$ is the cross-sectional area of the cell. The conductivity of the suspension is measured in a bottle-type conductance cell[55] with gray platinum electrodes, using an a.c. Wheatstone bridge network operating at radio frequency. The conductivity must be measured at the same temperature as that used for the measurement of the electrophoretic velocity.

The effective cross-sectional area $q$ at the point where the velocity is measured of a cylindrical cell may be obtained directly from the geometry of the cell, e.g., by measuring the length of a thread of mercury of known weight. In a rectangular cell, the cross-sectional area is best obtained by calibration using a particle of known electrophoretic mobility (see Section 5.4).

## 5.5. Optical Arrangement

The recommended optical system consists of a microscope, focused by calibrated movement of the objective fitted with a 40× objective and a 10× focusing eyepiece carrying a 1-mm crosshatch graticule. This combination gives an overall magnification of 400×. The insertion of an angled eyepiece makes for a more relaxed body position during observation and also increases the magnification to 600×. It is preferable that the objective and condenser should be of water immersion type; this improves definition.

With rectangular observation chambers, phase-contrast optics may be used with advantage; phase-contrast optics cannot be used with cylindrical cells. The principle of ultramicroscopy may be applied to suspensions containing particles of colloidal size. The suspension is placed in a cylindrical cell with the incident light beam at right angles to the optical axis of the microscope.

The source of light is a high-intensity source, which is brought to focus on the backplate of the condenser. This must be correctly aligned so that the light beam passes through the cell and straight up the optical axis of the microscope; incorrect alignment gives rise to uneven illumination and often spectral colors. The contrast between the particles and the suspension medium can be improved by closing down the iris diaphragm (on the condenser or lamp) or by interposing a colored filter in the light path.

The overall working distance of a 40× objective is 1.6 mm, so that it is possible to focus right through a rectangular cell in which two optical flats (0.5 mm thick) are separated by 0.5 mm. In a cylindrical cell of 2-mm internal diameter, with a ground flat on one surface, it is possible to focus to the axis of the tube only.

The fine adjustment of a microscope has many turns, but on account of its construction and wear, the movement of the objective as a result of one complete revolution may vary with the number of turns along the drive. It is therefore necessary to investigate the uniformity of response of this adjustment by measuring the depth of a standard-thickness coverslip (the depth of the observation chamber is very convenient) starting from different initial positions of the fine adjustment (i.e., at far end, then 1, 2, 3, ... revolutions in). In use, the calibrated fine adjustment must be used only over the range of uniform response. The depth of the cell must be determined each time it is required to focus on the stationary level. The micrometer adjustment should be tightened sufficiently to prevent mechanical drift during an observation.

To determine the depth of a rectangular cell, subsequent to locating the stationary level, it is necessary to focus on particles adhering to the top and bottom inside surfaces, when the cell is filled with water. The selection of a particle in focus is a subjective matter and, once established, must always be adhered to both at a glass surface and in free suspension. From the readings on the calibrated fine adjustment of particles in focus on the top and bottom glass surfaces, the depth of the cell may be determined, and hence the microscope can be adjusted to be in focus at the stationary level.

In the cylindrical cell, focusing on the stationary level is complicated by the planoconcave cylindrical lens (the tube is ground plane on the outside) (see Section 7.1). By geometrical optics, Henry[39] showed that the actual point of focus of the objective and the focal point of the same objective are related by

$$\frac{n_1}{pr - n_1 kr/n_2} = \frac{n_3}{qr - kr} + \frac{n_2 - n_3}{r} \tag{30}$$

where $n_1$, $n_2$, and $n_3$ are the respective indices of refraction of the surrounding fluid (air or water), the glass of the chamber, and the suspending medium. If $r$ is the radius of the tube, then $kr$ is the minimum thickness of the wall at the optical flat, $qr$ is the actual distance to the stationary level, and $pr$ is the distance through which the objective must be moved from the optical flat to focus on particles in the stationary level. In air, assuming $n_2 = 1.47$, the Henry correction amounts to 0.12; however, for a similar cylindrical chamber immersed in water, the correction is less than 0.012 and in most cases can be neglected.

For use in a water thermostat, the condenser and objective must be waterproofed.

## 5.6. Thermostat

It is desirable to refer all mobility values to a standard temperature, usually 25°C. Various workers[56–58] have shown that the relationship

$$\eta_1 \bar{v}_1 = \eta_2 \bar{v}_2 \tag{31}$$

where $\eta_1$ and $\eta_2$ are the coefficients of viscosity and $\bar{v}_1$ and $\bar{v}_2$ the electrophoretic mobility values at two temperatures, is valid over a range of temperatures. Thus, Gittens[57] showed that, within the limits of experimental error, the electrophoretic velocity of bacteria in suspension is independent of temperature in the range 20–30°C, providing that the measurements are made at constant current. This means that, provided that the velocity measurements and the conductance measurements [for $\kappa$ in equation (29)] are made at the same temperature, then the mobility value can be corrected to 25°C.

It is advisable, however, to thermostat the cell at a definite temperature, as this also maintains the constancy of temperature during an observation by reducing heating effects from electrical heating or from the light source, thereby reducing the convection currents within the chamber. With some biological cells, e.g., spermatozoa[59] and bacteria that are normally motile, it may be necessary to make measurements of the electrophoretic velocity at low temperatures to eliminate the effects of the motility. Most modern designs of apparatus now incorporate a water thermostat around the cell. Although it is more difficult to arrange than an air thermostat, a water thermostat has many distinct advantages, e.g., greater dissipation of heat.

## 6. Experimental Assembly of the Rectangular Cell

### 6.1. Design and Construction

The complete apparatus is shown diagrammatically in Figure 9 and is illustrated in Figure 10. The glassware is available from E. C. Linskey, 23, Tavistock Place, London, WC1. The cell, in this apparatus mounted in the horizontal position, is constructed by fusing two optically flat Hysil glass plates (40 × 25 × 0.5 mm thick) around the edges using spacers to give a cell depth of approximately 0.5 mm (width/depth = 50). The sidearms of 10-mm-bore Pyrex tubing are sealed directly onto the cell and bent as shown to allow the cell to be immersed in the water thermostat. Spherical ground glass joints sealed onto these sidearms are used for connection to the electrode compartments. The overall length between the joints is 270 mm, and the drop of the cell below the arms is 25 mm. Solid glass strengthening struts, fused to the sidearms around the cell and in a plane with it, give additional cell rigidity. This allows the cell to be handled with less risk of breakage.

The electrode compartments are constructed, as shown in Figure 9, by sealing sidearms carrying taps of vacuum quality (i.e., hollow) onto 10-mm Pyrex tubing carrying a no. 2 porosity sintered glass disk (maximum pore size 40–50 $\mu$m). The sintered glass disk is used to prevent excessive diffusion of potassium chloride from the electrode compartment into the suspension

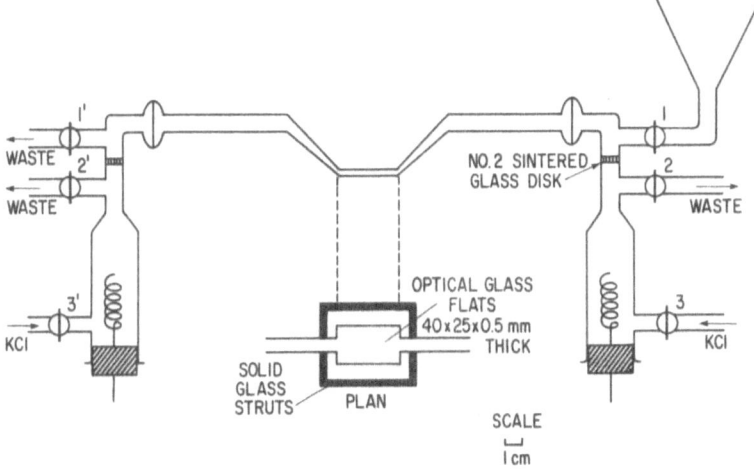

Figure 9. Electrophoresis assembly using rectangular cell mounted in horizontal position.

under investigation. The physical boundary between the two liquids forms in the disk. The sidearms are sealed as close as possible to the disk to prevent the accumulation of pockets of liquid or air bubbles, which may be difficult to flush out. For a similar reason, the bends at the top of the electrode compartments should be as smooth as possible. A funnel, fused on the right-hand sidearm, facilities filling; the corresponding exit arm on the left-hand electrode compartment is attached to a reservoir to collect the suspension as it passes from the cell. This reservoir can, with advantage, be located below bench level, so that filling and washing of the cell can be accomplished by gravity feed. The inlet taps, 3, 3′, are connected by nylon or rubber tubing to two different 500-$cm^3$ separating funnels filled with

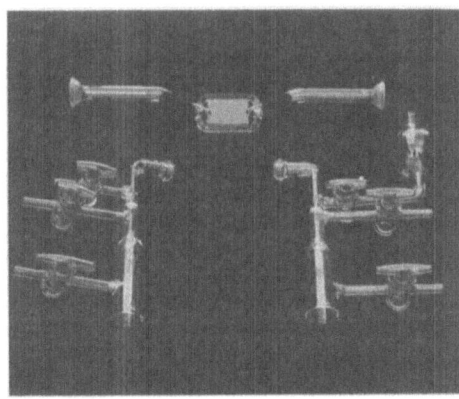

Figure 10. Complete glassware of rectangular cell assembly.

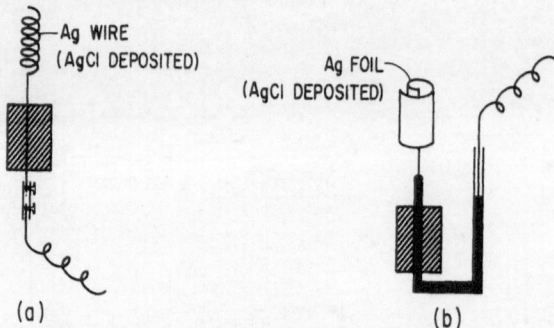

Figure 11. Types of electrode for use with electrophoresis apparatus in Figure 9: (a) anodized silver wire, suitable for measurements at low current densities; (b) anodized silver foil of large surface area, suitable for measurements at high current densities.

$3.5 \, \mathrm{mol \, dm^{-3}}$ potassium chloride solution, and the outlet taps, 2, 2′, are connected by tubing to separate waste reservoirs, again located below the bench.

For work in which only low current densities are to be used, the electrodes are best constructed of a 15-cm length of pure silver wire (14 SWG) wound into the form of a spiral on a round former. One end of this can be forced through the rubber bung that seals the electrode compartment (Figure 11a). For work at high ionic strengths and high current densities, the electrode area must be large. This is best achieved by using pure silver foil ($80 \times 30 \times 0.15$ mm thick) rolled into a cylinder and welded to 30 mm of 0.5-mm-diameter platinum wire. The wire is sealed into a length of bent glass tubing, filled with mercury for electrical connection (Figure 11b), which passes through the rubber bung.

The electrodes should be silver plated before being anodized. This is accomplished by removing grease from the electrode with acetone, followed by cleaning with dilute nitric acid. This latter step improves adhesion of the plate, possibly by oxide removal or etching the silver itself. The electrodes are then washed thoroughly with distilled water and silver plated from a potassium argentocyanide solution until a coherent deposit of silver is formed.[55] It is important that all the platinum on the second type of electrode be covered with silver. The electrodes are now thoroughly washed and anodized.[55] The silver electrodes connected in parallel are made the positive electrode and a platinum wire the negative electrode in a cell in which the electrolyte is $0.1 \, \mathrm{mol \, dm^{-3}}$ hydrochloric acid. Electrolysis at a current density of about $1 \, \mathrm{mA \, cm^{-2}}$ is carried out for about 1 h, when a uniform purple deposit of silver chloride covers the complete electrode. All the electrode right up to the rubber bung must be anodized, otherwise the electrode will not behave as a silver–silver chloride electrode. Once prepared, the electrodes must always be stored in $3.5 \text{ mol dm}^{-3}$ potassium chloride solution and must never be touched. When it is necessary to reanodize the electrodes, the deposit of silver chloride may be stripped off using concentrated aqueous ammonia solution.

The microscope, in which the focusing is by the movement of the objective, is best fixed to a bench that is level and free from vibration caused by people working on the bench or by people moving about the laboratory. The normal stage of the microscope is replaced by a 10-mm-thick rectangular plate (240 × 140 mm) with a hole cut in the center of it large enough to permit the substage condenser to pass through it. This plate also carries (on either side) adjustable mounting clamps (Figure 12) that permit the leveling and other fine adjustments in the positioning of the cell. Each mounting clamp, made from two brass plates (75 × 25 × 5 mm thick), separated by four adjustable bolts, is bolted into lateral slots cut in the base plate. To the top plate is bolted a Perspex holder with a groove shaped to take the sidearm of the cell. The slots in the Perspex plates are perpendicular to those in the base plate; this enables accurate positioning of the cell in three dimensions. Foam plastic pads are used between the top and bottom Perspex plates to protect the sidearm from breakage. With suitable care, the cell can be rigidly fixed in the required position so that there is no possibility of relative movement of cell and microscope. The electrode compartments are held to the sidearms of the electrophoresis cell by clips on the spherical joints and supported either by clamps or by rubber bungs.

The thermostat tank for this apparatus consists of two parts. The large reservoir tank constructed from 15-mm-thick Perspex has an overall size of

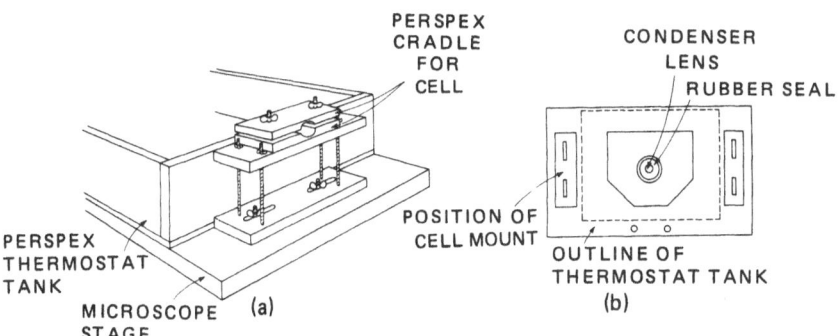

Figure 12. Base plate of microscope, showing mounting for sidearms of the electrophoresis cell and the position of the thermostat tank. (a) Perspective view of one mount, showing side of thermostat tank, (b) Plan of base plate.

$310 \times 310 \times 250$ mm deep. One corner of this is cut away to a depth of 70 mm to allow the attachment of the small tank ($180 \times 180 \times 40$ mm deep), constructed from 7.5-mm-thick Perspex (Figure 13), which is located on the modified stage of the microscope. A polythene tube is used to connect the circulating pump of the temperature control unit directly to the smaller tank. This allows a constant flow of water across the cell at any required temperature; for work at temperatures below ambient, it is necessary to incorporate a cooling unit in the larger reservoir. A hole is cut in the center of the base of the small tank large enough to allow the substage condenser to pass through; the top lens of the condenser should be at least 1 cm above the bottom of the thermostat. This hole in the Perspex is sealed with a thin rubber sheet bolted to the Perspex base between two aluminum O rings. A smaller hole is cut in the rubber sheet so that there is a watertight seal between the condenser and the tank; an improved seal is obtained by placing some rubber cement (e.g., Cow gum) between the rubber and the condenser. This rubber sheet must not be so taut that during the course of an observation the condenser lens is pulled away from the set position close to the far side of the observation chamber.

An alternative, and possibly more convenient, arrangement of the thermostat is to use two separate tanks, one mounted on the microscope stage as described, the larger one equipped with the control unit is located at a lower level. The water at the required temperature is pumped up into the small tank, across the cell, and returns by gravity through a tube, the level of which is adjusted to be just above the electrophoresis cell (Figure 14). This latter, more flexible, arrangement can be used when the focusing of the

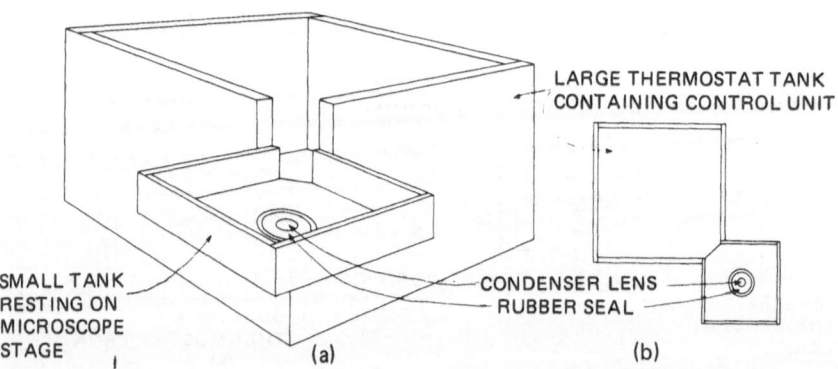

Figure 13. Thermostat tank for rectangular cell mounted in horizontal position. (a) Perspective view, showing smaller tank to fore supported on the base plate of the microscope, (b) Plan of tanks. These tanks may be separated and water circulated from the large tank located at a lower level using gravity return.

Figure 14. Rectangular cell assembled and mounted in thermostat tank on microscope stage; focusing is by movement of stage. (Courtesy of Dr. D. Cornick.)

microscope is by movement of the stage. Provision must then be made for the mounting of the electrode compartments onto the stage so that the cell, electrodes, and thermostat all move with the stage.

Depending on the dimensions and design of the microscope used, it may be necessary to mount the microscope on a thick block or to adjust the dimensions quoted. Additionally, it may be necessary to cut away the sides of the smaller tank to allow the cell arms to pass over. However, the water level in the smaller tank must be sufficient to give a coverage of the cell with about 25 mm of water.

## 6.2. Assembly and Preparation for Use

If the objective and condenser lenses are not waterproofed, then the edges between the glass of the lens and its metal mount must be proofed. This is best achieved by applying two or three coats of a good-quality nail lacquer, allowing time to dry between applications. Care must be taken not to allow the varnish to creep too far onto the glass, as this will cut down the light passing through.

The fine adjustment screw of the microscope is calibrated over the full number of turns, using a coverslip of constant thickness; the most uniform range is selected for subsequent use.

Before setting up, particles should be deposited on the inside surfaces of the chamber to aid location of these surfaces and hence the stationary level. Bacteria are very convenient for this purpose as they adhere well and are easily recognized. If an alcoholic suspension of bacterial cells is allowed to dry onto the glass at 37°C, then a fine deposit of bacteria will adhere to the surfaces. This should not be a thick deposit; only individual cells are required. A thick deposit will produce changes in the electroosmotic flow. Excess bacteria are washed out before assembly. For workers without a source of bacteria, the particles under study should be deposited; if a suspension in alcohol is allowed to dry, then a few particles will adhere. Excess particles are washed out before assembly.

The cell is now located in the Perspex cradles on the adjustable mounts on the stage of the microscope. All the vertical and horizontal adjustments are made until the cell is level and near the top surface of the substage condenser. When correctly positioned, the Perspex holders (with the protecting foam plastic) are carefully but firmly bolted together with wing nuts so that there is no possibility of movement of the cell relative to the microscope. A ready check that movement does not occur is obtained by the day to day observation of the position of the bacteria or particles in focus on the top and bottom surfaces.

The taps and barrels on the electrode vessels are carefully degreased with carbon tetrachloride and then adequately, but not excessively, greased. A thin, transparent, continuous film, with no streaks, indicates correct greasing. One simple method of greasing is to apply a very thin streak of petroleum jelly along the full length of the tap (not across the hole) and then warm gently in a luminous flame until it is just molten. If the tap is now pushed firmly into the barrel so that the holes in tap and barrel are in line and held stationary for a few seconds, the thin film of grease will creep around the tap. Unless the taps and barrels are numbered, all the taps should never be removed at the same time; incorrect assembly of tap and barrel can be responsible for leakages. The rubber bungs carrying the silver–silver chloride electrodes are now inserted into the electrode compartments. The spherical joints are greased (not excessively) and each electrode compartment attached to the sidearm of the observation chamber, the joints clipped, and the compartment supported.

The lower sidearm on each electrode vessel is connected by nylon or rubber tubing to a separate reservoir (500-cm$^3$ separating funnel) mounted about 30 cm above the apparatus, and the upper sidearm is connected by tubing to separate waste reservoirs located below the bench. The exit tube (tap 1′) is connected to another reservoir below the bench. The electrode compartments are filled from the reservoirs with 3.5 mol dm$^{-3}$ potassium chloride solution by opening taps 3 and 2 and 3′ and 2′ until some flows to waste; all air bubbles must be expelled from the electrode compartments. All these taps and those on the separating funnels are turned off.

The top part of the apparatus, including the observation chamber, must now be filled with distilled water. This is introduced through the funnel and tap 1. Air bubbles, which may become trapped in the bends at the top of the electrode vessels, may be removed by loosening the clip on the spherical joint and rotating the electrode vessel through 180°. (If the bends at the top of these vessels are smooth and round, air bubbles will not be so trapped.) Air bubbles trapped in the cell itself are best removed by applying alternate suction and pressure (using the mouth) through a wash-bottle used as a trap containing water connected on tap 1' in place of the reservoir. The exact procedure for the complete removal of the air bubbles from a particular setup is a matter of experience. Once the apparatus has been completely filled, care should be taken when introducing and renewing suspensions and solutions that further air bubbles are not admitted.

The thermostat tanks are filled with distilled (or deionized) water so that the observation chamber is covered with at least 25 mm of water. The substage condenser is then brought close to the bottom of the cell and the microscope lamp adjusted and focused until maximum definition of particles adhering to each glass surface is obtained. If phase contrast optics are used, it is further necessary to adjust the relative positions of the phase plates in the objective and condenser until they are correctly aligned. In adjusting the objective lens, extreme care should be taken, as the small working distance of a 40× objective leaves little clearance between the objective and the top outside face of the cell.

## 6.3. Mode of Operation

The procedure, established in the author's laboratory over a period of years, if rigorously followed, will ensure reproducibility of results.

1. Adjust the temperature of the thermostat and allow sufficient time for the establishment of temperature equilibrium.

2. Close shorting switch $S_1$ (Figure 8) for 1 min to eliminate electrode polarization; then open it.

3. Flush through each electrode compartment with 50 cm³ of the potassium chloride solution. With taps 1, 3, and 3' open and taps 1, 2, and 2' (Figure 9) closed, allow a small amount of potassium chloride solution to pass through the sintered glass disks into the region above. Finally, pass another 25 cm³ of potassium chloride solution through each compartment and close all the taps (including those on the separating funnels). It is necessary to perform this procedure at least twice each day.

4. Thoroughly wash the observation cell and all connections with glass-distilled water to remove all electrolytes (particularly the potassium chloride solution).

5. Flush the observation cell with 50 cm³ of the buffer solution to be used for the next suspension previously thermostatted to the working temperature.

6. Flush the experimental suspension (previously thermostatted) through the apparatus and close taps 1 and 1', thereby trapping a sample.

7. Adjust the eyepiece so that the graticule is in focus; there is no need to calibrate the graticule. Determine the cell depth in arbitrary units using the range of the micrometer that has been shown to have a uniform response. This is achieved by focusing on particles (or bacteria) adhering to the inside surfaces of the chamber. Always make the final focusing adjustment from below, i.e., objective moving upward, to avoid backlash in the fine adjustment screw. Calculate the position of the stationary level using the results previously obtained for the velocity–depth curve (see Section 6.5) and adjust the microscope accordingly, again moving up from particles in focus on the bottom of the chamber. Any particle now in focus (the same focus as used on the glass faces) is in the required level. It is preferable to work at the upper stationary level (see Section 6.5).

8. Check that the particles do not move systematically within the cell. Such movement may be due to badly fitting joints, to inadequately greased taps, to polarized electrodes, or to the nonestablishment of temperature equilibrium.

9. Switch on the current ($S_2$ and $S_3$, Figure 8) and adjust it to give a suitable excursion time of the particles (3–5 s) across a given distance in the crosshatch graticule. Select a particle in focus and, using a stopwatch reading to 0.01 s, record the time required for the particle to traverse a fixed distance on the graticule (i.e., a number $n$ squares of side $L$). Reverse the polarity of the electrodes (switch $S_3$) and measure the time in the reverse direction. To avoid the possibility of bias, the particle to be timed must be selected while stationary; then the current is switched on and the time recorded. Depending on the homogeneity of the particles in suspension, it is necessary to take between 20 and 50 separate timings on individual particles. Keep the current in either direction and throughout a set of readings constant with the resistance; record this value of the current. Differences in the excursion times in the two opposite directions should not exceed 1–2% of the average recorded time and should be random; any systematic difference larger than this means that there is a superimposed drift of liquid in one direction, possibly due to leaks. From the known polarity of the electrodes and the direction in which the particles are moving, the sign of the charge carried may be determined. (*Note*: The optical system reverses the actual direction of movement.)

10. Wash out the cell and leave filled with distilled water with tap 1 open when not in use to allow for volume changes occurring when the thermostat is turned off.

11. Determine the conductance of the suspension at the same temperature.

## 6.4. Calculation of Electrophoretic Mobility and Determination of Cell Constant

The electrophoretic mobility $\bar{v}/\text{m}^2\,\text{s}^{-1}\,\text{V}^{-1}$, i.e., the particle velocity $v/\text{m s}^{-1}$ per unit potential gradient in the stationary level, is given by

$$\bar{v} = \frac{v}{X} = \frac{nL}{t}\frac{q\kappa}{I} = \frac{nL}{t}\frac{q}{I}JG \tag{32}$$

where $nL/\text{m}$ is the distance traveled ($n$ is the number of squares of side $L/\text{m}$) in time $t/\text{s}$, $q/\text{m}^2$ is the cross-sectional area of the cell, $I/\text{A}$ is the current flowing, and $\kappa/\Omega^{-1}\,\text{m}^{-1}$ is the conductivity of the suspension, which is obtained from the measured conductance $G/\Omega^{-1}$ and the cell constant $J/\text{m}^{-1}$ of the conductance cell.

The values of $G$, $I$, and $t$ are obtained experimentally. It is not possible, however, to measure the cross-sectional area of a rectangular observation chamber. This difficulty is overcome by using a standard particle that has a known mobility $\bar{v}_s$, generally measured absolutely in a cylindrical cell, when suspended in a buffer solution of known pH and ionic strength. By timing this particle using the technique previously described, an apparatus constant $K$ for the apparatus (including the cell constant $J$ of the conductance cell) may be determined, where $K$ is given by

$$K = LqJ = \frac{\bar{v}_s t' I'}{nG'} \tag{33}$$

where the primed letters refer to the values for the standard particle. Thus, by combining equations (32) and (33), the mobility of the particle under study can be calculated:

$$\bar{v} = \frac{KnG}{It} \tag{34}$$

For use as a standard, a particle in suspension must be easily available and must have a reproducible mobility value, which for preference should not vary markedly with small changes in pH and ionic strength. The primary standard, which largely meets these criteria, is the human erythrocyte[38,50]; it has been used extensively. On account of difficulties encountered with obtaining and using fresh suspensions of erythrocytes daily, however, secondary standards have been more widely used (Table 1). Although the mobility values of red blood cells and bacterial cells have been shown to be constant under the specified conditions, it is not considered good practice to use such biological material, which is liable to variation, as a reference. It is only recently that a good reproducible nonbiological standard particle

*Table 1. Standard Particles Used for Calibration*

| Particle | Suspending medium | $I$/mol dm$^{-3}$ | pH | $10^8\bar{v}$/m$^2$ s$^{-1}$ V$^{-1}$ at 25°C |
|---|---|---|---|---|
| Red blood cells | Phosphate[45,60] | 0.172 | 7.35 | $1.31 \pm 0.02$ |
|  | Sodium chloride[38,50] | 0.145 | 7.0 | $1.07 \pm 0.02$ |
| *Klebsiella aerogenes* | Acetate–barbiturate[57] | 0.05 | 7.0 | $1.30 \pm 0.05$ |
| (NCTC 418) |  | 0.02 | 7.0 | $1.67 \pm 0.03$ |
| Micropaque[a] | Acetate–barbiturate[61] | 0.02 | 7.0 | $1.47 \pm 0.03$ |

[a] Supplied by Aspro-Nicholas Ltd., Slough, England.

has become available; this is the commercially available barium meal "Micropaque." It is a stabilized barium sulfate preparation that does not sediment out of suspension. At a concentration of about 0.1% (w/v), the suspension is sufficiently thick to observe individual particles and not too thick to cause abrasion and wear of the taps.

It is necessary to determine the apparatus constant $K$ either with a primary or a secondary standard particle every day. Initially, there may be slight day-to-day variation, but once the apparatus has settled down, any variations should decrease.

The number of particles timed depends on the homogeneity of the charge on the particles in suspension. A minimum of 20 particles should be timed to give an average value, but it may be necessary to increase this number, depending on the standard deviation $S$, defined as

$$S = \left( \frac{\sum(x - \bar{x})^2}{n-1} \right)^{1/2} = \left( \frac{\sum(x^2) - (\sum x)^2/n}{n-1} \right)^{1/2} \qquad (35)$$

where $\bar{x}$ is the arithmetic mean value of all the $x$ values and $n$ is the total number of observations. A typical histogram showing the distribution of mobility values of cells of *Klebsiella aerogenes* is depicted in Figure 15a; the standard deviation is $0.11 \times 10^{-8}$ m$^2$ s$^{-1}$ V$^{-1}$ on 80 individual bacteria. The coefficient of variation $100 \times Sx^{1/2}$ is 4.7%, and the standard error of the mean $\pm S/n^{1/2}$ is $0.012 \times 10^{-8}$ m$^2$ s$^{-1}$ V$^{-1}$. However, mixed populations have been reported in biological systems[62,63]; Figure 15b shows a typical mixed population of cells of *K. aerogenes* in which the cells with the lower mobility values have surface appendages known as "fimbriae." A study of the change of shape of such histograms can yield important information about changes in the nature of surface components during the culture of bacterial cells.

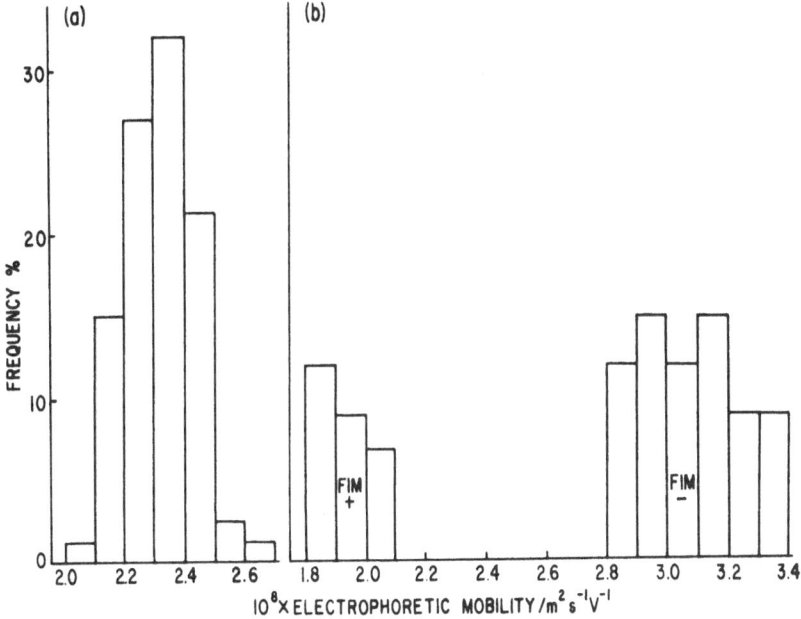

Figure 15. Histograms of mobility values for cells of *Klebsiella aerogenes* showing (a) normal population of homogeneous cap + strain, and (b) mixed population of cap − strain due to the presence of fimbriate (fim + ) and nonfimbriate (fim − ) cells.

## 6.5. Symmetry of the Observation Chamber

It is important that the symmetry of the cell be examined by determining the velocity–depth curve for a particle in suspension. For this purpose, the particles in suspension must have a homogeneous distribution of mobility values. In the experiment, the current flowing is kept constant, and the velocity of particles in suspension at different levels is measured. At least ten individual particles should be timed alternately in each direction at each of nine levels equally spaced throughout the cell. To simplify the calculation, it is advantageous to take readings of levels that are equally spaced on either side of the geometric central plane, but the position of the calculated level *must* always be approached from the bottom inside face of the cell (Section 6.3.7). In changing from one level to the next, it is essential that reference be made back to the particles in focus on the bottom inside face of the chamber. Figure 16 shows a typical velocity–depth curve measured for cells of *K. aerogenes* in phosphate buffer solution (pH = 7.0, $I = 0.012 \text{ mol dm}^{-3}$); since the current is constant throughout the experiment, it is only necessary to plot reciprocal times (i.e., it is not necessary to calculate a mobility value at each level).

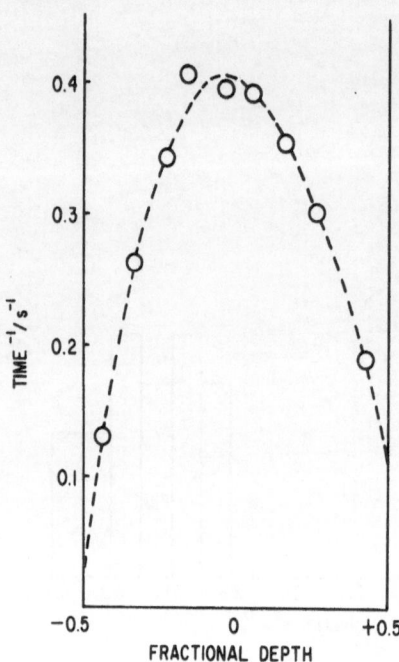

Figure 16. Typical velocity–depth curve obtained in rectangular cell. (Cells of *Klebsiella aerogenes*, in phosphate buffer solution pH = 7.0, $I = 0.012 \, \text{mol dm}^{-3}$.)

The equation to this curve is now calculated to establish the symmetry of the cell, the location of the two stationary levels, and hence the true electrophoretic velocity. The equation to any line may be expressed in the form

$$y = a + bx + cx^2 + \cdots \tag{36}$$

where $a, b, c$, etc., are the constants to be determined. In this instance, $y$ is the observed velocity of the particle and $x$ is the fractional depth of the cell measured from the geometric center. If $N$ readings have been made, then

$$\sum y = Na + b \sum x + c \sum x^2 + \cdots \tag{37}$$

$$\sum xy = a \sum x + b \sum x^2 + c \sum x^3 + \cdots \tag{38}$$

$$\sum x^2 y = a \sum x^2 + b \sum x^3 + c \sum x^4 + \cdots \tag{39}$$

Thus, from a knowledge of $\sum y, \sum xy, \sum x^2 y, \ldots, \sum x^4$, which are obtained from the experimental readings, the constants in simultaneous equations (37)–(39) may be determined. If the readings are made at levels spaced equally on either side of the center, then $\sum x = \sum x^3 = 0$. For the experimental data shown in Figure 16, the equation to the best curve calculated as above is

$$v_0 = 0.407 + 0.0523x - 1.494x^2 \tag{40}$$

The small coefficient of $x$ confirms the symmetry of the cell. Rewriting equation (18) with the now known variation of $v_0$ with depth gives

$$v_p = 2 \int_0^{0.5} (0.407 + 0.0523x - 1.494x^2) \, dx \qquad (41)$$

whence $v_p = 0.296$. Substituting this value of $v_p$ into equation (40) and solving for $x$ gives the positions of the stationary levels as fractional depths of 0.290 and 0.256 from the center, i.e., 0.210 and 0.756 from the top inside face. The position of the upper stationary level is in agreement with that predicted (Section 5.1); the position of the lower stationary level is, however, far removed from the theoretical level. Several workers[47,64,65] have observed that over a period of time the position of the lower stationary level tended to change, while that of the upper level remained constant at the predicted value; the symmetry of the cell was unchanged. This has been attributed to a change in the nature of the bottom cell surface and the concomitant change of the electroosmotic velocity. Since this change occurs very rapidly[57] after the initial cleaning of the cell, it is most probably due to the deposition of particulate matter, bacterial cell debris, etc. onto the bottom surface of the cell. Cleaning of the cell with acid results in the shift of the lower stationary level back to its theoretical position, but such a method cannot be used routinely, as the debris reaccumulates very rapidly. For this reason, it is recommended that all measurements should be made at the upper stationary level.

## 6.6. Other Sources of Error

For a given suspension, the main sources of error involved in the measurement of the electrophoretic mobility are the accurate location of the stationary level, the correct and reproducible focus of particles in suspension, polarization of the electrodes, starting and stopping the stopwatch, the distance traveled, Brownian movement of the particles, the preparation of the suspension, and, for biological samples in particular, variation within the sample. The Brownian shift for a spherical particle of 2-$\mu$m diameter amounts to 2 $\mu$m with a transit time of 10 s over a distance of 50 $\mu$m.

These errors are most conveniently investigated by a statistically designed experiment involving the use of a Latin square.[66] The experiment, as depicted in Table 2, can be designed to show any differences between the electrophoretic velocity when determined by two observers (A and B) or one observer on two occasions ($A_1$ and $A_2$) over different distances in the eyepiece graticule (10 or 20 divisions) in both directions of motion ($\leftrightarrows$) and at the two stationary levels for different applied field strengths (recorded as the current flowing in milliamperes, $i_1, i_2, \ldots$).

The positions of the stationary levels are the experimentally determined positions for the cell (Section 6.5). The readings are taken according to the numerical order indicated in the squares. All the usual procedures, e.g., approaching the stationary level from beneath, of Section 5.3, are followed, with the exception that five individual readings of different particles are taken with the current flowing in one direction (i.e., no reversal of polarity between readings as is normal practice).

In a typical set of observations (made on particles suspended in phosphate buffer solution, $I = 0.012$ mol dm$^{-3}$, pH = 7.0), the data shown in Table 3 were obtained. Analysis of the deviance between columns, rows, and indices (currents) (Table 4) shows that there is no significant difference between observations made at the different values of the currents but that there is a significant difference between rows and between columns.

A factorial analysis of the variance between the columns to find the effect of observers and distances traveled reveals a highly significant difference between the observers. There is no significant difference between the values for the different distances traveled or interaction between them and the observers. The significant difference between the observers makes it imperative that all measurements in a given series be made by one observer and that control experiments always be made so that comparison of mobility values between two observers can be made. No such difference between columns was observed when the Latin square was repeated with one observer taking all the readings on the same suspension at different times.

Factorial analysis of variance for the rows reveals a significant difference between the observations in the two directions but not between the stationary levels; there is no interaction of the two. The difference

*Table 2. Design of Latin Square*

| Level | Observer A | | Observer B | |
|---|---|---|---|---|
| | 10 divisions | 20 divisions | 10 divisions | 20 divisions |
| Upper stationary | 1  $i_4$ $\longrightarrow$ | 5  $i_3$ $\longrightarrow$ | 9  $i_1$ $\longrightarrow$ | 13  $i_2$ $\longrightarrow$ |
| | 2  $i_3$ $\longleftarrow$ | 6  $i_4$ $\longleftarrow$ | 10  $i_2$ $\longleftarrow$ | 14  $i_1$ $\longleftarrow$ |
| Lower stationary | 3  $i_2$ $\longrightarrow$ | 7  $i_1$ $\longrightarrow$ | 11  $i_3$ $\longrightarrow$ | 15  $i_4$ $\longrightarrow$ |
| | 4  $i_1$ $\longleftarrow$ | 8  $i_2$ $\longleftarrow$ | 12  $i_4$ $\longleftarrow$ | 16  $i_3$ $\longleftarrow$ |

*Table 3. Experimental Results for a Latin Square*[a]

| Level | Observer A | | Observer B | | Totals |
|---|---|---|---|---|---|
| | 10 divisions | 20 divisions | 10 divisions | 20 divisions | |
| Upper stationary ⟶ | 2.939 (1.1) | 3.075 (0.9) | 2.849 (0.5) | 2.855 (0.7) | 11.716 |
| ⟵ | 2.859 (0.9) | 2.624 (1.1) | 2.641 (0.7) | 2.359 (0.5) | 10.483 |
| Lower stationary ⟶ | 2.951 (0.7) | 2.950 (0.5) | 2.676 (0.9) | 2.776 (1.1) | 11.353 |
| ⟵ | 2.773 (0.5) | 2.794 (0.7) | 2.635 (1.1) | 2.616 (0.9) | 10.816 |
| Totals | 11.520 | 11.443 | 10.801 | 10.604 | 44.368 |

[a] In this table, the number in each square is the sum of the $n/it$ values for five timings, where $n$ is the number of squares traversed in $t$/s, and $i$/mA is the current indicated by the number in parentheses.

between readings in the two directions was shown to be an artifact of the experiment and was due to electrode polarization produced by taking five successive measurements all in one direction. This is overcome in the procedure (Section 6.3) by taking readings in alternate directions and by the use of the shorting switch at the end of measurements on one suspension to remove any residual polarization.

This analysis shows that the electrophoretic mobility is independent of applied field strength; it may be tested over a wider range in a separate experiment.

## 6.7. Determination of Reproducibility

This analysis of observations made on a single suspension gives a measure of the confidence limit of a single mean. It is of importance as an index of the reliability of the operator, and it also provides the limits for that operator outside which the difference between the mobility value of the particles of two suspensions can be considered significant.

In its simplest form, readings are made on three consecutive days on the same suspension at the upper stationary level—three sets of 30 individual readings on the first 2 days and five sets of 30 readings on the third day. If all the observations are taken at the same applied field strength, then a direct analysis of the timings may be made. For a typical set of data (Table 5), it is

Table 4. *Analysis of Deviance of Data in Table 3*

| | Degrees of freedom | Deviance | Variance | $F^a$ |
|---|---|---|---|---|
| Between columns | 3 | 0.158 | 0.0527 | 5.12 |
| Between rows | 3 | 0.226 | 0.0753 | 7.31 |
| Between currents | 3 | 0.020 | 0.0067 | 1.54 |
| Error | 6 | 0.062 | 0.0103 | — |
| | 15 | | | |

$^a$ $F$ = variance/variance of error; $F$ for 3/6 degrees of freedom at 5% probability level = 4.76; $F$ for 3/6 degrees of freedom at 1% probability level = 9.78.

apparent that, at $p = 0.05$, there is no significant difference between sets of readings taken on the same or different days.

The confidence limit of a single mean is obtained by averaging the variances of all the readings, calculating the standard deviation from the average, and including the student $t$ test to allow for the size of sample. The average value of the standard deviation of the mean $\sigma_{\bar{x}}$ for 30 readings is 0.05, and $t$ for 30 readings at 0.05 level is 2.04; hence, the general reproducibility on 30 readings is ±0.10 on average of 3.28. Thus, the error is 3.14%, and the average value is quoted as $3.28 \pm 0.10$. This means that, in this particular instance, any two mobility values differing by at least 10% are significantly different.

A slight improvement in the confidence limit may be obtained by reducing the applied field strength (e.g., to half), which has the effect of increasing the time, thereby reducing errors inherent in starting and stopping the watch. The time measured must not be extended too much, otherwise inaccuracies due to particles falling out of focus, etc., will have a pronounced effect.

Although it has been suggested[38] that, at times less than 5 s, errors involved in starting and stopping the watch become significant, it would seem from the data presented here that this situation does not obtain. The sources of error, listed at the beginning of Section 6.6 that pertain to the experimental procedure described can be investigated and, where necessary, eliminated by the testing procedures of Sections 6.6 and 6.7.

## 6.8. Lateral Mounting of the Rectangular Cell

This method of mounting in which the shorter edge of the cell is in the vertical plane[51] is of great advantage for suspensions in which the particles sediment out rapidly. In the horizontally mounted cell, such particles will fall out of the stationary level during timing, while in the laterally mounted cell, the same particles will sediment under gravity and still remain in the

Table 5. *Analysis of Readings on a Standard Suspension*

| Statistical quantity | Day A | | | Day B | | | Day C | | | | |
|---|---|---|---|---|---|---|---|---|---|---|---|
| | Set 1 | Set 2 | Set 3 | Set 1 | Set 2 | Set 3 | Set 1 | Set 2 | Set 3 | Set 4 | Set 5 |
| $\sum x$ | 100.42 | 97.07 | 99.58 | 98.95 | 97.15 | 95.81 | 100.51 | 98.78 | 97.49 | 100.22 | 97.85 |
| $\bar{x}$ | 3.35 | 3.24 | 3.32 | 3.30 | 3.24 | 3.19 | 3.35 | 3.29 | 3.25 | 3.34 | 3.25 |
| $\sum x^2$ | 340.48 | 315.72 | 332.44 | 329.12 | 316.76 | 302.08 | 339.58 | 328.21 | 318.10 | 336.61 | 321.06 |
| Variance, $S^2$ | 0.1496 | 0.0562 | 0.0655 | 0.0948 | 0.0744 | 0.0375 | 0.0983 | 0.1021 | 0.0445 | 0.0624 | 0.0641 |
| $\sigma_{\bar{x}}{}^a$ | 0.07 | 0.05 | 0.05 | 0.05 | 0.05 | 0.04 | 0.06 | 0.06 | 0.04 | 0.05 | 0.05 |
| $\sigma_{\bar{x}}$ | | 0.03 | | | 0.03 | | | | 0.022 | | |
| Limits | | $3.30 \pm 0.06$ | | | $3.24 \pm 0.06$ | | | | $3.30 \pm 0.05$ | | |

$^a$ $\sigma_{\bar{x}}$ = standard deviation of mean = $S/n^{1/2}$.

stationary level. Under an applied electric field, the particles appear to move in a diagonal fashion across the field of view; only the horizontal component of the velocity is measured.

The cell (Figure 17a), connected to the electrode compartments previously described (Figure 11), is now immersed in a small bath through which water is circulated. The microscope is in the horizontal position, and provision is made for both the objective and the condenser to pass through rubber seals in the sides of the thermostat tank (Figure 17b). The stage of the microscope is adapted to hold the thermostat and the mounts for clamping the cell rigidly in place. During mounting, it is essential that the cell be positioned correctly in the vertical plane to ensure that when particles settle out, they fall in a plane parallel to the glass walls and hence remain subject to the same liquid flow within the cell.

The method of preparation, operation, and calibration of the cell and the testing of reproducibility are exactly as described in the previous sections for the horizontal cell. Particles are timed between vertical cross lines on the cross-hatch graticule; no notice is taken of any vertical component of the velocity. In this mounting, the position of the stationary levels is more constant, since any sediment collecting is further removed from the position of observation, and further, it is without effect on the electroosmotic flow.

## 7. Experimental Assembly of the Cylindrical Cell

### 7.1. Design and Construction [38,50,67]

The essential features of the cylindrical cell are shown diagramatically in Figure 18. The tube consists of a cylindrical precision capillary of Veridia glass in which the internal diameter is uniform to ±0.01 mm throughout the whole length. The internal diameter chosen depends upon the working distance of the objective; 2.0–3.0 mm diameter is suitable for use with a 40× objective. At smaller diameters, the velocity–depth curve becomes very steep in the region of the stationary level, while at larger diameters, thermal convection effects become significant on account of the greater current flowing per unit applied field strength. A tube of length 120 mm is convenient both for the accurate leveling of the cell and for reducing the risk of contamination of the suspension with electrolyte from the electrode compartments (Figure 18a) or the products of electrolysis (Figure 18b). The wall thickness of the capillary should be less than 0.6 mm to permit adequate temperature control during a determination.

To facilitate observation, an optical flat, 20 mm long, is ground centrally on the outer wall of the tube using a machined steel block in which the surface is ground to ±0.001 mm. Two aluminum rings clamped 20 mm

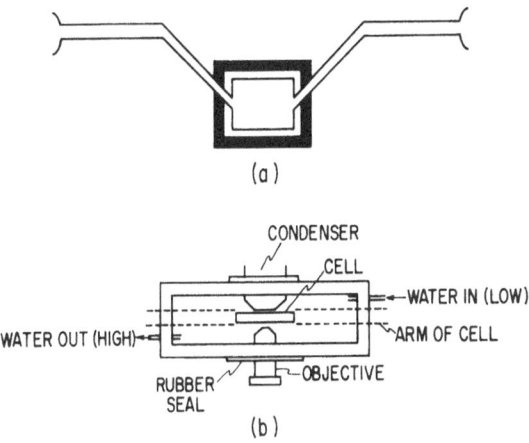

Figure 17. (a) Rectangular cell for mounting in lateral position, (b) Plan of thermostat tank for rectangular cell in lateral position.

apart along the tube act as guides; grinding is achieved with aqueous suspensions of 100-, 320-, and 600-mesh carborundum in succession to remove quantities of glass a (0.30 mm), b (0.15 mm), and c (0.05 mm), as shown in Figure 19. The resulting finely ground plane glass surface is polished with rouge. During the grinding and polishing, care must be taken to ensure that uniform pressure is maintained between the grinding or polishing tool and the glass; otherwise, the resulting plane surface will not be parallel with the axis of the tube. This must be checked with a micrometer screw gauge during preparation.

The internal diameter of the tube is determined by measuring the length of a thread of mercury at different places along the tube and then weighing the mercury. This value for the diameter should be confirmed optically at several points along the optical flat by direct observation with a traveling microscope of the near and far surfaces of the tube while filled with water colored with methylene blue to aid location of the surfaces.

For the cell with silver–silver chloride electrodes,[67] two female B7 Quickfit joints are sealed on at each end, one coaxial with the tube and the other at right angles and in the plane of the optical flat, for filling purposes (Figure 18a). The electrode compartments comprise a B7 socket (to accommodate the electrode) fused into a Pyrex glass tube, which carries a male B7 joint (for connection to the cell) fused centrally and at right angles to it. Sintered glass disks (no. 3 or 4 porosity) are cemented onto each male joint and ground to a thickness of about 0.5 mm. The disk is chosen to give a final pore size of 10–20 $\mu$m; this gives optimal conditions for minimal leakage of potassium chloride from the electrode compartment and minimal

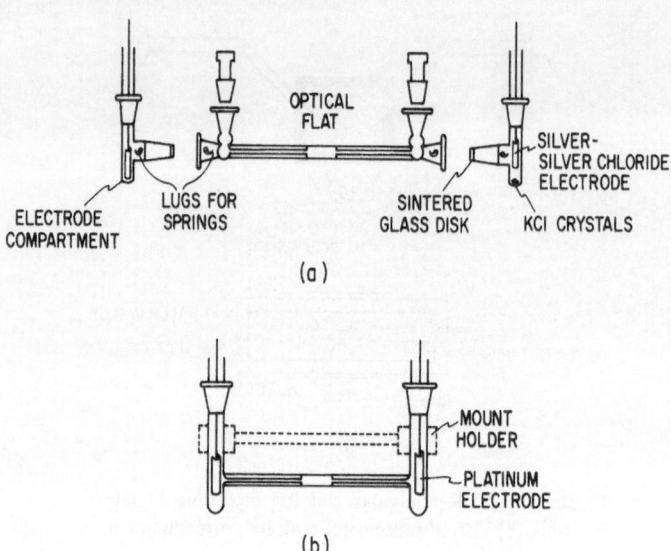

Figure 18. Essential features of cylindrical electrophoresis chambers (a) incorporating Ag, AgCl, KCl electrodes, (b) incorporating platinum electrodes. Figure 18a reprinted from Seaman,[38] p. 9, courtesy of J. and A. Churchill Ltd.

resistance. The silver electrodes, consisting of a cylinder of silver foil $(3 \times 2 \times 0.013$ cm thick) welded onto 1-mm-diameter silver wire, are sealed into a male B7 joint with a sealing wax. The silver electrodes are cleaned, silver plated, and anodized as described previously (Section 6.1). The electrode compartments are filled with a saturated solution of potassium chloride, and a layer of potassium chloride crystals is kept at the bottom of the tube to ensure that there is no appreciable change of concentration. The joint carrying the electrode is now inserted and made a tight fit. The joints are adequately greased with a thin film of vacuum grease, and the electrode compartments are connected to the capillary tube and held in place by stainless steel springs. The capillary tube filled with water is now ready for mounting. This cell has the advantage that it only requires very small amounts of suspension $(0.8-1.4$ cm$^3$, depending on the bore and length of the tube).

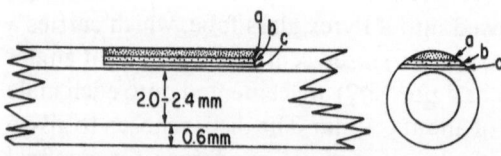

Figure 19. Preparation of optical flat on a glass tube to be used as a cylindrical electrophoresis chamber: (a) 0.30 mm thick, (b) 0.15 mm thick, and (c) 0.05 mm thick. Reprinted from Seaman,[38] p. 8, courtesy of J. and A. Churchill Ltd.

For the cell with platinum electrodes,[50] small electrode compartments are fused directly onto the capillary tube; these terminate (at the top) in a B10 Quickfit socket (Figure 18b). The platinum electrodes, which consist of a cylinder of platinum foil ($12.5 \times 20$ mm) welded onto platinum wire, are mounted in a B10 male glass joint; electrical connection is made through a mercury reservoir. The electrodes are first blackened and then converted to the gray form[55] by heating to a cherry red color (700–800°C) in a coal gas flame. The large size of the electrodes reduces the current density and so minimizes the risk of polarization.

A standard microscope tube (n) is attached horizontally to the steel crossbar (o), which is fixed across the glass thermostat tank (p) (Figure 20). The fine adjustment screw (q) is linked mechanically to a larger zero dial test indicator (r),which reads to 0.002 mm. This bears on the crossbar and gives an accurate and rapid reading of the position of the objective. The microscope lamp is mounted behind the thermostat tank, coaxially with the objective and at right angles to the optical flat of the chamber. The microscope objective (s) is waterproofed and inserted into the water bath through a watertight flexible rubber gasket. The cylindrical electrophoresis chamber, suitably mounted in a holder (w), is also attached to the crossbar through the vertical traverse (t). This vertical traverse, which is calibrated, allows the chamber to be raised or lowered by a measured amount; this permits the exact location of the horizontal diameter of the chamber in the field of view of the microscope. Once in the correct position, the vertical traverse is locked (u). The capillary tube is mounted in the holder by clamps at each end; this facilitates the rapid and symmetrical heat exchange across the capillary wall. The tank and cell are thermostatted in a well-stirred bath

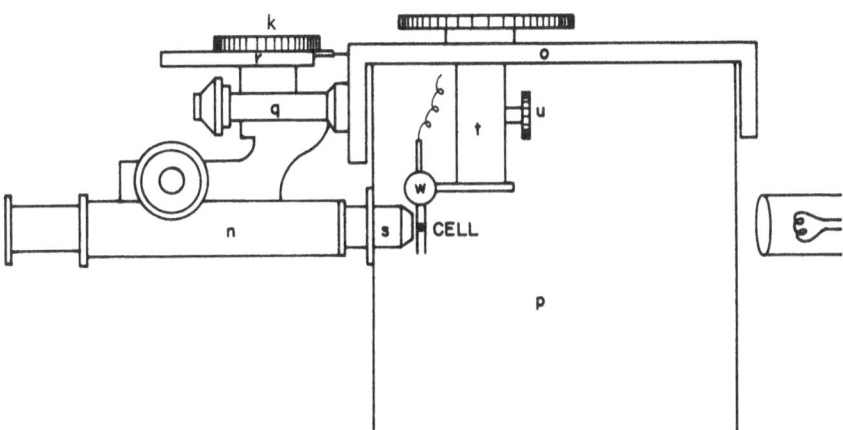

Figure 20. Cylindrical electrophoresis, showing complete assembly and the mounting of the cell and microscope. Reprinted from Bangham *et al.*,[50] p. 642, courtesy of Macmillan and Co. Ltd.

controlled to ±0.01°C. In mounting the cell holder onto the traverse (t), the tube is set so that its capillary is parallel to the surface of the water in the bath, i.e., it is truly in the horizontal plane.

The microscope is fitted with a 10× focusing eyepiece that incorporates a crosshatch graticule with squares of 0.25-mm edge. The optical system, with a 40× objective, gives a magnification of 400×. The chessboard graticule is calibrated under this magnification against a standard 1-mm standard micrometer scale ruled to 0.01 mm. The depth of focus is about 0.005 mm. Since the tube and objective lens are immersed in water, the effects of refraction by the planoconcave cylindrical lens are reduced, and the Henry correction[39] is negligible.

The electrical circuit recommended for use with this type of apparatus is as shown in Figure 8.[38] In the cell with platinum electrodes, the field strength may be obtained directly by measuring the applied potential with a high-resistance voltmeter and dividing this reading by the distance between the electrodes or indirectly from a knowledge of the current flowing using the relationship of equation (29). In practice, the two methods give results in excellent agreement.[50] For the cell using silver–silver chloride electrodes, it is preferable to use the latter method, as it has to be established that the effective electrical length of the electrophoresis chamber between the sintered glass disks corresponds to the physical length.

## 7.2. Assembly and Preparation for Use

The crosshatch graticule is calibrated at the working magnification against a standard micrometer scale. The objective is waterproofed. The fine adjustment screw and the test-dial indicator are calibrated over their full working distance, using a coverslip of constant thickness; the most uniform range is then selected for subsequent use. The diameter of the capillary (in the region of the optical flat) is determined both by the mercury thread method and also by direct observation. The electrode compartments and electrophoresis cell are filled with potassium chloride solution and water, or water, as appropriate to the apparatus; the joints are greased and the component parts joined together (and held with springs), and the whole is mounted in the cell holder. The holder is now attached to the vertical traverse (t) and adjusted so that the optical flat is in the vertical plane at right angles to the axis of the microscope and the capillary is horizontal, as indicated by a spirit level. The thermostat tank is filled with distilled water to cover the observation chamber to a depth of about 3–4 cm. The microscope lamp is adjusted so that it is along the optical axis of the microscope; care should be taken when adjusting the position of the objective, as the clearance between the lens and the optical flat is small.

## 7.3. Mode of Operation

1. Adjust the temperature of the thermostat and allow sufficient time for the establishment of equilibrium.

2. Check that the optical flat is perpendicular to the axis of the microscope. This is most easily accomplished by focusing on small scratches or dust on the optical flat and moving the cell in the vertical plane past the objective. Provided that the flat itself is plane, these scratches, etc., will remain in focus during the traverse.

3. Locate the position of minimum thickness of the optical flat (i.e., the true horizontal diameter of the cell) by filling the cell with a dilute solution of methylene blue and focusing at the top and bottom of the capillary. The distance between these two extremes is measured on the micrometer attachment of the vertical traverse. Set the position of minimum thickness of the optical flat as the midpoint of the vertical traverse. This setting may be checked by actual measurement of the thickness of the flat (using the dial test indicator, k) at this point and comparing it with the thickness when the cell is displaced vertically small distances above or below this point.

4. Close the shorting switch $S_1$ (Figure 8) to eliminate electrode polarization and then open it.

5. Thoroughly wash the observation chamber and all connections with glass-distilled water. The chamber is filled either by means of a Pasteur pipette or by pouring from a very small beaker. Remove the water from the cell by means of a polythene canula attached to a water pump.

6. Flush the chamber with 50 cm$^3$ of the buffer solution to be used for the next suspension, previously thermostatted at the working temperature.

7. Fill the chamber with the suspension, making sure that there are no air bubbles trapped anywhere in the apparatus. Replace the electrodes and/or stoppers and ensure that they fit tightly.

8. Focus on the front face of the optical flat and, by means of the dial test indicator, adjust the position of the objective so that it is in focus at the required level in the cell (usually the stationary level, which is 0.293 of the radius from the inside wall of the tube).

9. Check that the particles in suspension do not move systematically along the cell; vertical movement under the gravitational field is expected. Such horizontal movement may be due to badly fitting joints or to the nonestablishment of temperature equilibrium.

10. Switch on the current ($S_2$ and $S_3$), and adjust the current with the potential divider to give a suitable excursion time across a given distance in the crosshatch graticule (a time of about 5 s for a potential gradient of 2–5 V cm$^{-1}$ is suitable). Select a particle in focus that is still stationary, and using a stopwatch reading to 0.01 s, record the time for the particle to

traverse a fixed distance in the graticule. Reverse the polarity of the electrodes (switch $S_3$) and measure the time in the reverse direction. The number of particles timed depends on the homogeneity of charge on the particles; a minimum of ten should be timed. Record the current flowing and the applied voltage in each direction. Differences in the excursion time in the two opposite directions should not exceed 1–2% of the average recorded time and should be random; any systematic difference larger than this means that there is a superimposed movement of liquid in one direction due to leaks. From the known polarity of the electrodes and the direction in which the particles are moving, the sign of the charge carried can be determined.

11. Wash out the cell and leave filled with distilled water.

12. Determine the conductance of the suspension at the same temperature.

### 7.4. Calculation of the Electrophoretic Mobility

The mobility is calculated using equation (32) if the current flowing and the conductance of the suspension are measured or using the modified form

$$\bar{V} = \frac{v}{X} = \frac{nL}{t} \frac{l_e}{E} \qquad (42)$$

where $E/V$ is the measured applied voltage and $l_e/m$ is the effective electrical length of the cell (i.e., the distance between the electrodes). Since the diameter of the cell can be measured and the distance moved across the graticule is known, the values of $q$ and $L$ are known absolutely, and so the direct substitution of the experimental data in equations (32) or (42) gives the absolute mobility value of the particles. The mobility value of a standard reference particle (Table 1) should be determined as a check on the procedure.

### 7.5. Symmetry of the Observation Chamber and Location of the Stationary Level

This symmetry is normally achieved[50] by measuring the mobility of human erythrocytes in suspension (0.1% v/v) in 0.145 mol dm$^{-3}$ aqueous sodium chloride solution. The parabolic velocity–depth profile is determined by the electroosmotic flow of liquid along the walls of the tube returning through the center of the tube (Section 5.1). Since the $\zeta$-potential of the red blood cells is constant over the pH range 5–9 while the $\zeta$-potential of the glass surface varies with pH over this range, it follows that the velocity–depth curves measured at different pH values will intersect at the stationary level. On either side of this level, the curves will diverge. The straight lines of mobility against the square of the displacement from the axis will be of different slopes but will all intersect at the position of the stationary level. The difference between the curves can be enhanced by suspending the cells in dilute solutions of hemoglobin, which is readily adsorbed onto the capillary wall. Mobility–depth parabolas can be determined on either side of the isoelectric point of hemoglobin, so that in one instance the wall is positively charged and in the other negatively charged.

Experimentally, the velocities of the particles or erythrocytes in suspension at two pH values are measured at known depths within the cell; at least 10 individual particles should be timed alternately in each direction at as many levels as permitted by the working distance of the objective. The graph of the mobility against the square of the displacement from the axis of the tube should be linear, and point of intersection of two such lines (at different pH values) will be the position of the stationary level (Figure 21). It is advisable to calculate the equations to each of the straight lines by the method of least squares [using equations (37) to (39) to determine the constants $a$ and $b$] and hence to calculate the point of intersection.

Because of the limited working distance of the objective, it is usually only possible to obtain the first half of the parabola, and the method of plotting assumes that the cell is symmetrical. Treatment of the results as

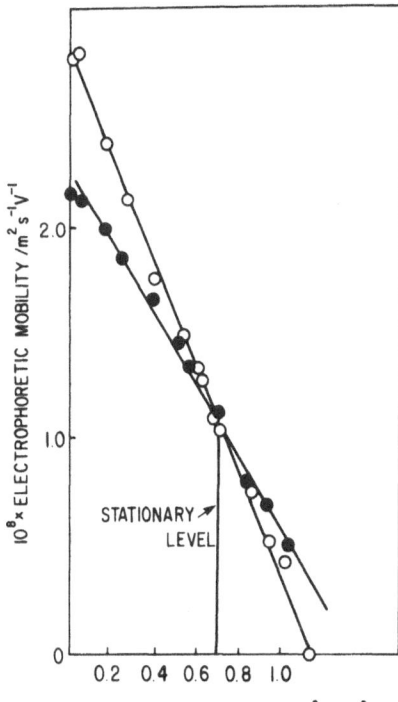

Figure 21. Graphical location of the stationary level in a cylindrical chamber using human red blood cells at pH 9.0 (○) and at pH 6.5 (●). Reprinted from Bangham *et al.*,[50] p. 644, courtesy of Macmillan and Co. Ltd.

indicated in Section 6.5 will give the equation to the first half of the parabola and will give some indication of the symmetry of the cell about its geometric axis.

## 7.6. Other Sources of Error and Reproducibility

These are most conveniently investigated by the methods outlined for rectangular cells in Sections 5.6 and 5.7.

## 8. Additional Precautions Necessary in Determination of Electrophoretic Mobility

### 8.1. Cleaning of the Electrophoresis Chamber

Cleaning should be accomplished by the least drastic method possible to avoid destruction of the glass surface or the adsorption of extraneous material, either of which can produce changes in the electroosmotic flow of liquid across the surface. If the cell has been maintained and used in the correct manner, flushing with large quantities of glass-distilled water or 6% sodium chloride solution to remove proteins should suffice. Cleaning agents that include chromate ions are *not* recommended when biological systems are being studied, since these ions, which are polyvalent and highly toxic, are very strongly adsorbed to glass and can only be removed by washing in flowing water for 24 h. If a detergent is used, the cell must be thoroughly washed first with a dilute solution of hydrochloric acid and then with large volumes of water to ensure the complete removal of the detergent.

Some suspensions, e.g., barium sulfate, present special problems, as the particles settle rapidly under gravity and adhere firmly to all glass surfaces. The cleaning procedure for cells used for this type of suspension must be established on an individual basis.

The outer surfaces of the optical flats in either type of cell tend to get dirty; this dirt is easily removed by a tissue or chamois leather. It is essential to keep these outer surfaces clean, otherwise there will be a considerable reduction in the intensity of light passing up the microscope tube.

### 8.2. Suspending Electrolyte

It is essential that the particles are suspended in a solution of known pH and ionic strength for mobility measurements. The addition of such substances as sucrose or sorbitol to increase the osmotic pressure should be avoided, since, in addition to altering the viscosity and relative permittivity of the solution, they may be preferentially adsorbed onto the particle surface. The actual surface then differs from that under study. The effect and range of concentrations over which adsorption occurs, if any, can usually be investigated.[68] The interpretation of experimental results of measurements in the presence of such nonelectrolytes is made more difficult. If it is considered necessary to increase the tonicity of a solution, it is best done by increasing the electrolyte concentration.

A standard buffer solution with the same ionic species present and a constant ionic strength over a wide range of pH values is advantageous. One used widely is that first described by Michaelis,[69] which consists of a mixture of sodium acetate, sodium barbiturate, and sodium chloride, the pH of which can be adjusted by the addition of hydrochloric acid in the range pH 2.6–9.6. The stock solution ($I = 0.5$ mol dm$^{-3}$) contains 154.635 g sodium diethylbarbiturate (0.15 mol dm$^{-3}$), 102.0675 g hydrated sodium acetate (0.15 mol dm$^{-3}$), and 58.450 g sodium chloride (0.2 mol dm$^{-3}$) dissolved in glass-distilled water and made up to 5 dm$^3$. An aqueous 1.0 mol dm$^{-3}$ solution of hydrochloric acid is added to a measured portion of this stock solution to give the required pH, and the whole diluted to the required ionic strength (0.02 to 0.05 mol dm$^{-3}$ is suitable for most purposes). The pH of the diluted buffer solution should be measured to allow for the effect of dilution. An acetate–barbiturate buffer solution of $I = 0.173$ mol dm$^{-3}$ is isotonic with blood.

The range of the acetate–barbiturate buffer solution can be extended to pH 11.0 by dilution to the required ionic strength and the addition of sodium hydroxide solution of the same ionic strength to give the required pH. At pH values 1.5–2.5, a mixture of hydrochloric acid and sodium chloride, and at 11.0–12.5, mixtures of sodium hydroxide and sodium chloride are satisfactory.

Solutions of 0.145 mol dm$^{-3}$ sodium chloride adjusted to the required pH by the addition of 0.145 mol dm$^{-3}$ hydrochloric acid or sodium hydroxide have been used extensively as the suspension medium for erythrocytes.[38,50] Such solutions are not buffer solutions, and in consequence are liable to changes of pH during a determination of mobility.

The Sørensen phosphate buffer solutions[55] are prepared from mixtures of the following solutions:

*Solution A*: 23.88 g Na$_2$HPO$_4$ · 12H$_2$O per 1 dm$^3$ solution   (0.0667 mol dm$^{-3}$)

*Solution B*: 9.074 g KH$_2$PO$_4$ per 1 dm$^3$ solution   (0.0667 mol dm$^{-3}$)

A mixture containing 8 volumes of solution A and 2 volumes of solution B, with a pH of 7.38 and $I = 0.172$ mol dm$^{-3}$, is the buffer solution used originally[45] for the measurement of the mobility of human red blood cells. Other mixtures of these two phosphate solutions have different ionic strengths, since solutions A and B are salts of the triprotic phosphoric acid.

Any calculation of charge density of particles in suspension in such an electrolyte requires an exact knowledge of the concentration of each ionic species present for substitution in equation (9) (in which $\psi_\delta$ is replaced by $\zeta$).

## 8.3. Preparation of Suspensions

Great care must be exercised in the preparation of suspensions for electrophoresis to avoid either the adsorption of material or the removal of surface components. The particles must be equilibriated with the suspending electrolyte solution; this is achieved during the washing process. Washing the particles involves suspension in the electrolyte, centrifugal sedimentation, followed by resuspension. It may be necessary to repeat this unit process several times; each new material for study must be tested by the determination of the mobility of the particles in suspension after one, two, three, or more washings until a constant mobility value is attained.

The establishment of a washing procedure is of particular importance in the study of the electrical properties of biological cells, bacteria, etc.[70] It is essential, on the one hand, that any adsorbed components from the growth medium (e.g., amino acids, polypeptides, proteins) be removed, while on the other, that loosely bound surface components are not washed off the surface.

In the suspension of particles in buffer solutions of varying pH, a check must be made that at the extremes of pH irreversible surface degradation is not occurring. Again, this is of particular importance with biological cells, which may suffer surface disorganization at extremes of pH. This may be readily checked by determining the mobility of the cells after washing and suspension at the extreme pH value, followed by washing and resuspension of the cells in a buffer solution of neutral pH. If the mobility is the same as that of cells directly suspended in buffer solution of neutral pH, then it is safe to assume that surface disorganization has not occurred as a result of suspension at the extreme pH.

The actual suspension should not be too thick, otherwise the possibility of particle–particle interaction could lead to erroneous results. From the practical viewpoint, a very thick suspension cuts down the intensity of light and makes the selection of a particle in focus at the stationary level very difficult. The particle will be seen moving against a gray background which is moving at a different speed. Suitable concentrations are: for washed human erythrocytes, 0.1% (v/v); bacteria, about $10^8$ organisms $cm^{-3}$; and for other particles, about 0.1% (w/v). As the observer becomes more experienced, it is possible to reduce the concentration of the suspension considerably.

## 9. Other Designs of Apparatus

The literature contains a large number of references to various modifications of the basic apparatus described previously.[37,38] In recent years, interest has been shown in the measurement of the $\zeta$-potential of suspensions in liquids of low relative permittivity. Suspensions in organic media show true electrophoretic phenomena, attributed to the selective adsorption of ions from solution.[71] Van der Minne and Hermainie[72] have discussed the stringent requirements for the electric character of cells for use with media of low relative permittivity. The capillary tube cell made of quartz described by these authors has the undesirable feature of the use of cement for assembly. Parreira[71] has described a rectangular cell made entirely of quartz joined to the electrode compartments by Teflon-sleeved ground glass joints. It is necessary in such cells to incorporate a pair of narrow capillary tubes (one on either side of the cell) to equalize the pressure differential while at the same time maintaining the principle of a closed cell. An additional difficulty using organic solvents is the field-induced polarization phenomenon; this occurs at high applied electric fields when the particle velocity is no longer a function of the field strength. It is thus necessary to make a series of measurements over a wide range of applied fields to establish the extent of the linear relationship. The symmetry of such cells is best tested as described previously (Section 6.5) using a 0.001% carbon black suspension in benzene containing calcium dioctyl sulfosuccinate.

Most of the electrophoresis cells suffer from the difficulty of dismantling and reassembly; this means that cleaning procedures must be carried out *in situ*. With certain types of suspension, particularly those that sediment out rapidly, this is not always a satisfactory or convenient procedure. An easily demountable electrophoretic cell has been described[73]; in this apparatus, the cell is a 1-mm spectrophotometer cell 90 mm long and 25 mm wide constructed without ends. The electrode compartments are machined from solid Perspex blocks; the glass cell fits into a slot in each compartment and is held in position by bolts and an O ring. Platinum electrodes are used. This cell can be used in either the horizontal or lateral position. The position of the stationary level agrees with that predicted by the Komagata equation [equation (26)] for a cell in which $k = 10$.

Several commercial instruments are available. The Zeiss Cytopherometer has been developed in collaboration with Ruhenstroth-Bauer and his associates. This apparatus uses a rectangular cell (width/depth ratio = 20) of fused glass mounted in the lateral position. The cell is surrounded by a water jacket for temperature control. Reversible copper–copper sulfate electrodes are used with gelatin and plaster plugs to prevent the diffusion of the polyvalent ions from the electrode compartments into the observation chamber. The electrode compartments are located above the level of the observation cell, in marked contrast to all other designs. The preparation and assembly of the electrodes is unnecessarily complicated and tedious; the use of silver–silver chloride electrodes would improve the setup. The cell assembly is rigidly mounted on the microscope stand. The optics provided give either normal light or phase contrast microscopy, with a magnification of up to 800×. A photographic attachment serves to document the experimental results. The power supply consists of an electrically regulated transformer and rectifier; it stabilizes the desired current independent of changes of resistance in the electrophoresis circuit. This instrument has been used extensively, in the Max Planck Institut für Biochemie, Munich, for the study of the surface properties of blood and other mammalian cells.

The Particle Electrophoresis Apparatus, Mark II, manufactured by Rank Brothers, Cambridge, is based essentially on the design of Bangham *et al.*[50] The instrument can be used with either a Pyrex cylindrical cell or a fused silica rectangular cell (width/depth ratio = 10) mounted in the lateral position. Either cell can be thermostatted up to 80°C. Illumination is dark field only in the case of the cylindrical cell, but can be light field, dark field, or phase contrast for the flat cell. Electrodes of platinum, palladium, or silver–silver chloride are supplied from a 100 V d.c. power supply, which is capable of giving fields of up to $0.1 \text{ V m}^{-1}$. Also included in the apparatus are a voltmeter, ammeter, and electromagnetic digital timer. The manufacturers claim that, using quartz–iodine illumination, polystyrene particles of $0.2 \ \mu m$ are visible, while using a neon laser, polystyrene particles of $0.09 \ \mu m$ are visible.

The Zeta-meter (Zeta Meter Inc., 1720 First Avenue, New York) has been used in the study of the stability and the dipserion of colloidal systems. The cell consists of a cylindrical tube of nonbreakable clear plastic (glass Teflon tubes are available for use with corrosive liquids) of diameter 4.4 mm. This large size compared with the 0.5–1.0 mm depth of normal cells makes it possible to study flocculated particles of moderate size. The tube is mounted between plastic electrode compartments. The electrodes used depend upon the conductivity of the sample (which can be measured *in situ*) and the need to prevent gassing, which would cause false migration. The anode compartment, with either a platinum–iridium electrode (for solutions of low conductivity) or a molybdenum electrode (for solutions of high conductivity), is filled with the sample and closed. The use of the readily oxidizable molybdenum prevents the formation of gaseous oxygen by the formation of a dense adherent film of oxide. A strip platinum cathode is used; the cathode compartment is open so that the hydrogen evolved passes directly into the atmosphere. There is provision for the thermostatting of the cell. The tube and electrode compartments (Figure 22) are mounted on the stage of a stereoscopic microscope. A thin beam of intense white light projected upward at an angle of 50° is scattered by the particles; this gives rise to a dark field. Overall magnification of up to 160× enables indirect viewing of particles well below $0.5 \ \mu m$ diameter. The microscope is fitted with attachments to enable accurate location of the stationary levels; the large depth of focus and the large particles studied must give rise to errors in

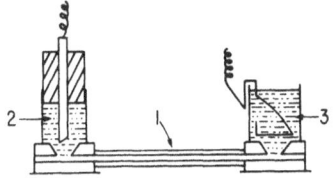

Figure 22. Diagrammatic representation of Zeta-meter assembly: (1) Perspex cell, (2) platinum–iridium anode, (3) strip platinum cathode.

the exact location of particles in focus at these levels. The instrument is provided with a power unit, incorporating a reversing switch, voltage control, voltmeter, ammeter, and electric timer. It is claimed that this instrument is easier to use and that it is more robust than those using glass cells, while at the same time being capable of handling a greater size range of particles suspended in solvents with a wide range of relative permittivities.

In the Model 1202 Electrophoretic Mass Transport Analyzer (Micromeritics Instrument Corp., 800 Goshen Springs Road, Norcross, Georgia), the mobility of particles is measured by the rate at which they migrate into or out of a sample cell under a known applied field strength. The demountable plastic mass transport chamber (Figure 23) consists of a reservoir (b), a sample cell (c), reversible zinc electrodes (a), and a shutter (d). One electrode is accessible to the bulk suspension and the other is at the closed end of the smaller sample cell. The cell is normally rotated at 30 rpm to counteract sedimentation of the larger particles. A constant potential gradient is maintained in the cell during a measurement, and at a predetermined time, the change in the mass concentration in the sample cell is determined. Since, in general, the particles have a different density from the suspending electrolyte, this is most conveniently determined gravimetrically. The instrument can be used to study a wide range of industrial dispersions and suspensions with concentrations of up to 50% by volume without dilution (compare the less than 1% concentrations used in conventional particulate electrophoresis cells); at these high concentrations, the effects of particle–particle interaction on the mobility are not negligible. In the short time (5 min) for a normal test, the ionic environment of the

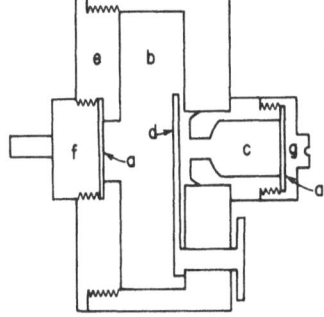

Figure 23. Diagrammatic representation of demountable mass transport chamber: (a) electrodes, (b) reservoir, (c) sample cell body, (d) shutter, (e) reservoir cover, (f) electrode retainer, (g) sample cell cap.

particles will not change appreciably due to electrolysis; this situation will not apply to the longer times required for measurements on more dilute suspensions. There is no provision for thermostat control of the mass transport chamber; at high applied potential gradients for long periods of time, electrical heating will result.

## 10. Applications of the Particulate Electrophoresis Technique

It is important to realize that the absolute mobility or $\zeta$-potential of a particle in a given suspension of defined pH and ionic strength does not serve to characterize the nature of the surface. The charge on a surface is governed by (1) the nature and number of ionogenic groups on the surface and (2) the properties of the suspending electrolyte; e.g., pH, ionic strength, presence of surfactants.

### 10.1. Measurement of Mobility in Solutions of Constant pH and Ionic Strength

Any change of surface properties with time can be followed by determining the mobility of the particles, in suspension in buffer solution of fixed pH and ionic strength, at regular intervals. Examples include the variations observed during the growth cycle of bacteria,[70,74-76] during the regeneration of tissue cells,[77,78] and during the periodic illumination of *Chlorella* cells.[79] In addition, surface changes occurring as a result of reaction at the surface can be studied, e.g., the detection of antibody bound to bacterial cells,[80] the interaction between montmorillonite and bacteria,[81] the interaction of red blood cells with viruses[82] and with basic polyamino acids,[83] the infection of cells with bacteriophage,[84] and the physical treatment of insoluble inorganic suspensions.[85] Although such changes may be recorded, other methods must be employed to seek the actual cause of the change.[70]

From the mobility, the $\zeta$-potential and hence the surface charge density can be calculated, and from information on the area of the particles, an idea can be obtained as to the number of charges carried by the particle in suspension.

### 10.2. Variation of Mobility with pH in Solutions of Constant Ionic Strength

To characterize or identify a surface in more detail, it is necessary to record the variation of the electrophoretic mobility with a change in one of the properties of the suspending buffer solution. The most information can

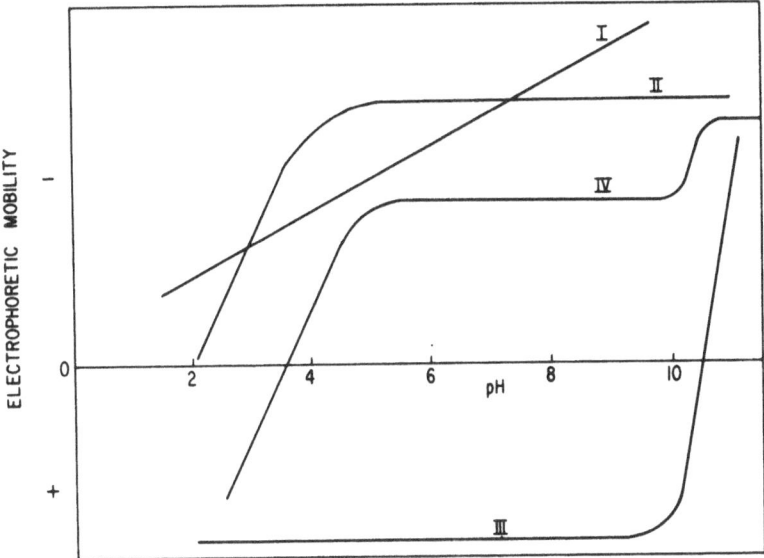

Figure 24. pH–mobility curves (at constant ionic strength) for (I) nonionogenic surface, (II) ionogenic surface with carboxyl surface groups, (III) ionogenic surface with surface amino groups, (IV) ionogenic surface with positively and negatively charged dissociating groups.

be obtained by determining the mobility of the particles in buffer solution of fixed ionic strength but varying pH; this is essentially a titration of surface charged groups and as such is strictly comparable with a potentiometric titration. In using this method, care must be taken to ensure that in suspensions of extreme pH the surface is not irreversibly damaged. Figure 24 shows typical diagrammatic pH–mobility curves which may be obtained for different surfaces.

Curve I is typical of a nonionogenic surface and represents the adsorption of anions and/or desorption of protons with increase of pH; such behavior has been reported for polystyrene latex particles,[20] Nujol oil droplets,[86] and suspensions of octadecanol.[87] Deviations from linearity are generally attributed to the presence of traces of surface impurities.

Curve II is typical of an ionogenic surface in which the charged group is either carboxyl, sulfate, or phosphate. Experimental evidence suggests that on such surfaces the charge due to ion adsorption is negligible.[57] At high pH values, the surface groups are fully dissociated, giving rise to a plateau mobility value; as the pH decreases, so association occurs until at the lowest pH the charge is zero. The p$K$ of the surface is the pH at which the mobility value is half the plateau value; the stronger the acid group the lower the p$K$ value. The pH of the surface pH$_s$ of a particle is related to the pH of the bulk solution $pH_b$ by the equation[88]

$$pH_s = pH_b + A\bar{v} \qquad (43)$$

where $A$ is a constant (0.217 at 25°C), for particles 1 to 2 $\mu$m in diameter. This correction is largest at low concentrations of electrolytes and, as the concentration increases, $pH_s$ tends toward $pH_b$.

pH–mobility curves of this type II have been reported for cells of *Klebsiella aerogenes*[57,75] due to the carboxyl groups of glucuronic acid, for red blood cells[60,89–91] due to the carboxyl groups of $N$-acetylneuraminic acid, for ungerminated spores of *Bacillus subtilis*[92] due to an acidic polysaccharide, for inert hydrocarbon droplets coated with sugar,[93] for barium sulfate particles coated with natural or synthetic stabilizing agents[94] due to the dissociation of the acidic carboxyl groups, and for droplets of octadecanoic acid.[87]

Curve III is characteristic of an amino-type surface, which is positively charged at low pH values due to the presence of the $NH_3^+$ group; at higher pH values, the proton is lost and the surface eventually becomes negatively charged, due presumably to anionic adsorption onto the polar nitrogen atom. Such behavior is shown by suspensions of octadecylamine droplets.[87]

Curve IV is a composite curve of a surface containing both positively and negatively charged dissociating groups. Such behavior is commonly exhibited by bacterial surfaces, e.g., *Streptococcus pyogenes*[95] and spores of *Bacillus megatherium*,[92] in which there are more carboxyl than amino groups. In suspension at low pH value the surface is positively charged ($COOH$, $NH_3^+$); at higher pH values, the negative charge increases due to the increased dissociation of the carboxyl group to give a plateau mobility value where $COO^-$ and $NH_3^+$ exist. At still higher pH values (generally above 9), the proton is lost and the charge is due to $COO^-$ alone, resulting in a higher plateau mobility value. From the plateau mobility values, it is possible to calculate the ratio of $COO^-/NH_3^+$.[95] The more complicated pH-mobility curves reported for cells of methicillin-sensitive strains of *Staphylococcus aureus*[96,97,97a] are due to the additional presence of surface teichoic acid, which undergoes conformational changes with pH. The shape of the pH-mobility curve of cells of *Staphylococcus aureus* can be used to distinguish between methicillin-sensitive and resistant cells.[97,97b,c,d] The position of the pH-mobility curve depends on the ionic strength of the buffer solution used for suspension[57,60]; the general shape of the curve is, however, unaffected by the ionic strength.

Further confirmation of the nature of ionogenic groups can be obtained by specific chemical treatments of the surface followed by a study of the pH–mobility curves of the modified surface. Thus, diazomethane esterifies surface carboxyl groups, reducing the negative charge carried by bacteria to zero[98,99] and also by aldehyde-treated erythrocytes.[100] This treatment suffers from the disadvantage that it must be carried out in organic solvents on surfaces previously exposed to acid conditions, which often cause irreversible surface damage. Fluordinitrobenzene, benzene sulfonyl chloride, and *p*-tosyl chloride block surface amino groups and hence increase the negative charge carried[95,101,102]; this is a mild chemical treatment carried out in aqueous or aqueous alcoholic suspension. Treatment of red blood cells with formaldehyde or acetaldehyde stabilizes the cells and permits the use of other chemical or enzymic treatments.[100,103] Reaction of cells of *Staphylococcus aureus* with sodium metaperiodate results in the destruction of the teichoic acid and the formation of a typical carboxyl surface; the cells are much more susceptible to lysis.[97b,104]

The treatment of biological surfaces with enzymes has been used extensively; any change in the mobility or pH–mobility curve resulting from such treatment may be the result of one or more of the following: (1) enzymic reaction which leads to the removal and/or production of ionizable groups in the plane of shear; (2) changes in the distribution of ions in solution near the interface; (3) shape and orientation changes, e.g., reorientation of groups, swelling, lysis, surface denaturation; (4) irreversible adsorption of the enzyme or other component present in the enzyme system; (5) changes in cell porosity (Section 2). Changes due to enzyme reactions are probably the most important; if sufficient control experiments, including analytical procedures for the assay of liberated material, are made, changes due to the other factors can be largely eliminated. The main advantages of enzymic treatments of biological surfaces are that, in general, the conditions are mild, the enzyme is specific, and only small amounts of enzyme are required.

Cells of *Streptococcus pyogenes* treated with hyaluronidase[70] lose hyaluronic acid, which partially covers the surface; subsequent treatment with proteolytic enzymes results in the loss of surface protein antigens and the production of cells that are electrokinetically and serologically identical, irrespective of their original type.[95] Treatment of bacterial cells with specific amino-acid decarboxylases has been used to establish the identity of surface carboxyl groups arising from amino acid residues.[105] Spores of *Bacillus subtilis* are affected by lysozyme, hyaluronidase, and polyglutamylpeptidase; spores of *Bacillus megatherium* by glutamylpeptidase and lipase; and spores of *Bacillus cereus* by none of these enzymes[106]; these observations are consistent with the three surfaces being composed of differently oriented layers of the same hexosamine peptide. Lipase reduces the sensitivity of cells of *Staph. aureus* to sodium tetradecyl sulfate,[107] suggesting that lipid is a normal cell wall component.

Sialomucopeptides are released from erythrocytes by proteolytic enzymes; as a result, there is a marked reduction of the negative mobility value of the cells.[90,100] Bull sperm incubated with neuraminidase releases neuraminic acid, and the mobility of the sperm cells decreases.[108] The change of shape and position of the pH–mobility curves of tumor and tissue cells after treatment with neuraminidase has given valuable information about the surface of such cells. Ehrlich ascites tumor cells[109,110] were shown to contain three types of ionogenic groups, one of which was the carboxyl group of sialic acid, and it was observed that mesenchymal tumor cells have a higher surface negative charge than carcinoma cells because of their higher sialic acid content.[111]

## 10.3. Variation of Mobility with Ionic Strength in Solutions of Constant pH

The variation of the charge carried by a particle with ionic strength depends on the valence type of the electrolyte and on the presence of ions that may undergo specific interaction or adsorption with the surface. Since all surfaces give a very similar pattern of variation of mobility with concentration of electrolyte, this is not a very good method for surface classification.

In general, with increasing concentration of electrolyte, there is an increase of the negative $\zeta$-potential (of a negatively charged surface), which passes through a maximum value and thereafter decreases. The initial increase of charge is often referred to as the "charging process." The decrease of the $\zeta$-potential is determined by the nature and concentration of the ion of opposite sign to that of the particle (i.e., the cation for a negatively charged surface). The charge and nature of the anion has insignificant effect on the shape or position of the curves. Thus, the concentration of electrolyte required for the particle to attain a zero charge is greatest for a univalent ion (usually this is an extrapolated value) and decreases with increasing valence (Figure 25). Thus, the sign of the charge carried by a negatively charged particle may be completely reversed in the presence of a very low concentration of a thorium salt solution; this is the charge reversal concentration.

The method of plotting "charge reversal spectra" has been used to classify, and in some instances to identify, the charged groups on biological surfaces. Essentially, the method is to determine the bulk concentrations of salts of a variety of different valence cations with simple anions required to reduce the mobility of the particle, and hence the net charge, to zero. When the results are plotted in the form of spectra, Bungenberg de Jong[112] showed that a variety of naturally occurring substances can be differentiated as carboxyl, sulfate, or phosphate colloids according to the nature of their principal surface ionized groups. Within any one group, spectra differ in detail from one compound to another. Apart from the very large number of measurements required over a wide range of concentrations, it is often necessary to extrapolate the $\zeta$-potential–concentration curve to concentrations well above the highest concentration used in order to obtain the

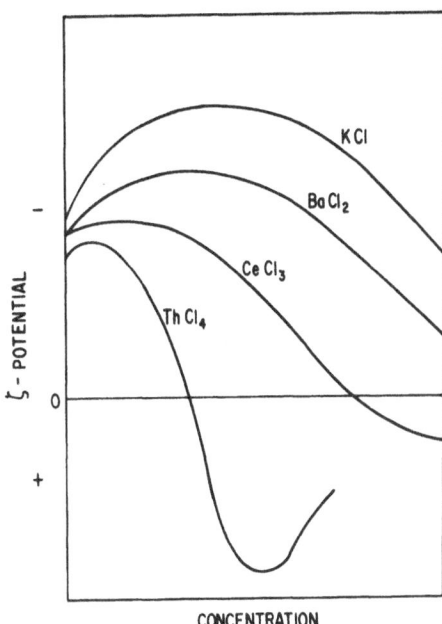

Figure 25. Typical $\zeta$-potential–concentration curves for electrolytes of different valence type.

charge reversal concentration; this gives rise to considerable errors. Some of the electrolytes, particularly those containing polyvalent ions, undergo hydrolysis; this results in a change of pH and also possible change in the character of the ionic species present in solution.

Using this method, several yeasts were considered to be typical carboxyl colloids[113]; the surface of cells of *Escherichia coli* was classified as an acidic polysaccharide, possibly an arabinate[18]; the surface of cells of *Mycobacterium phlei* was shown to be partially covered with phosphatidic fat[114]; and the principal surface groups of spores of *Bacillus megatherium*, *Bacillus cereus*, and *Bacillus subtilis* were characterized as carboxyl, possibly in the form of a polysaccharide–peptide complex.[115] By comparison with model systems, the dominant group on sheep erythrocytes was shown to be phosphate, whereas that on the polymorphonuclear leukocyte was carboxyl.[89] The cation spectra for lymphocytes and platelets were similar to one another and closely resembled egg lecithin, suggesting that phosphate and amino were the predominant groups on the surface.

There are many reports in the literature of the variation of the mobility of insoluble inorganic compounds with the nature and concentration of the suspending electrolyte. The main objective has been the identification of the charge-determining group, e.g., thoria,[116,117] silver halides,[118–120] silica,[121] lead sulfate,[85] barium sulfate.[122,123] The mobility of polystyrene latex particles in different electrolytes has been widely studied[124] in an attempt to explain the stability of such suspensions.

## 10.4. Effect of Surface Active Agents on Mobility of Particles Suspended in Buffer Solutions of Fixed pH

The presence of small amounts of anionic or cationic surface active agents in the buffer solutions used as suspending media is of assistance in detecting lipidlike material at the surface of biological cells.[92,107,125–131] The concentration of these agents must be very low, otherwise cell lysis and nonspecific irreversible adsorption of the surfactant occurs; either of these will interfere with the interpretation of the results. The negative charge of a surface containing lipid is increased in the presence of anionic agent, e.g., sodium dodecyl sulfate, due to the attraction of the nonpolar residue to the lipid by van der Waals forces. Since the polar sulfate group is oriented toward the polar medium, there will be an increased number of negative charges at the surface (Figure 26). In contrast, the charge at a nonionogenic nonlipid surface is unaffected at the low concentrations involved. With a cationic agent, e.g., cetylpyridinium bromide, there is, for all surfaces, a reduction of the negative mobility to zero followed by the establishment of a positively charged surface; this is due to the large organic ion entering the Helmholtz layer. In the presence of lipidlike material, the charge reversal concentration is lower than that for a nonlipid surface, again due to the interaction of the hydrophobic residue of the surfactant with the lipid molecules.

## 11. Nomenclature and Units

The recent introduction of the Système International d'Unités (SI) and its use in this article may cause confusion for some readers. Although detailed information on the use of these units and their relation to cgs units is widely available,[132,133] a brief account relevant to the quantities discussed in this article may clarify the position.

The nomenclature adopted in the tables and in equations arises from the relation

$$\text{physical quantity} = \text{numerical value} \times \text{unit}$$

thus, a concentration of say $2.5 \text{ mol m}^{-3}$ may be written as

$$c = 2.5 \text{ mol m}^{-3} \quad \text{or} \quad c/\text{mol m}^{-3} = 2.5$$

Adopting this method, it is possible to indicate the units so that the correct numerical value can be used or ascertained. This is the only recommended use of the slash (/); thus, writing a concentration as $2.5 \text{ mol/m}^3$ is discouraged. The letter s is never added to a symbol to indicate a plural form; s is the symbol for second, the basic SI unit of time.

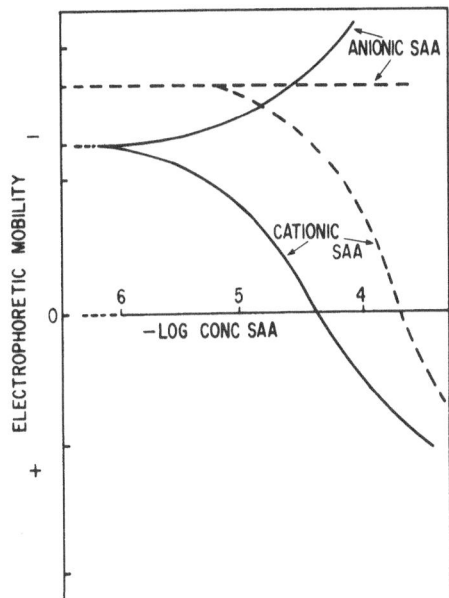

Figure 26. Variation of mobility with concentration of surface active agent for different types of surface: (——) lipid surface, (– – –) nonlipid surface.

Since the meter, m, is the basic unit of length, the basic unit of volume is the cubic meter, $m^3$, but $dm^3$ (equivalent for all practical purposes to a liter) and $cm^3$ (equivalent to cc or ml) are used as working units of volume. Concentrations are expressed in the working units of $mol\ dm^{-3}$; this is numerically the same as concentrations expressed as $mol\ l^{-1}$ or molarity, $M$. (In SI units, the word "molar" has a different connotation.) Ionic strength defined as $\frac{1}{2}\sum c_i z_i^2$, where $c_i$ is the concentration and $z_i$ the valence of the $i$th ionic species, are quoted in the working units of $mol\ dm^{-3}$, which again is equivalent to that expressed as $mol\ l^{-1}$. However, when the concentration or ionic strength is used in an equation based on SI units, then it must be expressed as $mol\ m^{-3}$.

Using the cgs values of the physical constants, equation (5) becomes

$$\kappa/cm^{-1} = (0.327 \times 10^8)\,(I/mol\ l^{-1})^{1/2} \tag{5a}$$

Equation (14) for the calculation of $\zeta$ becomes

$$\zeta/mV = 12.85x \tag{14a}$$

where the measured mobility $x$ has the units of $10^{-4}\ cm^2\ sec^{-1}\ V^{-1}$, and equation (15) for the calculation of charge density becomes

$$\sigma_D/esu\ cm^{-2} = (3.52 \times 10^4)\,(c/mol\ l^{-1})^{1/2} \sinh \frac{\zeta/mV}{51.3} \tag{15a}$$

In equations (32)–(34), if the distance traveled is expressed in cm, the specific conductance in $\text{ohm}^{-1}\,\text{cm}^{-1}$, and the cell constant in $\text{cm}^{-1}$, the units of mobility will be $\text{cm}^2\,\text{sec}^{-1}\,\text{V}^{-1}$. The conversion factor for mobility in SI units from the value in cgs units is $10^{-4}$.

On the SI scale, the pH is still approximately given by the Sørensen definition:

$$pH = -\log\left(c_{H^+}/\text{mol dm}^{-3}\right)$$

ACKNOWLEDGMENTS

The author thanks all his research students over the years for their help and assistance in establishing and exploiting the technique of particulate electrophoresis. In particular, he is indebted to Drs. Alan W. Hill, David T. Pechey, and Roger J. Simmonds for their helpful comments and criticism during the preparation of the manuscript. Finally, I thank my wife Mary for her help and encouragement over the years and for the correction of the manuscript.

## References

1. O. Stern, *Z. Elektrochem.* **30**, 508 (1924).
2. H. von Helmholtz, *Ann. Physik. Wiedemann* **7**, 337 (1879).
3. J. Perrin, *J. Chim. Phys.* **2**, 607 (1904).
4. G. Gouy, *J. Phys.* **9**, 457 (1910); *Ann. Phys.* **7**, 129 (1917).
5. D. L. Chapman, *Phil. Mag.* **25**, 75 (1913).
6. J. T. Davies and E. K. Rideal, *J. Colloid Sci. Supp.* **1**, 1 (1954).
7. P. J. Anderson, *Trans. Faraday Soc.* **55**, 1421 (1959).
8. D. A. Haydon, *Proc. Roy. Soc.* **258A**, 319 (1960).
9. H. A. Abramson, L. S. Moyer, and M. H. Gorin, *The Electrophoresis of Proteins and the Chemistry of Cell Surfaces*, Reinhold, New York (1942).
10. P. Debye and E. Hückel, *Physik. Z.* **25**, 97 (1924).
11. D. C. Henry, *Proc. Roy. Soc.* **133A**, 106 (1931).
12. H. A. Abramson, *J. Phys. Chem.* **35**, 289 (1931).
13. H. A. Abramson and L. Michaelis, *J. Gen. Physiol.* **12**, 587 (1929).
14. M. Mooney, *J. Phys. Chem.* **35**, 331 (1931).
15. C. C. Brinton and M. A. Lauffer, in *Electrophoresis* (M. Bier, ed.), p. 427, Academic Press, New York (1959).
16. D. C. Henry, *Trans. Faraday Soc.* **44**, 1021 (1948).
17. J. T. Davies and E. K. Rideal, *Interfacial Phenomena*, Chap. 3, Academic Press, New York (1961).
18. J. T. Davies, D. A. Haydon, and E. K. Rideal, *Proc. Roy. Soc.* **145B**, 375 (1956).
19. G. J. Gittens and A. M. James, *Biochim. Biophys. Acta* **66**, 250 (1963).
20. M. N. A. Carter, Ph.D. thesis, London University, London (1966).
21. M. H. Wright, Ph.D. thesis, London University, London (1971).
22. M. H. Wright, G. E. Nichols, and A. M. James, *Kolloid-Z. Z. Poly.* **245**, 525 (1971).

23. H. Fricke and H. J. Curtis, *J. Phys. Chem.* **40**, 715 (1936).
24. E. J. W. Verwey and J. Th. G. Overbeek, *Theory of the Stability of Lyophobic Colloids*, Elsevier, New York (1948).
25. D. A. Haydon, *Biochim. Biophys. Acta* **50**, 450, 457 (1961).
26. E. N. Da C. Andrade and C. Dodd, *Proc. Roy. Soc.* **204A**, 449 (1951).
27. D. C. Grahame, *J. Chem. Phys.* **18**, 903 (1950).
28. B. E. Conway, J. O'M. Bockris, and I. A. Ammar, *Trans. Faraday Soc.* **47**, 756 (1951).
29. D. A. Haydon, *Recent Progr. Surface Sci.* **1**, 94 (1964).
30. L. G. Longsworth, in *Electrophoresis* (M. Bier, ed.), pp. 91, 137, Academic Press, New York (1959).
31. F. O. Howitt, *Biochem. J.* **28**, 1165 (1934).
32. R. Ellis, *Z. Phys. Chem.* **78**, 321 (1911).
33. L. S. Moyer, *J. Bacteriol.* **32**, 433 (1936).
34. M. von Smoluchowski, *Handbuch der Elektrizität und des Magnetismus*, Vol. 2, p. 366, Barth, Leipzig (1921).
35. S. Komagata, *Res. Electrotech. Lab. Tokyo*, No. 348 (1933).
36. H. Lamb, *Phil. Mag.* **25**, 52 (1888).
37. A. M. James, *Progr. Biophys. Biophys. Chem.* **8**, 96 (1957).
38. G. V. F. Seaman, in *Cell Electrophoresis* (E. J. Ambrose, ed.), p. 4, J. and A. Churchill, London (1965).
39. D. C. Henry, *J. Chem. Soc.*, 997 (1938).
40. M. E. Smith and M. W. Lisse, *J. Phys. Chem.* **40**, 399 (1936).
41. A. J. Rutgers, L. Facq, and J. L. van der Minne, *Nature* **166**, 100 (1950).
42. G. A. Dworkin, *Biofizika* **3**, 610 (1958).
43. J. D. Hamilton and T. J. Stevens, *J. Colloid Interface Sci.* **25**, 519 (1967).
43a. C. J. van Oss, R. M. Fike, R. J. Good, and J. M. Reinig, *Anal. Biochem.* **60**, 242 (1974).
44. J. H. Northrop and M. Kunitz, *J. Gen. Physiol.* **7**, 729 (1925).
45. H. A. Abramson, *J. Gen. Physiol.* **12**, 469 (1929).
46. L. S. Moyer, *J. Bacteriol.* **31**, 531 (1936).
47. G. J. Gittens and A. M. James, *Analyt. Biochem.* **1**, 478 (1960).
48. M. Mooney, *Phys. Rev.* **23**, 396 (1924).
49. A. E. Alexander and L. Saggers, *J. Sci. Instr.* **25**, 374 (1948).
50. A. D. Bangham, R. Flemans, D. H. Heard, and G. V. F. Seaman, *Nature* **182**, 642 (1958).
51. R. S. Hartman, J. B. Bateman, and M. A. Lauffer, *Arch. Biochem. Biophys.* **39**, 56 (1952).
52. G. F. Fuhrmann and G. Ruhenstroth-Bauer, in *Cell Electrophoresis* (E. J. Ambrose, ed.), p. 22, J. and A. Churchill, London (1965).
53. D. E. E. Loveday and A. M. James, *J. Sci. Instr.* **34**, 97 (1957).
54. R. Neihof and S. Schuldiner, *Nature* **185**, 526 (1960).
55. A. M. James and F. E. Pritchard, *Practical Physical Chemistry*, Longmans, London (1974).
56. J. Powney and L. J. Wood, *Trans. Faraday Soc.* **36**, 420 (1940).
57. G. J. Gittens, Ph.D. thesis, London University, London (1962).
58. H. W. Douglas and F. Parker, *Biochem. J.* **68**, 94 (1958).
59. A. D. Bangham, *Proc. Roy. Soc.* **155B**, 292 (1961).
60. D. H. Heard and G. V. F. Seaman, *J. Gen. Physiol.* **43**, 635 (1960).
61. A. M. James and G. H. Goddard, *Pharm. Acta Helv.* **46**, 708 (1971).
62. A. M. James and C. F. List, *Biochem. Biophys. Acta* **112**, 307 (1966).
63. G. A. Maccacaro and M. Turri, *Giorn. Microbiol.* **7**, 21 (1959).
64. P. White, *Phil. Mag.* **23**, 811 (1937).
65. J. R. Wright and R. R. Minesinger, *J. Colloid Sci.* **18**, 802 (1963).
66. K. A. Brownlee, *Industrial Experimentation*, H.M.S.O., London (1957).

67. G. V. F. Seaman and D. H. Heard, *Blood* **18**, 599 (1961).
68. J. M. Gebicki and A. M. James, *Biochim. Biophys. Acta* **59**, 167 (1962).
69. L. Michaelis, *Biochem. Z.* **234**, 139 (1931).
70. M. J. Hill, A. M. James, and W. R. Maxted, *Biochim. Biophys. Acta* **66**, 264 (1963).
71. H. C. Parreira, *J. Colloid Interface Sci.* **29**, 432 (1961).
72. J. L. van der Minne and P. H. J. Hermainie, *J. Colloid Sci.* **7**, 600 (1952).
73. H. L. Shergold, O. Mellgren, and J. A. Kitchener, *Trans. Inst. Mining Metallurgy* **75C**, 331 (1966).
74. J. H. B. Lowick and A. M. James, *Biochim. Biophys. Acta* **17**, 424 (1955).
75. D. T. Plummer and A. M. James, *Biochim. Biophys. Acta* **53**, 453 (1961).
76. A. M. James, D. E. E. Loveday, and D. T. Plummer, *Biochem. Biophys. Acta* **79**, 351 (1964).
77. S. Ben-Or, S. Eisenberg and F. Doljanski, *Nature* **188**, 1200 (1960).
78. S. Eisenberg, S. Ben-Or and F. Doljanski, *Exp. Cell. Res.* **26**, 451 (1962).
79. S. Lukiewicz and W. Korohoda, in *Cell Electrophoresis* (E. J. Ambrose, ed.), p. 171, J. and A. Churchill, London (1955).
80. M. J. Hill, A. M. James, and W. R. Maxted, *Nature* **202**, 187 (1964).
81. K. C. Marshall *J. Gen. Microbiol.* **56**, 301 (1969).
82. E. Straub, in *Cell Electrophoresis* (E. J. Ambrose, ed.), p. 125, J. and A. Churchill, London (1955).
83. A. Nevo, A. de Vries, and A. Katchalsky, *Biochem. Biophys. Acta* **17**, 536 (1955).
84. J. Kleczkowski and A. Kleczkowski, *J. Gen. Microbiol.* **21**, 308 (1959).
85. T. Morimoto, Y. Hirata, and S. Kittaka, *Bull. Chem. Soc. Japan* **38**, 566 (1965).
86. H. W. Douglas and D. J. Shaw,*Trans. Faraday Soc.* **54**, 1748 (1958).
87. J. N. Mehrishi and G. V. F. Seaman, *Trans. Faraday Soc.* **64**, 3152 (1968).
88. G. S. Hartley and J. W. Roe, *Trans. Faraday Soc.* **36**, 101 (1940).
89. A. D. Bangham, B. A. Pethica, and G. V. F. Seaman, *Biochem. J.* **69**, 12 (1958).
90. G. V. F. Seaman and D. H. Heard, *J. Gen. Physiol.* **44**, 251 (1960).
91. G. M. W. Cook, D. H. Heard, and G. V. F. Seaman, *Nature* **191**, 44 (1961).
92. H. W. Douglas, *Trans. Faraday Soc.* **51**, 146 (1955).
93. H. W. Douglas, *Trans. Faraday Soc.* **46**, 1082 (1950).
94. G. H. Goddard, Ph.D. thesis, London University, London (1970).
95. M. J. Hill, W. R. Maxted, and A. M. James, *Biochim. Biophys. Acta* **75**, 402 (1963).
96. A. M. James and J. E. Brewer, *Biochem. J.* **108**, 257 (1968).
97. A. W. Hill, Ph.D. thesis, London University, London (1971).
97a. N. J. Marshall, J. H. Hewitt, and A. M. James, *Microbios* **4**, 241 (1971).
97b. A. W. Hill and A. M. James, *Microbios* **6**, 157 (1972).
97c. A. W. Hill and A. M. James, *Microbios* **6**, 169 (1972).
97d. A. M. James and S. M. S. Al-Salihi, *Microbios Lett.* **1**, 177 (1976).
98. G. A. Maccacaro and A. M. James, *Biochem. Biophys. Acta* **36**, 279 (1959).
99. G. J. Gittens and A. M. James, *Biochim. Biophys. Acta* **66**, 237 (1963).
100. G. V. F. Seaman and G. M. W. Cook, in *Cell Electrophoresis* (E. J. Ambrose, ed.), p. 48, J. and A. Churchill, London (1965).
101. S. S. Cohen, *J. Exp. Med.* **82**, 133 (1945).
102. H. W. Douglas, *Trans. Faraday Soc.* **55**, 850 (1959).
103. D. H. Heard and G. V. F. Seaman, *Biochem. Biophys. Acta* **53**, 366 (1961).
104. A. M. James and J. E. Brewer, *Biochem. J.* **107**, 817 (1968).
105. M. J. Hill, A. M. James, and W. R. Maxted, *Biochim. Biophys. Acta* **71**, 740 (1963).
106. H. W. Douglas and F. Parker, *Biochem. J.* **68**, 99 (1958).
107. M. T. Dyar, *J. Bacteriol.* **56**, 821 (1948).
108. E. Bey, in *Cell Electrophoresis* (E. J. Ambrose, ed.), p. 142, J. and A. Churchill, London (1965).

109. G. M. W. Cook, D. H. Heard, and G. V. F. Seaman, *Exp. Cell Res.* **28**, 27 (1962).
110. P. D. Ward and E. J. Ambrose, *J. Cell Sci.* **4**, 289 (1969).
111. P. S. Vassar, *Lab. Investigation* **12**, 1072 (1963).
112. H. B. Bungenberg de Jong, in *Colloid Science II* (Kruyt, ed.), Elsevier, Amsterdam (1949).
113. H. E. Jansen and F. Mendlik, in *Proceedings of the European Brewery Convention, Brighton, England*, p. 59, Elsevier, Amsterdam (1951).
114. D. M. A. Adams and E. K. Rideal, *Trans. Faraday Soc.* **55**, 185 (1959).
115. H. W. Douglas and F. Parker, *Trans. Faraday Soc.* **53**, 1494 (1957).
116. T. Morimoto and M. Sakamoto, *Bull. Chem. Soc. Japan* **37**, 720 (1964).
117. O. Griot, *Trans. Faraday Soc.* **62**, 2904 (1966).
118. R. H. Ottewill and M. C. Rastogi, *Trans. Faraday Soc.* **56**, 880 (1960).
119. R. H. Ottewill and R. F. Woodbridge, *J. Colloid Sci.* **19**, 606 (1964).
120. G. R. Weiss, R. H. Ericson, and A. H. Herz, *J. Colloid Interface Sci.* **23**, 277 (1967).
121. O. Griot and J. A. Kitchener, *Trans. Faraday Soc.* **61**, 1032 (1965).
122. A. S. Buchanan and E. Heyman, *Nature* **161**, 649 (1948).
123. T. Morimoto, *Bull. Chem. Soc. Japan* **37**, 386 (1964).
124. R. H. Ottewill and J. N. Shaw, *J. Colloid Interface Sci.* **26**, 110 (1968).
125. M. T. Dyar and E. J. Ordal, *J. Bacteriol.* **51**, 149 (1946).
126. K. McQuillen, *Biochim. Biophys. Acta* **5**, 463 (1950).
127. J. H. B. Lowick and A. M. James, *Biochem. J.* **65**, 431 (1957).
128. M. J. Hill, A. M. James, and W. R. Maxted, *Biochim, Biophys. Acta* **75**, 414 (1963).
129. N. J. Marshall and A. M. James, *Microbios* **4**, 217 (1971).
130. D. T. Pechey and A. M. James, *Biomed. Express* **19**, 127 (1973).
131. D. T. Pechey, A. O. P. Yau, and A. M. James, *Microbios* **11**, 77 (1974).
132. M. L. McGlashan, *Physicochemical Quantities and Units*, R.I.C. Monograph for Teachers 15 (1968).
133. M. N. Hughes, A. M. James, and N. R. Silvester, *SI Units and Conversion Tables*, Machinery Publishing Company Ltd., London (1970).

# 5

# Methods of Producing Ultrahigh Vacuums and Measuring Ultralow Pressures

## J. P. Hobson

## 1. Introduction

Ultrahigh vacuum (UHV) or ultralow pressure is here defined as a pressure range below $10^{-9}$ Torr (1 Torr = 1 mm Hg = 133 Pa = 133 N m$^{-2}$). The decade of the sixties saw ultrahigh vacuum technology develop from a field of active research whose outlines were only dimly visible to one whose major components became commercially available with an accompanying decline in fundamental research activity. Today, all the components necessary for producing and measuring UHV may be purchased. Systems in which UHV has been produced range from space simulation chambers the size of a small house,[1] through 2 km of 6-in. tubing in the CERN intersecting storage ring accelerator,[2] to small laboratory systems with a volume of a few liters.[3,4] The volume of the universe at UHV far exceeds the volume at higher pressures. Indeed, it has been suggested that the universe itself originated as a spontaneous event in vacuum[5]! The development of UHV has stimulated a rapid growth, both experimental and theoretical, in the fundamental study of solid surfaces.[6] The great bulk of the vacuum market still rests above the UHV range, but it may be expected that UHV will play an expanding role in the fabrication of surface-sensitive devices, where extreme purity is important.

It is not the purpose of this article to review these widespread developments[7] but to examine in a modest way the general problem of the

*J. P. Hobson* • Radio and Electrical Engineering Division, National Research Council, Ottawa, Ontario, K1A OR8 Canada.

creation and measurement of UHV along with some of the devices, often quite simple, that are used for these two problems. Emphasis will be placed on operating capabilities rather than on the physical basis of operation, the latter aspects having been examined earlier.[8]

## 2. General Problem of UHV

An UHV system consists in general of a vacuum envelope, some means of reducing the pressure, some means of measuring the pressure, some means of admitting gases, and some apparatus for which the UHV is needed. The final item is not our concern here, but it is necessary that it be consistent with the rest of the system and not destroy the environment that has been prepared for it.

The ultimate partial pressure ($P$ Torr) for a given gas in a volume is given by

$$P = Q/S \qquad (1)$$

where $Q$ is the leak rate of the gas into the volume in Torr liters per second, $S$ is the pumping speed for the particular gas in liters per second. For conventional vacuum systems using, for example, a rotary mechanical pump and a trapped diffusion pump, the value of $Q$ will be such that the pressure will limit in the range $10^{-6}$–$10^{-7}$ Torr. $Q$ in a tight UHV system, in the absence of gas introduced deliberately, comes from gas evolved from the internal surfaces. Real leaks from the atmosphere can be reduced to negligible levels by careful sealing techniques, such as welding, brazing, or glass-blowing. Real leaks develop, of course, at the UHV level, and there is a temptation to seal them with various sealants on the market (Vacseal, Gevac, etc.) after full processing of the system. The author has made this choice on occasion and has almost always regretted it. Complete sealing at the UHV level with a sealant was rarely achieved. Thus, the level to which evolution from internal surfaces can be reduced is an important fundamental property of UHV systems. Consider first the vacuum envelope. For stainless steel (normally type 304) or equivalent, a value of $10^{-12}$ Torr liters sec$^{-1}$ cm$^2$ after a bake under vacuum at 400°C for 12 hr is a reasonable value.[9] Values several orders of magnitude less than this have been reported[10] but should be treated with caution because of experimental difficulties in making these measurements. Most of this gas load is hydrogen. Vacuum bakes at higher temperatures and for longer periods reduce this figure but are not always practical. Many laboratory systems have internal surface areas of about 1000 cm$^2$, making $Q \sim 10^{-9}$ Torr liters sec$^{-1}$. Pumps for hydrogen with $1 < S < 1000$ liters sec$^{-1}$ are quite practical, yielding base pressure in the range $10^{-12} < P < 10^{-9}$ Torr, and thus, stainless steel envelopes have become standard for UHV. It has been reported that aluminum has a lower outgassing rate than stainless steel,[11,12] but aluminum envelopes are relatively rare. The pressure in Pyrex glass envelopes is normally limited by the permeation of atmospheric helium through the walls, which at room temperature with a wall thickness of 1.5 mm has a value about $10^{-14}$ Torr liters $cm^2 \, sec^{-1}$.[13] For 1000 $cm^2$ of surface area, $Q = 10^{-11}$ Torr liters $sec^{-1}$. Helium pumps of 0.1 liters $sec^{-1}$ are quite practical, yielding an equilibrium pressure of $10^{-10}$ Torr of helium. The helium permeation can be drastically reduced by metallic envelopes or by the use of aluminosilicate glass,[14] which, however, is more difficult to work than Pyrex glass. Extensive measurements have been made on the permeation and outgassing of vacuum materials.[15] Equation (1) is useful for general discussion but gives an oversimplified view of reality. Many UHV systems contain devices having hot filaments or areas of heavy particle bombardment causing local heating, thus generating additional local gas sources.[10,16,17] Internal preoutgassing of parts in such regions is normally employed to heat the affected regions well above their ambient operating temperatures. This outgassing can cause the local deposition of thin metallic films that pump some residual gases by adsorption (called "gettering"). Thus, local sources and sinks of gas are commonplace in UHV systems, and strictly speaking, equation (1) can be applied only to regions in which the pressure is not a strong function of position. A pressure measurement at one point in a UHV system is frequently not indicative of the pressure throughout the system. This spatial effect is worst for the chemically active gases such as carbon monoxide and hydrogen and frequently negligible for the inert gases such as helium or argon.[18] As a representative example, results for $S$, $Q$, and $P$ are given in Table 1 for a carefully processed UHV system made from a combination of

Table 1. Leak Rates (Q) and Speeds (S) for the Residual Gases in an UHV Laboratory System of Stainless Steel and Aluminosilicate Glass Pumped only by Gauges[a]

| Gas | $Q$ (Torr liters $sec^{-1}$)[b] | $S$ (liters $sec^{-1}$) | $Q/S$ (calculated) (Torr) | $P$ (measured) (Torr) |
|---|---|---|---|---|
| Hydrogen | $4.2 \times 10^{-10}$ | 4.7 | $8.9 \times 10^{-11}$ | $1.0 \times 10^{-10}$ |
| Helium | $1.7 \times 10^{-15}$ | 0.0074 | $2.3 \times 10^{-13}$ | $2.6 \times 10^{-13}$ |
| Methane | $1.8 \times 10^{-13}$ | 0.34 | $5.3 \times 10^{-13}$ | $5.3 \times 10^{-13}$ |
| Carbon monoxide[c] | $<5.5 \times 10^{-12}$ | 1.9 | $2.9 \times 10^{-12}$ | $5.3 \times 10^{-11}$ |
| Nitrogen[c] | $<1.7 \times 10^{-11}$ | 0.40 | $4.2 \times 10^{-11}$ | $3.9 \times 10^{-11}$ |
| Argon | $1.2 \times 10^{-15}$ | 0.012 | $1.0 \times 10^{-13}$ | $1.3 \times 10^{-13}$ |

[a] System had been subjected to prolonged baking and outgassing.
[b] These are equivalent nitrogen values based on the assumption that all gases have the ionization probability of nitrogen.
[c] The values of $Q$ for carbon monoxide and nitrogen are given as upper bounds, because small leaks of these gases had been introduced deliberately either during or before the measurements.

stainless steel 304 and aluminosilicate glass in which no pumps were used except those appearing spontaneously as a result of system preparation and pressure measurement. Note that for the gases hydrogen, helium, methane, nitrogen, and argon, equation (1) is verified, whereas for carbon monoxide, it is grossly in error as a result of some of the processes noted above. Hydrogen also frequently violates equation (1).

System processing to reduce the spontaneous outgassing rates from internal surfaces is perhaps the most characteristic feature distinguishing UHV practice from conventional vacuum practice. The key step is usually a bake of the entire system for 10–20 h at a temperature of 300–500°C. The detailed procedure and physical processes involved are still the subject of research.[19] Figure 1 shows the residual gas pressures measured for the system of Table 1 at various stages in processing. It may be seen that the primary effect of the bake was the reduction of the partial pressure of water vapor (mass 18) relative to the other gases, which declined without major relative change. Note that, after processing step 8, the total pressure measured by a Bayard–Alpert gauge appeared lower than the partial pressure of hydrogen measured by a residual gas analyzer, another example of nonuniform pressures in an UHV system. It is believed that Figure 1 is representative of most UHV systems, but the gradual disappearance of water vapor should not be interpreted as a simple desorption of adsorbed water. An UHV system is a center of catalytic activity in which the behavior of water is still poorly understood. An illustration of the complex catalytic behavior of an UHV system is provided by Figure 2, which shows results obtained inside a system when water was admitted deliberately at a very low pressure. The main gas seen internally was hydrogen. This behavior is interpreted as a combination of oxygen with carbon on the internal surfaces of the system, with the subsequent release of hydrogen into the gas phase. The peaks at mass 1 and 17 are interpreted as coming from water adsorbed on the molybdenum grid of the mass spectrometer. Water vapor itself was a minor gaseous component throughout. A variety of reactions of this type involving chemically active gases can occur in UHV systems and illustrate the necessity of having a mass spectrometer on any system in which the gas composition is important. In general, an UHV system has a "memory" for the gases introduced to it. For the chemically reactive gases (hydrogen, nitrogen, carbon monoxide, oxygen, carbon dioxide, etc.) this memory resides at or near the chemically active surfaces, while for the inert gases (helium, neon, argon, methane, etc.), the memory resides primarily in the ion pumps or ion gauges. For gaseous amounts up to a monolayer on a surface, the memory can frequently be erased by the evaporation of a metallic film, e.g., titanium on top of the gas layer. This process is widely used in gettering and/or ion pumping.[20]

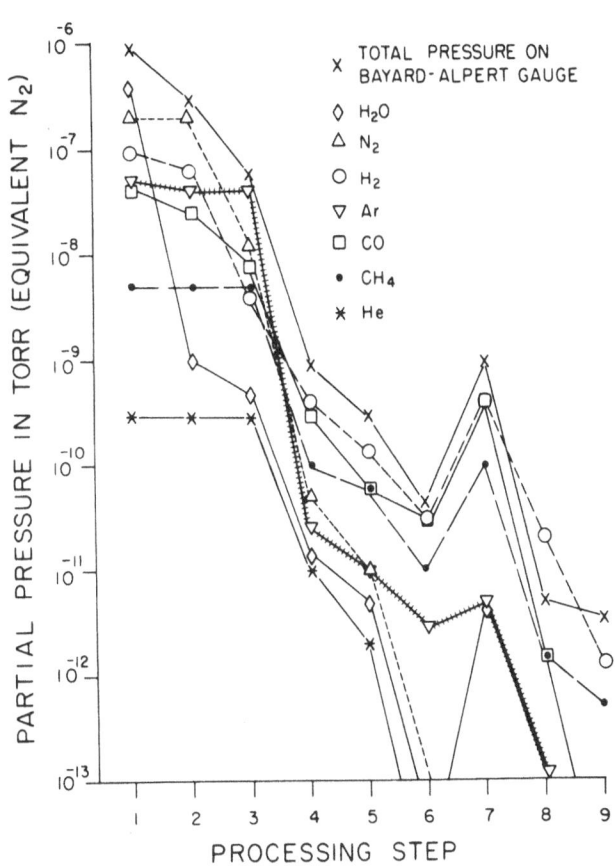

Figure 1. Partial pressures of residual gases after various processing steps in system of stainless steel and aluminosilicate glass; lowest measurable partial pressure, $10^{-13}$ Torr. Processing steps: (1) system pumped by external ion pump before any baking (room temperature); (2) same pumping condition as step 1 after whole system had been baked at 500°C for 16 hr (room temperature); (3) 10 min after closing valve to ion pump (note that the system pressure tends to fall even though no pumps are being used) (room temperature); (4) after 1 hr of heavy internal outgassing and activation of getter and getter–ion pumps inside the baked region (room temperature); (5) 16 hr after step 4 (room temperature); (6) 1 week later after further internal outgassing and pump activation (room temperature); (7) whole system at 50°C pumped by internal pumps only; (8) after 6 days at 50°C—system now at room temperature again (note that the total pressure gauge reads lower than the sum of the partial pressures as measured on a mass spectrometer; this is not unusual in an ultrahigh vacuum system) (room temperature); (9) several days after step 8 (note that only hydrogen and methane remain measurable).

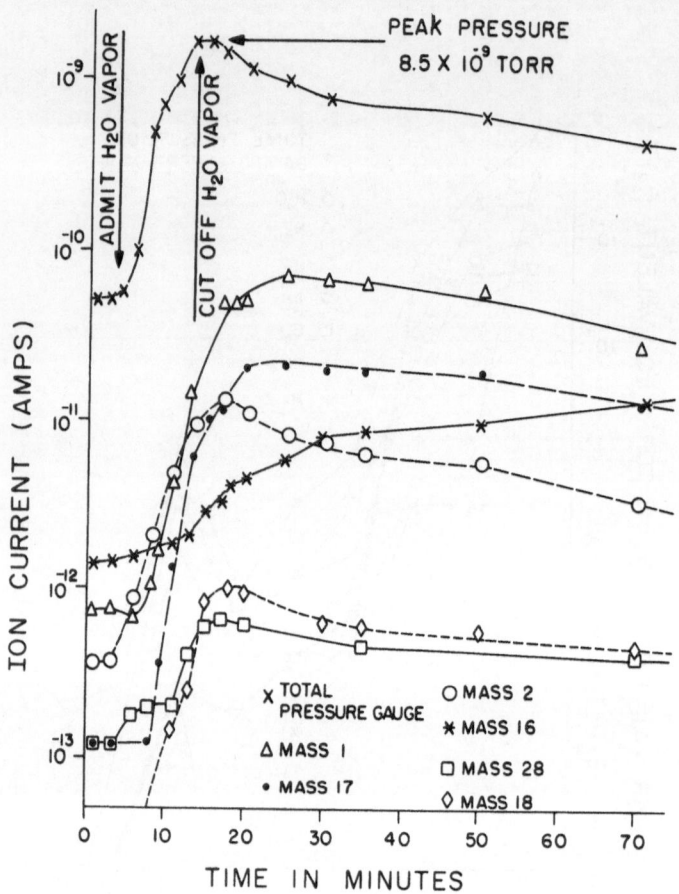

Figure 2. Admission of water vapor to an UHV system of stainless steel and aluminosilicate glass. Water vapor admitted at $t = 3$ min and cut off at $t = 14.5$ min. Points to note: (1) the first peak to rise sharply was mass 2 (hydrogen gas), which remained the dominant gaseous component; (2) next came peaks at mass 1 and mass 17 approximately in the ratio 3:1 (these are probably due to adsorbed water on the mass spectrometer grid), they responded only slowly after water was cut off; (3) the peak at mass 16 (almost certainly $O^+$ desorbed from the grid by electron bombardment) rose steadily, showing little relationship to the cut-off of water vapor, and became the dominant peak at $t = 150$ min; (4) mass 18 (water vapor itself) was always a minor component of the spectrum but did respond to the cutoff of the input vapor; (5) mass 28 (mainly carbon monoxide) remained at a low level throughout.

## 3. Pumps for UHV

There is a variety of pumps that have a high enough pumping speed and a low enough internal value of $Q$ to reach the UHV range.[21,22] They can be used in various combinations. In general, these UHV pumps require other pumps on the high-pressure side operating either continuously or during initial evacuation.

## 3.1. Diffusion Pumps

The major advantage of a diffusion pump is its ability to pump large quantities of gas continuously (so-called throughput) and to remove this gas permanently from the system. In modern oil diffusion pumps, half the molecules crossing the input plane are pumped without return to the low-pressure region. These pumps have a maximum throughput of more than 1 Torr liter $sec^{-1}$ per kilowatt of power, can tolerate 0.5 Torr at the high-pressure end, and have a low-pressure limit without traps of $1 \times 10^{-9}$ Torr.[23] With suitable traps it has been claimed that this limit can be reduced to $10^{-14}$ Torr [24] at some cost in pumping speed and throughput. Mercury diffusion pumps have also been used successfully at UHV,[25] but here trapping is even more essential, since the vapor pressure of mercury even at room temperature is $10^{-3}$ Torr. Diffusion pumps are necessarily appendage pumps and at UHV require precautions against trap warm-up or forepump failure.

## 3.2. Ion Pumps

Ion pumps[26-30] operate by the ionization of the gas to be pumped by electrons, with the subsequent driving of the ions into a surface (normally metallic) where some of the ions lodge permanently as neutral atoms or molecules. The cathode surface may also be sputtered to another location, where the sputtered film may pump chemically active molecules by gettering. Neutralization of ions near the cathode may occur, with subsequent trapping of these energetic neutrals in nearby electrodes. The detailed pumping mechanisms are complex, but no traps are required and no fluids of any kind are inside the vacuum system. The pumped gas is stored within the vacuum system, and an inherent disadvantage of these pumps is that this stored gas is potentially available for desorption later into the vacuum. When approximately one monolayer of gas ($\sim 10^{15}$ molecules $cm^{-2}$ or $3 \times 10^{-4}$ Torr liters $cm^{-2}$) has been pumped, the pumping surface loses its pumping capability and must be regenerated by the deposition of additional metallic material. There are exceptions when the pumped gas dissolves in the cathode, e.g., hydrogen in titanium. The full storage capacity of a relatively large ion pump is about 2000 Torr liters.[26] Two popular forms of ion pump have been developed for UHV. The sputter ion pump,[27,28] containing two electrodes (diode) and three electrodes (triode), utilizes a cold cathode discharge (3–6 kV) confined by a magnetic field ($\sim 1$–2 kG). Discharge currents, which may also be used as a rough measure of pressure, are typically 10–1000 A Torr$^{-1}$. Speeds range from 0.1 to $10^5$ liters sec$^{-1}$. Highest pressures are in the $10^{-3}$ Torr range, with lowest pressures well into the UHV range, although loss of pumping speed does occur in this range depending on design.[29] The fresh pumping film is generated continuously by sputtering at the cathode. Only a single insulated electrical connection is required. The second type of ion pump is the electrostatic getter ion pump,[26] which has no magnetic field, provides long electron orbits in a cylindrically symmetric electric field, and has continuous titanium sublimation. Highest pressure is $1 \times 10^{-1}$ Torr and lowest pressure is $10^{-12}$–$10^{-11}$ Torr. Active gas speeds are higher and inert gas speeds lower than for sputter ion pumps. Ion pumps normally are used as appendage pumps, but there is no reason in principle why they cannot be built in to operating equipment, making use of particular situations such as local magnetic fields.[31]

## 3.3. Getter Pumps

Getter pumps consist simply of metals or alloys either in the form of solids[32] or of deposited thin films,[33] which may or may not be cooled. Getter pumps incorporate no electric discharge, and their action results entirely from the solubility of the gas to be pumped in the solid or from the chemical affinity between the gas and the film. Thus, an outstanding advantage of a getter pump is its simplicity and flexibility in a variety of experimental situations. However getter pumps do not pump the inert gases such as helium, neon, and argon, and some other means must be supplied for pumping these gases, even though in many situations they do not present a major gas load. For the active gases, hydrogen, water vapor, carbon monoxide, carbon dioxide, oxygen, and nitrogen, on getter films, a sticking probability of $\sim 0.5$ is readily achieved on several metals. Thus, speeds of 5 liters sec$^{-1}$ cm$^{-2}$ can be developed. Pump areas of 1000 cm$^2$ can easily be utilized yielding getter pump speeds of 5000 liters sec$^{-1}$. There is no major diminution of speed at low pressures, but renewal of a film is necessary when pumped amounts approach a monolayer on the surface. For titanium, a film capacity of one pumped molecule per titanium atom evaporated can be approached. An instructive discussion of a 60-liter vacuum system pumped mainly with a titanium film is given by Shen.[34]

## 3.4. Turbomolecular Pumps

These pumps comprise high-speed rotors (12,000–42,000 rpm) with suitably mounted blades to propel gas by transferring momentum by collision to the molecules toward the downstream end of the pump. Rotor surface speeds must approach molecular speeds ($\sim 5 \times 10^4$ cm sec$^{-1}$). Pumps have been designed with both horizontal[35] and vertical[36] shafts. Pumping speeds vary between about 70 and 9000 liters sec$^{-1}$, and pressures in the UHV range can be reached without traps. Upper limit of pressure is about $10^{-1}$–$10^{-2}$ Torr. Since the gases pumped are exhausted from the system, memory effects are not troublesome. A single source of conventional power is all that is required. Against these advantages are rather poor pumping of hydrogen (low mass), some vibration and noise, relatively high initial cost, and potentially easy severe damage to the pump by introduction of solid objects. For particular applications, a liquid-nitrogen trap for water vapor[35] and a titanium getter pump for hydrogen[36] have been suggested as useful additions to turbomolecular pumps.

## 3.5. Cryopumps

Cryopumps represent a class of pumps in which molecules striking surfaces are held there (pumped) by the forces responsible for physical adsorption and condensation.[37,38] Generally, the surfaces are cooled below room temperature, often to liquid-nitrogen or liquid-helium temperatures.

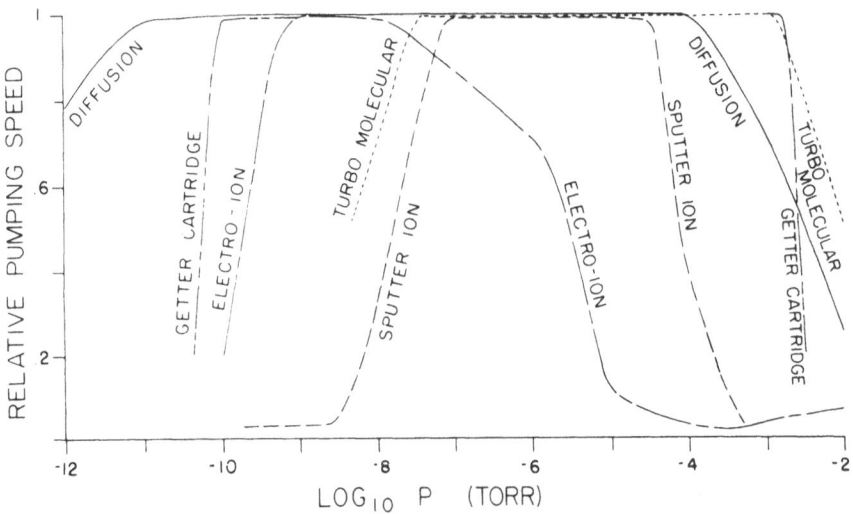

Figure 3. Pumping speed versus pressure for various ultrahigh vacuum pumps. Data taken from following sources: diffusion pump, gas nitrogen[23]; getter-cartridge pump, gas hydrogen[32]; electroion pump (getter ion), gas air[26]; turbomolecular pump, gas air[36]; sputter ion, gas hydrogen.[26,29] In addition to the pumps shown, evaporated getter pumps,[29] cryogenic pumps, and cryotrapping pumps[37] can be made to produce flat curves at unity over the whole range shown for particular gases, but being surface pumps, their performance depends somewhat more critically than the pumps shown in the figure on the quantity of gas pumped. The performance of all pumps depends upon operating conditions, but all pumps are characterized by an upper and lower operating pressure.

In the range where less than a monolayer of adsorbed gas is pumped, they are termed "cryosorption pumps," and their equilibrium behavior is controlled by the physical adsorption isotherm. In the range where more than a monolayer of gas is pumped, they are termed "condensation pumps," and their equilibrium behavior is controlled by the vapor pressure of the adsorbate. Condensation coefficients of 0.5 can be obtained and, hence, speeds of 5 liters $sec^{-1}$ $cm^{-2}$. The low-pressure limit can in principle be indefinitely low, since the equivalent of backstreaming can be reduced to zero at sufficiently low surface temperature. The high-pressure limit is at least atmosphere, although no single cryopump is suitable for atmosphere to zero pressure. The surfaces vary with application and range from high-area molecular sieves ($\sim 1000$ $m^2$ $gm^{-1}$) to bare flat surfaces. A relatively recent type of surface is a condensed layer of a second, more easily condensed gas.[39,40] Cryopumps have seen application as specialized custom built pumps, as well as routine off-the-shelf items. The gas is pumped within the system, and periodic regeneration of the pumps is necessary.

Figure 3 gives pumping speed versus pressure for one example of each type of pump discussed. No significance is to be given to the absolute values of the speeds. Each pump has a lower limit of pressure at which it can be used, determined by a physical process particular to each pump. For the getter pump and cryopump, this can be indefinitely low.

## 4. Gauges for UHV

All the important gauges used for measuring pressure at UHV are ionization gauges; i.e., electrons are used to ionize the gas in the vacuum and the resulting ions are measured as a flux, which is converted to an equivalent pressure. An important point not often noted explicitly is that the ion current is actually a measure of gas density not gas pressure; i.e., the ion current is insensitive to the temperature of the gas in the vacuum. Frequently this qualification is not quantitatively important, since most gauges operate near room temperature. General reviews of gauges for UHV have been given by Lafferty[41] and Lange.[42] Three broad categories can be identified—hot-cathode total pressure gauges, cold-cathode total pressure gauges, and partial pressure gauges usually employing hot cathodes. It is worth noting here that modern ultrahigh vacuum was born when it was realized that the ion current corresponding to $10^{-8}$ Torr in an early hot cathode ionization gauge (Figure 4a) was not coming from the gas phase but from instrumental effects due to x rays. To overcome this problem, the Bayard–Alpert gauge[43] was proposed (Figure 4b), which lowered the instrumental limit to about $10^{-11}$ Torr, making the pressure range between $10^{-8}$ and $10^{-11}$ Torr subject to quantitative measurement.

To a greater or lesser degree, all UHV ion gauges are sources of gas, as are pumps. This is a complex subject that should be examined in each particular application.

### 4.1. Hot-Cathode Total Pressure Gauges

These gauges all contain a hot thermionically emitting cathode, an anode or grid collecting the electrons after a flight as long as possible through the vacuum during which ionization occurs, and an ion collector measuring the ions. A quantity $K$ (sensitivity) is defined as

$$K = \frac{I^+}{I^- P} \quad \text{Torr}^{-1} \tag{2}$$

where $I^+$ is the ion current, $I^-$ is the electron current, and $P$ is the pressure in Torr. For a standard Bayard–Alpert gauge (Figure 4b), $K \sim 25 \ \text{Torr}^{-1}$ for nitrogen and varies for the other gases as the ionization probability at about $100 \ \text{eV}$, i.e., $S \sim 4$ for helium, $\sim 8$ for hydrogen, $\sim 50$ for xenon. Values of $I^-$ are typically $0.1$ to $10 \ \text{mA}$. Thus, for $I^- = 10 \ \text{mA}$, the observed ion current is multiplied by 4 to yield the pressure in Torr of nitrogen, generally taken as the standard gas. These ionization gauges are linear in pressure between about $10^{-3} \ \text{Torr}$ and $n \times 10^{-11} \ \text{Torr}$, where the non-pressure-dependent residual current in the gauge begins to become quantitatively important (Figure 5). It is this residual current and not any inability to measure small ion currents that represents the low-pressure limit of the hot-cathode ionization gauge. Hot-cathode total pressure gauges with values of $K = 10^4$ Torr$^{-1}$,[44] $K = 10^5 \ \text{Torr}^{-1}$ [45] (Figure 4e), and $K = 10^7 \ \text{Torr}^{-1}$ [46] (Figure 4f) have been reported, but their additional complexity has outweighed their increased sensitivity for routine use.

The residual current in a hot-cathode gauge arises, in the main, when electrons strike the electron collector, where they ionize adsorbed gas, these ions being measured as an ion current, or release x rays, which in turn release photoelectrons from the ion collector, these photoelectrons giving a current in the same sense as arriving ions. The general effect of the residual current in hot-filament ionization gauges is to give an overestimate of the true pressure. Much effort has been invested in the attempt to reduce this residual current and hence to lower the low-pressure limit of the hot-cathode ionization gauge. Some of the resulting gauge designs are shown in Figure 4 (c–f). Table 2 shows the degree to which these designs have been successful.[50] In the modulated Bayard–Alpert gauge, the residual current is measured and subtracted from the observed current to yield the ion current arising in the gas phase. In choosing between these various designs,

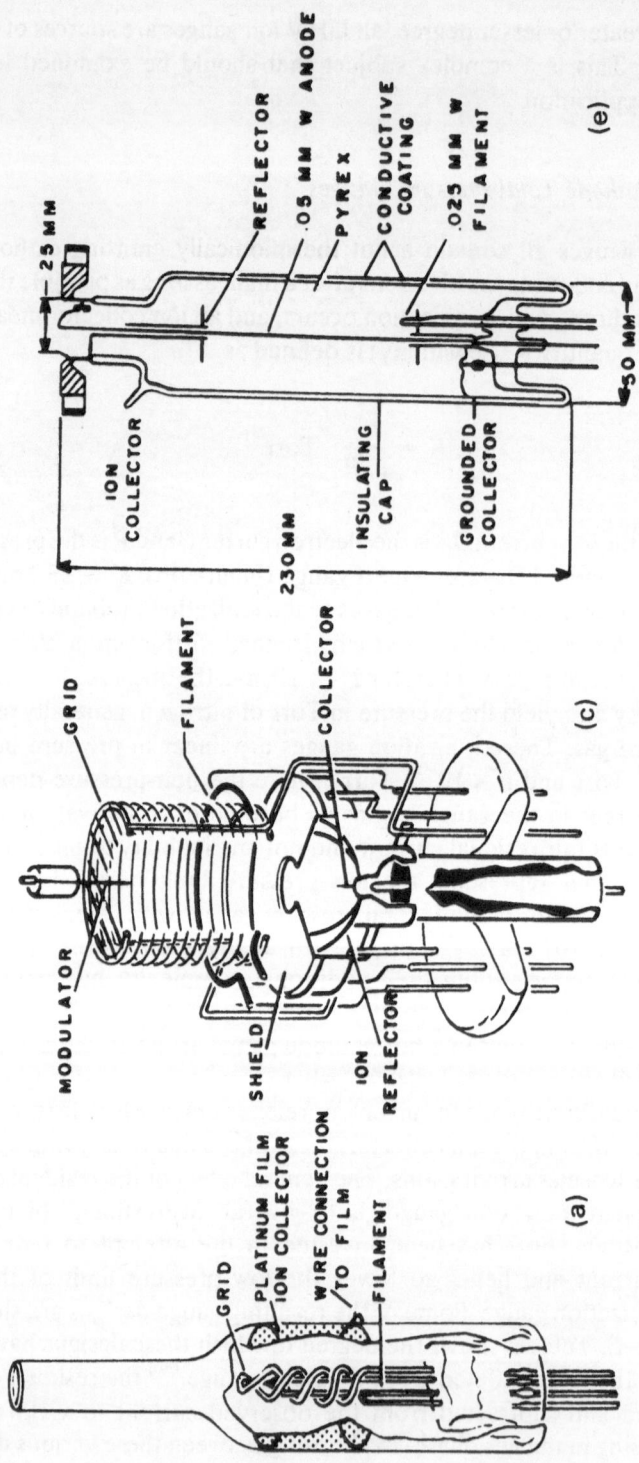

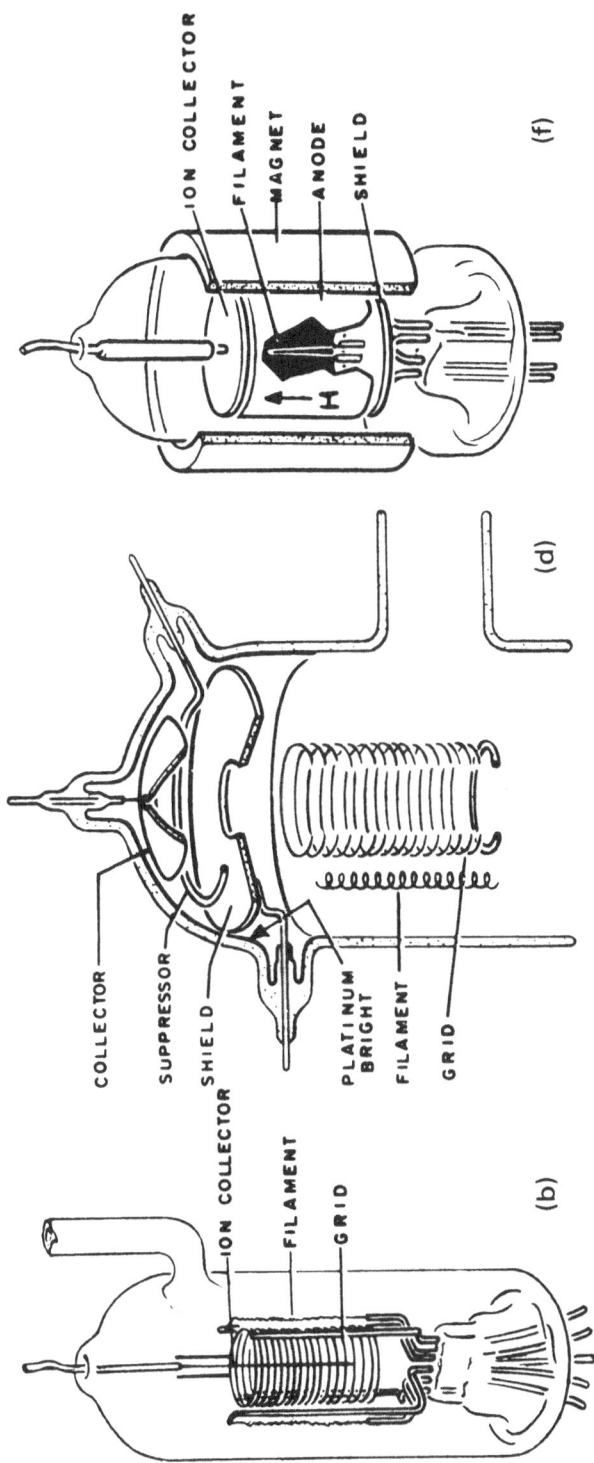

Figure 4. Various types of ultrahigh vacuum hot-cathode pressure gauges: (a) early triode gauge, (b) Bayard–Alpert gauge, (c) extractor gauge, (d) suppressor gauge, (e) orbitron gauge, (f) hot-cathode magnetron gauge (from Lafferty[41]).

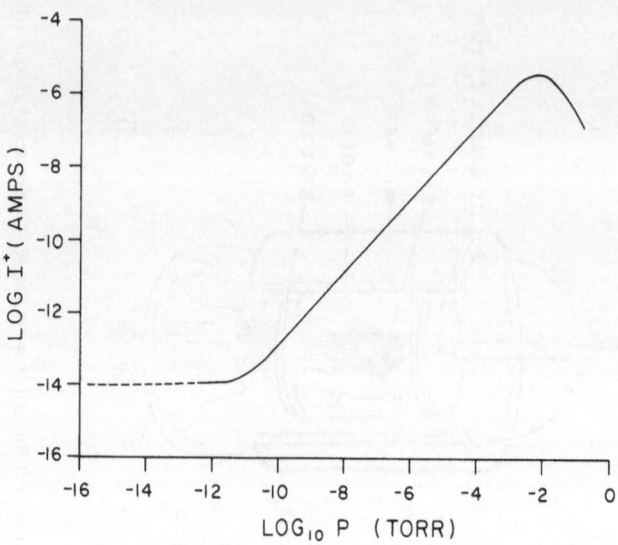

Figure 5. Measured ion current $I^+$ versus pressure $P$ for a suppressor gauge. $I^- = 80 \, \mu A$; argon gas; for the linear region, $K = 16 \, \text{Torr}^{-1}$. Gauge similar to that shown in Figure 4d.

Lange[42] recommends the modulated Bayard–Alpert gauge or the extractor gauge. Various modes of operation of the modulated Bayard–Alpert gauge have been examined,[61] and those involving no power changes at the modulator appear to be the most desirable. A Bayard–Alpert gauge and an extractor gauge have been directly compared in the same vacuum environment, and the latter has been proven to be less sensitive to ion desorption from the grid.[62] Several hundred modulated Bayard–Alpert gauges have been used to measure pressures of $1 \times 10^{-10}$ Torr in the CERN intersecting storage ring.[63] For total pressure measurements below $10^{-12}$ Torr, Blechschmidt[60] has described a hot-cathode gauge similar in general structure to an extractor gauge but with a channel electron multiplier as detector. This multiplier has a gain of $10^8$, giving the whole gauge a value of $K = 2 \times 10^9 \, \text{Torr}^{-1}$ with a low-pressure limit below $10^{-15}$ Torr.

## 4.2. Cold-Cathode Total Pressure Gauges

Three types of cold-cathode total pressure gauge—the inverted magnetron,[64] the magnetron,[65] and the trigger discharge[66]—have seen application at UHV. All have cylindrical geometry, one electrode being a cylinder. This electrode is the cathode in the inverted magnetron and the anode in the magnetron and trigger discharge gauge. This cylinder is capped at each end by disks, which are the complementary anode or cathode, respectively. These disks are joined electrically by a central rod in both magnetrons. Along the cylindrical axis, there is a magnetic field of 1–2 kG, and there is a radial electric field between anode and cathode supplied by an external supply of 2–6 kV. The spontaneous discharge, initiated by an auxiliary filament in the trigger discharge gauge, confines electrons to spiralling paths around the central axis. The volume of the discharge region is less than 15 cm$^3$ and has recently been reduced to 0.5 cm$^3$,[67] although the latter gauge has not been fully tested at UHV. Typical curves of ion current versus pressure obtained with these cold-cathode gauges are shown in Figure 6.[68] It may be noted that they are nonlinear at pressures below $10^{-9}$–$10^{-10}$ Torr. This makes pressures obtained from linear extrapolation underestimates and has led to claims of pressures in the literature lower than those actually obtained. These gauges are also subject to sudden jumps in sensitivity associated with mode changes in the discharge.[70] In general, the pumping speed of cold-cathode gauges is high and is not controllable. This is a problem in some circumstances. However, despite these disadvantages, the great advantage of small power requirement with no requirement for filament power has led to the magnetron gauge (Figure 7) being the first pressure gauge to measure pressure on the moon.[71]

Table 2. Low-Pressure Limits of Hot-Filament Ionization Gauges$^{a,b}$

| Gauge | $P_x$ (Torr) | $P_{min}$ (Torr) | Reference |
|---|---|---|---|
| Bayard–Alpert | $3 \times 10^{-11}$ | $3 \times 10^{-10}$ ($P_x$ not measured) | (48) |
|  |  | $1 \times 10^{-10}$ ($P_x$ measured) |  |
| Fine collector (4 μm) | $1.5 \times 10^{-12}$ | $2 \times 10^{-11}$ | (49) |
| Bayard–Alpert | (est) |  |  |
| Modulated Bayard–Alpert | $3 \times 10^{-11}$ | $3 \times 10^{-11}$ (ε assumed zero) | (50) |
|  | $3 \times 10^{-11}$ | $3 \times 10^{-13}$ (ε measured) | (51) |
| Suppressor | $<2 \times 10^{-12}$ | $<2 \times 10^{-11}$ | (52) |
| Modulated suppressor | $\sim 10^{-14}$ | $<2 \times 10^{-12}$ (est $10^{-13}$) | (53) |
| Orbitron | $10^{-12}$ (est) | $<5 \times 10^{-11}$ | (54) |
| Klopfer | — | $10^{-11}$ | (55) |
| Mass spectrometer (90° sector) | $\sim 0$ (est) | $<10^{-15}$ (est $5 \times 10^{-17}$) | (56) |
| Hot-cathode magnetron | — | $<10^{-11}$ (est $4 \times 10^{-14}$) | (57) |
|  |  | without multiplier |  |
|  |  | (est $3 \times 10^{-17}$) |  |
|  |  | with multiplier |  |
| Extractor | $3 \times 10^{-13}$ (est) | $7 \times 10^{-13}$ | (58) |
| Bent beam | $1 \times 10^{-14}$ | $1 \times 10^{-13}$ (est) | (59) |
| Channeltron | $4 \times 10^{-17}$ | $2 \times 10^{-15}$ | (60) |

$^a$ Table based on Hobson and Redhead.[47]
$^b$ $P_x$ = pressure at which ion current equals x-ray photocurrent, $P_{min}$ = pressure at which estimated error is 10%, est = Estimated but not measured directly. Values are given for normal gauge operating conditions.

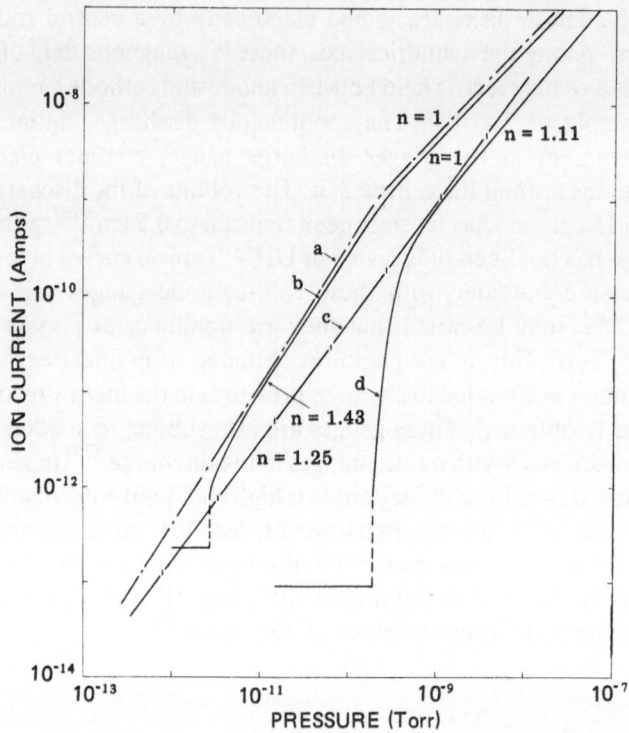

Figure 6. Ion current versus pressure for various cold-cathode, crossed-field gauges; $n$ is the exponent in the expression $I^+ = kp^n$. $B$ is the magnetic field, $V_a$ is the anode voltage. (a) Magnetron gauge, $B = 1$ kG, $V_a = 4.5$ kV, helium[69]; (b) magnetron gauge, $B = 1$ kG, $V_a = 4.8$ kV, helium[68]; (c) inverted magnetron gauge, $B = 2.1$ kG, $V_a = 6$ kV, helium[68]; (d) magnetron gauge (Kreisman type without shield electrodes), $B = 1$ kG, $V_a = 4$ kV, helium.[68]

## 4.3. Partial Pressure Gauges

There are methods of obtaining the partial pressures of the gases in an UHV system that depend upon the adsorption properties of the gases, but they are not widely used, and the discussion here will be restricted to devices that separate by some means the ions generated in an ionization source not unlike that of a total pressure gauge.

These devices are often called "partial pressure analyzers," "residual gas analyzers," or "mass spectrometers." They represent a modification of the principles of conventional mass spectrometry to the particular needs of UHV. These may be taken to be high sensitivity, since the original gas samples are at low pressure; modest resolution, since frequently the gases of interest do not have masses above 100; extremely low outgassing rates, since this property will influence the lowest pressures measurable; low background currents, since this will also influence the lowest pressures measurable; and an acceptably short time in which a spectrum can be completed. Pertinent reviews are available.[72-74] There are perhaps five types of mass spectrometers that have been developed to meet UHV requirements and that are commercially available—magnetic sector,[75] cycloidal,[76] radio frequency,[77] quadrupole–monopole,[78] and time of flight.[79] To meet the sensitivity requirements of UHV, these all need electron multipliers as detectors. Without multipliers they have sensitivities in the range of $5 \times 10^{-4}$ A Torr$^{-1}$. The multipliers tend to change their gain under various conditions. This is a distinct disadvantage of UHV partial pressure gauges. The sensitivity is in general dependent upon the mass, and quite careful calibration is required if corrections for this effect are important. Many modern partial pressure analyzers include a means of summing all the partial ion currents to yield a total pressure reading. It should not be assumed that the sum of the partial pressures will indeed be the total pressure. If this is important, it should be checked and methods developed to deal with any discrepancy. While several partial pressure analyzers have operated in the range of $10^{-12}$ Torr, outgassing facilities are not always adequate to ensure this, and indeed some commercial instruments have very poor outgassing facilities. It is a rare UHV partial pressure gauge that does not raise the

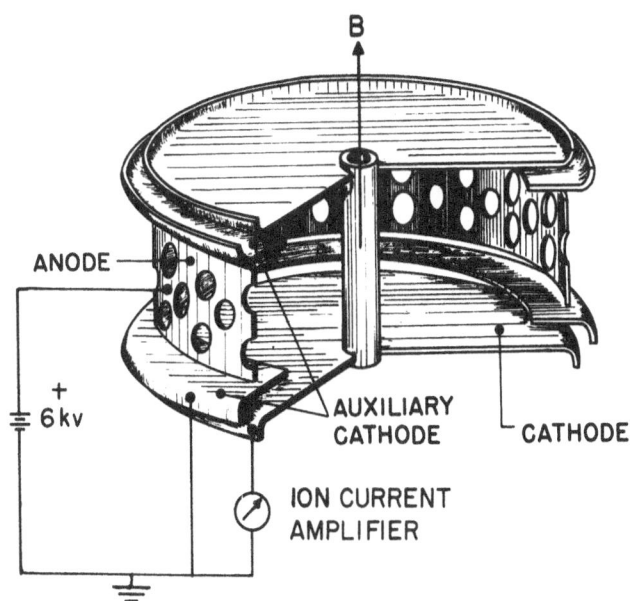

Figure 7. Schematic of cold-cathode magnetron gauge.[65]

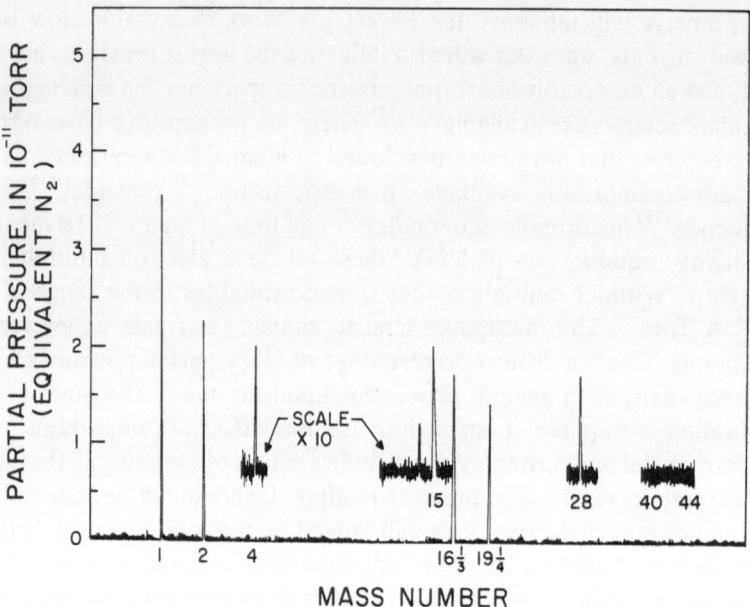

Figure 8. Background spectrum in an ultrahigh vacuum system with envelope of stainless steel and aluminosilicate glass, measured with 90° sector magnetic mass spectrometer. Of the four main peaks only mass 2 (hydrogen) is a true gas-phase peak. The others arise from ions released by electronic bombardment of the grid cage.

background pressure permanently in a system when it is turned on. It is necessary to take account of this in some types of measurements. The operating range of UHV partial pressure analyzers is from about $10^{-15}$–$10^{-4}$ Torr. While modern UHV partial pressure gauges have many deficiencies, as pointed out by van Oostrom,[80] they are nevertheless indispensable in many measurements and yield far more information than a total pressure gauge. Useful tables of cracking patterns of particular gases are available.[81] Figure 8 shows a residual spectrum in a system dominated by hydrogen outgassing from stainless steel. Figure 9 shows a series of spectra obtained with a 90° magnetic mass spectrometer under various conditions in an UHV system with envelope of stainless steel and aluminosilicate glass.

### 4.4. Gauge Calibration

Calibration of UHV gauges,[82] both total and partial, in terms of an absolute force measurement such as a torr has always presented a problem. The problem is twofold. Direct measurements of force, such as might be made with a McLeod gauge or with the deflection of a calibrated membrane,

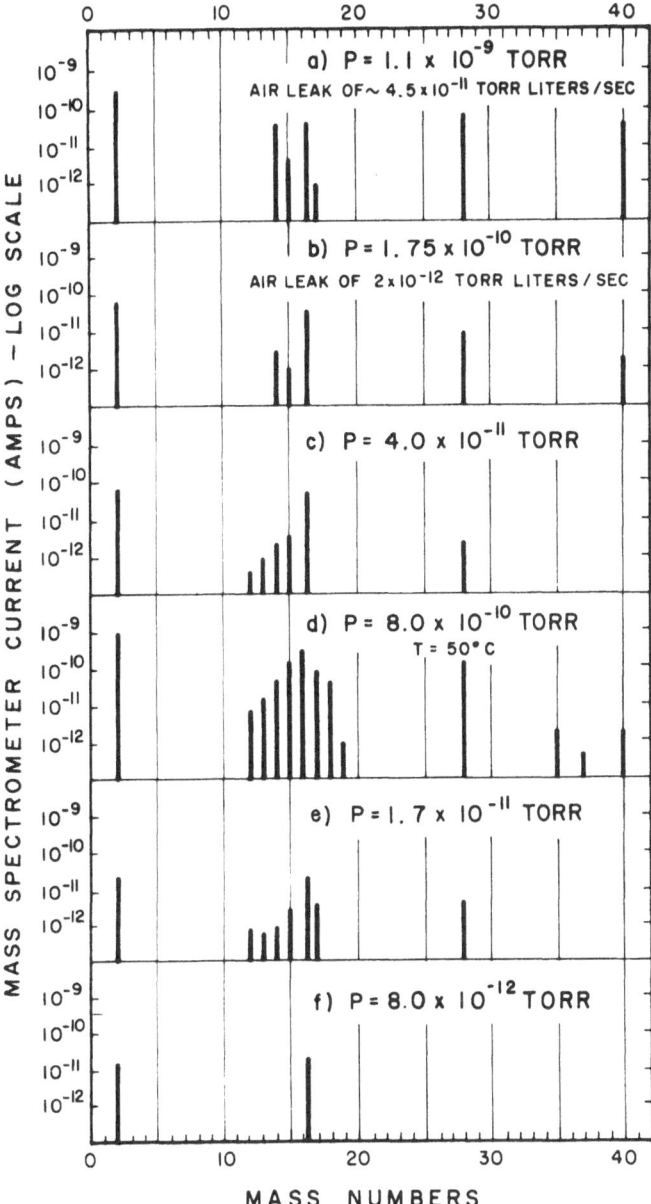

Figure 9. Residual gas spectra obtained in the same system as described in Figure 8 (except that mass 1 was not observable) for various system conditions: (a) large leak present, (b) leak partially sealed, (c) leak removed, (d) entire system in full operation being baked at 50°C, (e) immediately following bake in d, (f) cleanest spectrum achieved at room temperature. Same system as for Figures 1 and 2 and Table 1.

tend to become inaccurate below the range $10^{-5}$–$10^{-4}$ Torr. McLeod gauges accurate to lower pressures than this[83] tend to be still in the development stage. Thus, direct calibration of UHV pressure gauges against an absolute standard is at present not possible. The use of the equilibrium pressure of a gas dissolved in a solid, for example, hydrogen in erbium or niobium,[84,85] has been suggested as a secondary standard for gauge calibration but is restricted to certain gas–solid combinations and is not widely used at present. For the inert gases, the repeated use of static expansion may be used to reduce the pressure from an absolutely measured value to one arbitrarily low in principle. Background errors will, however, limit the range covered. Alternatively, the establishment of a fixed leak into an UHV system permits a pressure rising linearly with time to be established, provided ion pumping by the gauge is reduced to negligible levels. The latter can be achieved by turning on the gauge during measurements only. Hence, gauge readings as a function of time provide a measurement of the linearity of gauge response versus pressure. When the pressure has risen to a value where an absolute manometer may be used, a simultaneous reading with the ion gauge provides an absolute calibration. Should the number of decades of pressure covered make this measurement too time consuming, then a double-leak method yielding a pressure rising quadratically with time may be used.[86] These methods, however, are restricted to gases that are not pumped chemically in UHV systems, i.e., the inert gases. In the case of active gases, pressure dividers or calibrated conductances are used in systems with high-speed pumps.[87,88] Davis,[88] using a surface at liquid helium temperature as his final high-speed pump (cryosorption pump), calibrated gauges to $10^{-13}$ Torr, using a McLeod gauge as an absolute standard. Porous leaks[89] have been used to admit gases to systems for calibration purposes. Should an experiment require a measure of pressure more precise than about ±20%, then the nominal manufacturer's gauge calibration, either total or partial, should not be accepted, and a method for absolute calibration of the gauges should be built into the experiment.

## 5. UHV Hardware

UHV hardware, as distinct from pumps and gauges, includes such items as piping, flanges, gaskets, valves, experimental containers, electrical feedthroughs, controlled leaks, windows, and manipulators. A great variety of commercial equipment exists,[90] and the suppliers are truly world wide. General discussions of hardware are available.[91,92] Much commercial UHV equipment is made of stainless steel (often 304), which is nonmagnetic and relatively inert during exposure to the atmosphere at temperatures up to 500°C. Vacuum-tight welds are usually made by tungsten inert gas (TIG) welding, although electron-beam welding is sometimes used for bellows where the basic material is thin ($\sim$0.005 in.). A common gasket material is copper, but gold and aluminum are often used, and other malleable metals less frequently. UHV valves represent one of the most impressive achievements of modern ultrahigh vacuum technology. They can be opened and closed repeatedly with a torque of a few foot-pounds to a closed conductance less than $10^{-13}$ liters sec$^{-1}$ (a bicycle tire with a leak of this magnitude would take more than a million years to go flat). The seal is normally gold against stainless steel, but silver and copper are sometimes used as the deformable material.

Some UHV valves can be baked closed, usually at a reduced temperature of 250°C. UHV bellows are available off the shelf with internal diameters from 0.1 to 20 in. and of variable length, since standard sections can be welded together indefinitely. If a user wishes to incorporate a bellows into an existing apparatus, he should be careful to order a convenient termination to the bellows, since the thin material of the bellows itself is difficult to weld. Experimental chambers are normally selected for a particular purpose. The modern trend is toward chambers with many welded flanges of different diameters to which various gauges, pumps, mass spectrometers, windows, electrons guns, etc., can be attached, yielding great flexibility of design. The flexibility and complexity of manipulators, usually for targets in surface analytic apparatus, is steadily increasing. UHV manipulators giving $x$, $y$, $z$, and two rotary motions are currently available. Controlled gas leaks of $10^{-12}$ Torr liters sec$^{-1}$ are quite practical. Indeed, quite simple modifications to these leaks permit controlled leaks matching the spontaneous leaks shown in Table 1, at least for the inert gases.

## 6. Surface Cleaning

Strictly speaking, surface cleaning is not a part of the production and measurement of extremely low pressures; however, the reason for obtaining an extremely low pressure is often the study of clean surfaces. Hence, a brief discussion of surface cleaning may not be out of place in this chapter. Moreover, the cleaning of surfaces generally disturbs the pressure conditions in a system and, hence, surface cleaning may be regarded as a major obstacle to the achievement of very low pressures. By "surface cleaning" is meant the process of ridding a surface of all unknown contaminants coming either from the bulk of the sample or from its vacuum environment so that it is left, one hopes, in a chemically pure and possibly monocrystalline state on which meaningful experiments may be performed.

Figure 10 shows the general stages in the prevacuum preparation of bulk single crystal targets for use in UHV as outlined by Whitton.[93] If

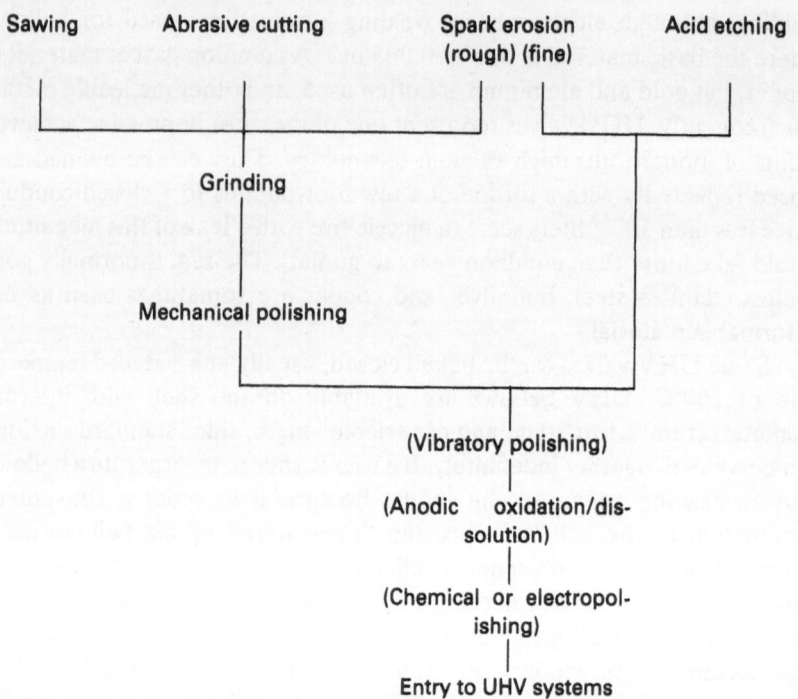

Figure 10. Stages in the prevacuum preparation of bulk single-crystal targets (from Whitton[93]).

polycrystalline surfaces or thin film surfaces are to be used, then the prevacuum procedures are likely to be less extensive.

Each operation in Figure 10 creates a certain depth of damage, which must be reduced by the next step in the chain until, at the point of entry to the UHV system, the surface is flat and sufficiently free from surface damage to permit the attainment of a perfect surface by postvacuum methods. For example, mechanical sawing introduces damage to a depth of about $10^6$ Å, while mechanical polishing with medium grit (600 grit silicon carbide) introduces damage to a depth of about $10^5$ Å, and with fine abrasive on cloth ($\gamma$-alumina) to a depth of about $10^3$ Å. The final stages of prevacuum preparation leave damage to a depth varying from 0 to 100 Å. Since x-ray examination of a surface by Laue back reflection extends to a depth of several thousand angstroms, such examination becomes insensitive in the final stages of target preparation.[93]

Following installation in an UHV system a bulk single crystal may be subjected to one or more of a number of additional preparative procedures. These are often combined with examination by UHV surface analytic instruments *in situ*. The number of analytic instruments is large and becoming larger. A survey has been given recently by the author,[94] and these instruments will not be discussed here except to note that each technique is sensitive to a different depth below the surface and yields its own particular type of information about surface order, surface chemical composition, or surface electronic states.

The postvacuum surface preparative procedures are now listed with a few comments on each. Further details may be found in the literature.[95–101]

## 6.1. Bulk Heating to High Temperature[95–98]

This may be done by resistive, induction, or radiation heating or by electron bombardment. The temperature is limited, of course, by the melting temperature of the sample. Only for a limited number of materials is the temperature at which gases readily desorb sufficiently far below the melting temperature. The method has been used with some success for tungsten, molybdenum, rhenium, tantalum, niobium, platinum, rhodium, nickel, chromium, germanium, silicon, graphite, and diamond. It is perhaps best suited to the refractory metals, where very high temperatures can be used (e.g. tungsten, $T = 2500$ K). However high temperatures enhance the diffusion of impurities from bulk to surface (e.g., carbon in tungsten, molybdenum, rhenium, tantalum, and nickel), with the resulting appearance in the gas phase of carbon monoxide, carbon dioxide, or methane. High temperatures may also cause etch pits, exposing undesired crystal faces at the surface (silicon at 1100°C), or whisker growth. Heating in Pyrex apparatus can cause deposition of boron from the walls onto a sample surface.

## 6.2. Chemical Reactions at Elevated Temperatures[95–97]

Samples may have surface contaminants of carbon, sulfur, or phosphorus, which can arise from diffusion from the bulk. Such samples have been cleaned by heating in oxygen to produce volatile oxides (e.g., nickel at 500°C in $10^{-6}$ Torr oxygen). However, it was found that adsorbed oxygen remained on the surface and was in turn removed by heating the nickel sample at 200°C in hydrogen at $10^{-7}$ Torr. The latter step may take place spontaneously during rebake of an entire stainless steel system. The effectiveness of these methods may depend on the particular crystal face studied.[100]

## 6.3. Ion Bombardment Followed by Annealing[95–98]

This method is quite general and is applicable to metals, insulators, and semiconductors. Ion bombardment sputters off surface atoms of a solid, thereby removing damaged layers and adsorbed contaminants but at the cost of creating its own type of damage and injecting atoms of the bombarding gas into the solid. The latter two disadvantages are counteracted by a stage of postbombardment annealing, which anneals the damage created by the bombardment and desorbs the trapped gas atoms. Inert gas atoms are clearly desirable to permit easy desorption, argon being commonly used at pressures less than $10^{-3}$ Torr at voltages of 200–600 V, and at positive ion current densities of 100 $\mu$A cm$^{-2}$. Following bombardment, the pressure is reduced to the UHV range, where a typical annealing cycle is a few minutes at an absolute temperature about one-half of the melting temperature of the solid. Several cycles of bombardment–annealing are generally necessary before clean, ordered surface conditions are obtained. Ion bombardment cleaning is usually preceded by a stage of bulk heating. Although widely used, there exist examples [e.g., carbon on nickel (111)[100]] where ion bombardment was found ineffective.

## 6.4. Electron Bombardment

Electrons striking a surface have a probability for desorption of adsorbed gas that can range as high as $10^{-4}$ atoms per electron[102,103] (oxygen on molybdenum). For this case, 10 mA cm$^{-2}$ electron current for 10 sec at about 100 eV is sufficient to remove a monolayer of adsorbed oxygen. Momentum transfer from electron impact is not sufficient to cause sputtering of a metal, and hence, electron bombardment is useful for removing adsorbed layers from a metal without disturbing the substrate. However, for some adsorbates [e.g., carbon monoxide on nickel (110)[104]] electron bombardment causes dissociation of the carbon moxoxide, leaving carbon on the surface. Desorption of neutral molecules is often accompanied by an ion component about 1% of the neutral component. The ion current is useful as a continuous monitor of the quantity of adsorbate, and its periodic measurement can yield useful information about contamination processes. It is this component that is responsible for the surface peaks in Figure 8. However, for some insulators, even low-level electron bombardment causes decomposition (e.g., potassium chloride[105]), and it has been difficult to make low-energy electron diffraction measurements on alkali halides.

## 6.5. Pulsed Laser Bombardment

A pulsed laser beam[99] can provide sudden localized heating of a surface without the introduction into the system of any auxiliary apparatus, since the beam can be introduced to the vacuum through a window. Bedair and Smith[99] found that adsorbed background gases such as carbon monoxide, hydrogen, water, and hydrocarbons could be desorbed without damage by 20 nsec pulses of 30 MW cm$^{-2}$. For oxygen on nickel (100), pulses of

40–120 MW cm$^{-2}$ produced increasing degrees of oxygen desorption but at the expense of increasing surface damage. On silicon (111), laser pulses in the range 10–40 MW cm$^{-2}$ removed the 7 × 7 surface structure, which could be regained by annealing at 1000°C for a few minutes. However, repeated pulses of 60 MW cm$^{-2}$ produced visible damage.

### 6.6. Cleaning by Ultraviolet Radiation[100]

This is not a highly developed method. Irradiation of a surface in UHV by 140 W of ultraviolet radiation does not appear to clean a gold surface but rather appears to break down or polymerize adsorbates. However, in the presence of 10$^{-4}$ Torr of oxygen,[100] the formation of volatile oxides was enhanced, which was interpreted as caused by the presence of ozone. Gold surfaces were cleaned to give the same coefficient of adhesion by this method and by ion bombardment.

### 6.7. Field Desorption

A method of specialized importance is that of field desorption from the tip in the field-emission or field-on microscope.[97,106] Adsorbed atoms or atoms of the substrate can be physically pulled from the surface by a field of $n × 10^8$ V cm$^{-1}$ when the tip potential is positive. The method preferentially removes protruding atoms, thus tending to leave flat monocrystalline areas on the microscope tips. Tips of many materials have been cleaned in this way, including the refractory metals, nickel, iron, platinum, copper, silicon, and germanium. The areas cleaned are very small, of order 10$^{-10}$ cm$^{-2}$, but the effects are immediately observable and can be photographed.

### 6.8. Vacuum Cleaving[95,96]

A surface of desired orientation and as clean as the impurity level of the bulk crystal can be created by vacuum cleaving. A new surface is created each time a knifelike cleaver removes a slice from a crystal. Cleavage has been used with a wide variety of materials,[96] including metals, mica, pyrolytic graphite, alkali halides, and many semiconductors. It cannot be assumed that the surfaces are always perfect, the degree of disorder depending on the material and the method of cleavage. In general, bursts of gas are generated at cleavage that might recontaminate the surface, particularly if they arise from the cleavage mechanism rather than the crystal itself.

### 6.9. Crushing[95,96]

This is a grosser form of vacuum cleaving wherein much larger cleaved surface areas are produced at the expense of uncontrolled crystal orientation and much less controlled surface disorder patterns. Surfaces of germanium, silicon, lead telluride, and lead sulfide have been studied in this way. As with all large area samples, sintering effects may play a role in measurements.

### 6.10. Wire Brush[98]

This technique has not been widely used. It was found[98] that a stainless steel wire brush with strands of 0.25-mm diameter rotating at a speed of 300 rps under no load contaminated specimens of tungsten and stainless steel but cleaned copper specimens. The interpretation was that stainless steel, being harder than copper, could remove copper from the surface. The copper surface, however, was highly abraded and extremely rough.

### 6.11. Evaporation[95,96]

This is a simple method for forming a clean surface, provided the evaporating source is sufficiently pure. A film from an evaporation source is deposited at a vapor pressure of about $10^{-4}$ Torr on a suitable substrate, which may have been subjected to any the cleaning methods above. Normally, a polycrystalline film is obtained, but the films may be single crystals or amorphous or whiskers, depending on the kind of material evaporated, the evaporation rate, the nature of the substrate, and the substrate temperature. It is also possible to obtain a clean surface by evaporation in a poor vacuum by deposition of a film faster than the ambient atmosphere can contaminate it.[107] This method has the disadvantage that measurements must be made during film deposition, since only during evaporation is a clean surface obtained.

### 6.12. Chemical Deposition (Vapor Plating)[97]

This method involves a suitable vapor phase chemical reaction in which the desired surface appears as a product that can be deposited, e.g.,

$$SiCl_4 + 2H_2 \xrightarrow{1100^\circ C} Si + 4HCl$$

Not all such reactions will be consistent with maintaining an UHV environment.

### 7. Conclusion

This chapter has reviewed the modern methods of producing ultrahigh vacuums and measuring ultralow pressures. Great advances have been made in these fields since the invention of the Bayard–Alpert gauge in 1950. Commercial equipment is now available for the pressure range $10^{-9}$–$10^{-12}$ Torr, with extensions to lower pressures in particular situations. A great variety of surface analytic instruments utilizing ultrahigh vacuum techniques are appearing on the commercial market. There is little doubt that these instruments represent techniques of growing importance in the study of surface physics and chemistry.

ACKNOWLEDGMENTS

The author would like to thank, in particular, his former colleague J. W. Earnshaw, with whose aid Table 1 and Figures 1, 2, 8, and 9 were made, as well as many other members of the Electron Physics Section of the National Research Council of Canada for discussions in the general area of this chapter.

## *References*

1. L. V. Omelka, *J. Vac. Sci. Technol.* **7**, 257 (1970).
2. E. Fischer, *J. Vac. Sci. Technol.* **9**, 1203 (1972).
3. P. A. Redhead, E. V. Kornelsen, and J. P. Hobson, *Can. J. Phys.* **40**, 1814 (1962).
4. G. A. Rozgonyi, W. J. Polito, and B. Schwartz, *Vacuum* **16**, 121 (1966).
5. E. P. Tryon, *Nature* **246**, 396 (1973).
6. Proceedings of the 1st International Conference on Solid Surfaces, *J. Vac. Sci. Tech.* **9**, 561–952 (1972).
7. The reader may find an account in the Proceedings of the 6th International Vacuum Congress and the 2nd International Conference on Solid Surfaces, held in Kyoto, Japan, March 25–29, 1974. Published in *Jap. J. Appl. Phys.*, Supplement 2, Parts 1 and 2 (1974).
8. P. A. Redhead, J. P. Hobson, and E. V. Kornelsen, *The Physical Basis of Ultrahigh Vacuum*, Chapman and Hall, London (1968).
9. R. Clader and G. Lewin, *Brit. J. Appl. Phys.* **18**, 1459 (1967).
10. L. De Chernatony, *Vacuum* **22**, 635 (1972).
11. G. Moraw and R. Dobrozemsky, Proceedings of the 6th International Vacuum Congress, Kyoto, Japan. Published in *Jap. J. Appl. Phys.*, Supplement 2, Part 1, 261 (1974).
12. J. R. Young, *J. Vac. Sci. Tech.* **6**, 398 (1969).
13. W. A. Rogers, R. S. Buritz, and D. Alpert, *J. Appl. Phys.* **25**, 868, 1954.
14. F. J. Norton, *J. Appl. Phys.* **28**, 34 (1957).
15. W. G. Perkins, *J. Vac. Sci. Technol.* **10**, 543 (1973).
16. A. Klopfer, *Vakuum Tech.* **10**, 113 (1961).
17. G. A. Rozgonyi and J. Sosniak, *Vacuum* **18**, 1 (1968).
18. J. P. Hobson and J. W. Earnshaw, *J. Vac. Sci. Technol.* **4**, 257 (1967).
19. G. Carter, D. G. Armour, and L. De Chernatony, *Vacuum* **22**, 643 (1972).
20. T. Tom, *Phys. Today* **20**, 32 (1972).
21. B. D. Power, *High Vacuum Pumping Equipment*, Reinhold, New York (1966).
22. J. Singleton, *J. Phys. E.* **6**, 685 (1973).
23. M. H. Hablanian and J. C. Maliakal, *J. Vac. Sci. Technol.* **10**, 58 (1973).

24. D. J. Santeler, *J. Vac. Sci. Technol.* **8**, 299 (1971).
25. A. Venema, *Vacuum* **9**, 54 (1959).
26. D. G. Bills, *J. Vac. Sci. Technol.* **10**, 65 (1973).
27. S. L. Rutherford, *Trans. Amer. Vac. Soc. Vac. Symp.* **10**, 185 (1963).
28. L. D. Hall, *J. Vac. Sci. Technol.* **6**, 44 (1969).
29. J. H. Singleton, *J. Vac. Sci. Technol.* **6**, 316 (1969).
30. M. V. Kuznetsov. A. S. Nasarov, and G. F. Ivanovsky, *J. Vac. Sci. Technol.* **6**, 34 (1969).
31. M. D. Malev and E. M. Trachtenberg, *Vacuum* **23**, 403 (1973).
32. J. Vaumoron, G. Gasparini, and G. Bertoli, *J. Vac. Sci. Technol* **9**, 932 (1972).
33. D. J. Harra and T. W. Snouse, *J. Vac. Sci. Technol.* **9**, 552 (1972).
34. L. Y. L. Shen, *Rev. Sci. Instr.* **43**, 1301 (1972).
35. G. E. Osterstrom and A. H. Shapiro, *J. Vac. Sci. Technol.* **9**, 405 (1972).
36. K. H. Mirgel, *J. Vac. Sci. Technol.* **9**, 408 (1972).
37. J. P. Hobson, *J. Vac. Sci. Technol.* **10**, 73 (1973).
38. C. Benvenuti, *J. Vac. Sci. Technol.* **11**, 591 (1974).
39. V. B. Yuferov, P. M. Kobzev, and B. V. Glasov. *Sov. Phys. Tech. Phys.* **15**, 457 (1970).
40. K. E. Templemeyer, R. Dawdarn, and R. L. Young, *J. Vac. Sci. Technol.* **8**, 575 (1970).
41. J. M. Lafferty, *J. Vac. Sci. Technol.* **9**, 101 (1972).
42. W. J. Lange, *Phys. Today* **20**, 40 (1972).
43. R. T. Bayard and D. Alpert, *Rev. Sci. Instr.* **21**, 571 (1950).
44. R. K. Fitch, T. Mulvey, W. J. Thatcher, and A. H. McIlraith, *J. Phys. E. Sci. Instr.* **4**, 553 (1971).
45. E. A. Meyer and R. G. Herb, *J. Vac. Sci. Technol.* **4**, 63 (1967).
46. J. M. Lafferty, *Trans. Amer. Vac. Soc. Vac. Symp.* **7**, 97 (1960).
47. J. P. Hobson and P. A. Redhead, *J. Vac. Sci. Technol.* **2**, 93 (1965).
48. D. Alpert, *J. Appl. Phys.* **24**, 860 (1953).
49. A. Van Oostrom, *Trans. Amer. Vac. Soc. Vac. Sump.* **8**, 443 (1961).
50. P. A. Redhead, *Rev. Sci. Instr.* **31**, 343 (1960).
51. J. P. Hobson, *J. Vac. Sci. Technol.* **1**, 1 (1964).
52. W. C. Schuemann, *Rev. Sci. Instr.* **34**, 700 (1963).
53. P. A. Redhead and J. P. Hobson, *Brit. J. Appl. Phys.* **16**, 1555 (1965).
54. W. G. Mourad, T. Pauly, and R. G. Herb, *Rev. Sci. Instr.* **35**, 661 (1964).
55. A. Klopfer, *Trans. Amer. Vac. Soc. Vac. Symp.* **8**, 439 (1961).
56. W. D. Davis, *Trans. Amer. Vac. Soc. Vac. Symp.* **9**, 438 (1962).
57. J. M. Lafferty, *Trans. Amer. Vac. Soc. Vac. Symp.* **9**, 438 (1962).
58. P. A. Redhead, *J. Vac. Sci. Technol.* **3**, 173 (1966).
59. J. C. Helmer and W. H. Hayward, *Rev. Sci. Instr.* **37**, 1652 (1966).
60. D. Blechschmidt, *J. Vac. Sci. Technol.* **10**, 376 (1973).
61. P. J. Szwemin, *J. Vac. Sci. Technol.* **9**, 122 (1972).
62. U. Beeck and G. Reich, *J. Vac. Sci. Technol.* **9**, 126 (1972).
63. B. Angerth, *Vacuum* **22**, 7 (1973).
64. J. P. Hobson and P. A. Redhead, *Can. J. Phys.* **36**, 271 (1958).
65. P. A. Redhead, *Can. J. Phys.* **37**, 1260 (1959).
66. J. R. Young and F. P. Hession, *Trans. Amer. Vac. Soc. Vac. Symp.* **10**, 234 (1963).
67. R. D. Woods, *J. Vac. Sci. Technol.* **10**, 433 (1973).
68. P. J. Bryant, W. W. Longley, and C. M. Gosselin, *J. Vac. Sci. Technol.* **3**, 62 (1966).
69. F. Feakes, F. L. Torney, and F. J. Brock, *NASA Report CR 167*, STAR Index N65-17126 (1965).
70. K. F. Poulter, *J. Phys. E* **5**, 267 (1972).
71. F. S. Johnson, J. M. Carroll, and D. E. Evans, *J. Vac. Sci. Technol.* **9**, 450 (1972).
72. W. K. Huber, *Vacuum* **13**, 399 (1963).

73. W. K. Huber, *Vacuum* **13**, 469 (1963).
74. E. W. Blauth, *Proc. 4th Intern. Vac. Congress*, p. 21, Institute of Physics and Physical Society, London (1968).
75. W. D. Davis and T. A. Vanderslice, *Trans. Amer. Vac. Soc. Vac. Symp.* **7**, 417 (1960).
76. W. J. Lange, *J. Vac. Sci.* **2**, 74 (1965).
77. A. Klopfer, *Vakuum Tech.* **10**, 113 (1961).
78. P. H. Dawson and N. R. Whetten, *Adv. Electronics Electron Phys.* **27**, 59 (1969).
79. D. C. Damoth, *Le Vide* **28**, 27 (1973).
80. A. van Oostrom, *Vacuum* **22**, 15 (1972).
81. R. Souchet, J. Sarrach, and G. Valdener, *Le Vide* **27**, 125 (1972).
82. Stanley Rutherberg, *J. Vac. Sci. Technol.* **9**, 186 (1972).
83. J. R. Miller III, *J. Vac. Sci. Technol.* **9**, 201 (1972).
84. R. A. Outlaw and R. E. Stell, *J. Vac. Sci. Technol.* **8**, 608 (1971).
85. C. G. Titcomb and W. F. Wallace, *J. Vac. Sci. Technol.* **9**, 1253 (1972).
86. D. Alpert and R. S. Buritz, *J. Appl. Phys.* **25**, 202 (1954).
87. J. R. Roehrig and J. C. Simons, Jr. *Trans. Vac. Symp.* **8**, 511 (1961).
88. W. D. Davis, *J. Vac. Sci. Technol.* **5**, 23 (1968).
89. W. W. Hultzmann and L. N. Krause, *J. Vac. Sci. Technol.* **11**, 889 (1974).
90. Exhibitions of commercial items are held annually in conjunction with the Annual Symposium of the American Vacuum Society and every three years in conjunction with the International Vacuum Congress, as well as at more restricted symposia.
91. W. R. Wheeler, *Phys. Today*, 52 (Aug. 1972).
92. W. F. Brunner and T. H. Batzer, *Practical Vacuum Techniques*, Reinhold Publishing Corp., New York (1965).
93. J. L. Whitton, *Proc. Roy. Soc.* **A311**, 63 (1969).
94. J. P. Hobson, *Jap. J. Appl. Physics*, Supplement 2, Part 1, 317 (1974).
95. J. A. Dillon, *Trans. Amer. Vac. Soc. Vac. Symp.* **8**, 113 (1961).
96. R. W. Roberts, *Brit. J. Appl. Phys.* **14**, 537 (1963).
97. R. W. Roberts, Report No. 67-C-087, General Electric Research and Development Center, Schenectady, New York (March 1967).
98. F. J. Brock. *Surface Cleaning Techniques in UHV*, Rep. NASA-CR-66273, National Research Corp., Cambridge, Massachusetts (1966).
99. S. M. Bedair and H. P. Smith, *J. Appl. Phys.* **40**, 4776 (1969).
100. A. M. Horgan and I. Dalins, *J. Vac. Sci. Technol.* **10**, 523 (1973).
101. R. R. Sowell, R. E. Cuthrell, D. M. Mattox, and R. D. Bland, *J. Vac. Sci. Technol.* **11**, 474 (1974).
102. T. E. Madey and J. T. Yates, *J. Vac. Sci. Technol.* **8**, 525 (1971).
103. J. H. Leck and B. P. Stimpson, *J. Vac. Sci. Technol.* **9**, 293 (1972).
104. H. H. Madden and G. Ertl, *Surface Sci.* **35**, 211 (1973).
105. P. W. Palmberg and T. N. Rhodin, *J. Phys. Chem. Solids* **29**, 1917 (1968).
106. E. W. Müller, *Science* **149**, 591 (1965).
107. J. N. Smith, Jr., and H. Saltsburg, *J. Chem. Phys.* **40**, 3585 (1964).

# 6

# Electron Probe Microanalysis

## Gudrun A. Hutchins

## 1. Introduction

Electron probe microanalysis is an analytical technique that may be used to determine the chemical composition of a solid specimen weighing as little as $10^{-11}$ g and having a volume as small as 1 $\mu m^3$. The primary advantage of electron probe microanalysis over other analytical methods is the possibility of obtaining a quantitative analysis of a specimen of very small size.

During the analysis, the selected area of the specimen is bombarded with a beam of electrons. The accelerating voltage of the electrons determines the depth of penetration into the specimen; the degree of beam focusing affects the diameter of the analyzed volume. The electron bombardment of the specimen causes the emission of an x-ray spectrum that consists of characteristic x-ray lines of elements present in the bombarded volume. The chemical analysis is accomplished by the dispersion of this x-ray spectrum and the quantitative measurement of the wavelength and intensity of each characteristic line. The wavelength identifies the emitting element, and the line intensity is related to the weight concentration of the corresponding element within the analyzed volume.

This chapter describes the analytical procedure and gives examples of the great variety of suitable specimens. It is written primarily for surface and material scientists who are interested in using electron probe microanalysis as a tool in research and development. An understanding of the analytical technique and its limitations is important even for those users who will never operate the instrumentation themselves. The knowledge gained will aid in

*Gudrun A. Hutchins* • Research and Development Laboratories, Sprague Electric Company, North Adams, Massachusetts 01247.

the design of optimum specimens and in the extraction of the maximum useful information from an analysis. Section 2 of the chapter gives a general description of measurement procedures and of the physical basis upon which electron probe microanalysis rests. Section 3 shows some bias toward surface and thin film characterization.

## 2. Fundamentals

### 2.1. Measurement Techniques

#### 2.1.1. The Instrument

Several good electron probe microanalyzer models are available commercially. They differ in the specific optical design, electron and x-ray resolution, the range of accessories available, and the price. The simpler instruments are quite suitable for most analyses. The more complex and more expensive instruments generally provide more rapid or semiautomatic data accumulation and a range of accessories for special techniques. Some of the newer instruments can double as transmission or scanning electron microscopes. Instrumental options are discussed in some detail by Beaman and Isasi[1] and by Duncumb[2]; only the most important aspects are described in this section.

The basic electron probe microanalyzer (EPMA) consists of an electron probe forming system, an x-ray analyzing system, and a specimen insertion and positioning system. The most common geometry is shown schematically in Figure 1. Some design compromises are necessary because of conflicting requirements for the optimization of the individual systems. The nature of the compromises accounts for most of the differences between commercial models.

*2.1.1.1. Electron Optics.* The purpose of the probe-forming system is to focus the maximum possible electron current into a probe of a given diameter at the specimen in order to achieve maximum x-ray output at a given level of spatial resolution. The conventional system consists of a self-biasing tungsten filament gun and two magnetic lenses.

In the gun, electrons are released by thermionic emission from a hot V-shaped tungsten filament held at a negative potential that may be varied between 5 and 30 kV. They are accelerated toward the anode held at ground potential. The thermionic emission current density $i_0$ increases rapidly with temperature and is given by

$$i_0 = AT^2 \exp(-B/kT) \tag{1}$$

where $A$ is a material constant, $T$ is the Kelvin temperature, $B$ is proportional to the work function of the filament material, and $k$ is the Boltzmann

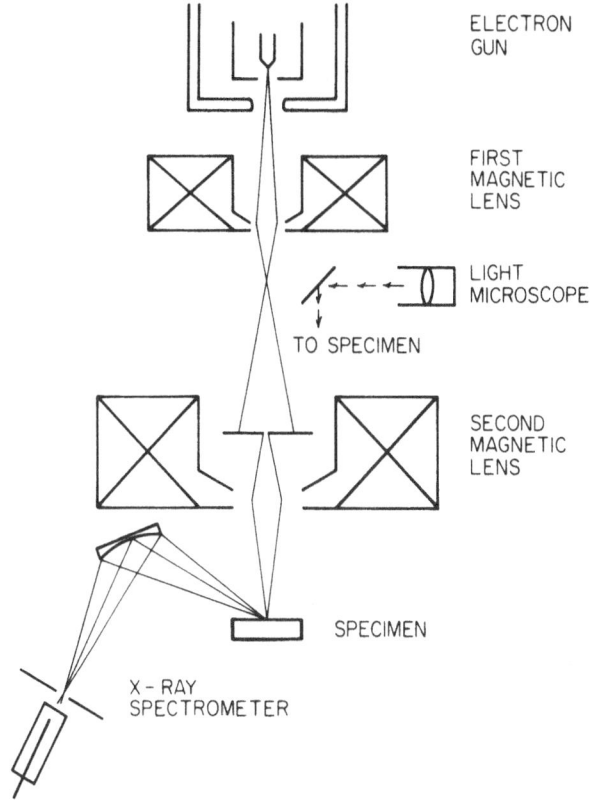

Figure 1. Schematic diagram of electron probe microanalyzer.

constant. The operating temperature of a 4-mil tungsten filament is typically 2700 K; the corresponding emission current density is approximately 2 A/cm$^2$. At higher filament temperatures, emission current densities over 10 A/cm$^2$ can be achieved,[3] but filament life rapidly decreases with increasing temperature due to tungsten evaporation. The Wehnelt cylinder or grid is maintained at a negative potential a few hundred volts greater than that of the filament through the use of a variable bias resistor. This self-biasing configuration has the following two separate functions: (1) it restricts emission to the filament tip, producing a small image that is subsequently demagnified by the magnetic lenses, and (2) the bias voltage produces a limiting field that makes the number of electrons passing through the Wehnelt cylinder opening into the electron optical column nearly independent of filament temperature above a certain value of $T$. Without this space charge limitation, a small fluctuation in $T$ would produce a larger fluctuation in $i_0$, and beam current stability would be a serious limitation. The effective value of $T$ for which the beam current saturates decreases with increasing bias voltage and increasing filament to Wehnelt cylinder distance. For optimum performance, an EPMA gun should be operated at the highest saturation temperature consistent with a small source size and filament life of approximately 40 hr.

The condenser lens (first lens after the electron gun) is used to give an initial demagnification of the source and to control the beam current at the specimen. A second condenser lens is used in the scanning electron microscopy mode of some combination instruments to provide a greater specimen current range and a second demagnification before the objective or probe-forming lens. The beam current is controlled by variation of the focal length of the condenser lens, which in turn varies the fraction of the beam passing through an aperture placed in the divergent beam on the image side of the condenser lens. As the focal length of the condenser lens is decreased (by an increase in lens current), the beam current and the image size are decreased. The effect of condenser lens aberrations on the final image is insignificant.

The objective lens further demagnifies the condenser lens image. The final beam diameter is larger than the geometrically reduced source because of spherical aberration in the objective lens. Spherical aberration arises because electrons farther from the optical axis in a divergent beam are focused more strongly. The magnitude of the effect increases with increasing beam divergence (aperture size) and increasing focal length (working distance). The minimum aperture size and working distance are both severely restricted by x-ray, specimen, and light optics requirements; the conventional EPMA objective lens is by necessity a "poor" lens in the electron optical sense. A relatively large aperture is needed to transmit enough electrons to produce a strong x-ray signal. A working distance of 1 cm or more between specimen and pole piece is required in many designs to obtain a sufficiently large x-ray emergence angle $\psi$. (This is the angle between the measurement direction of the x-ray spectrometer and the tangent to the specimen surface.) To minimize absorption and surface roughness effects, it is desirable to make $\psi$ as large as possible. For many instruments in which the electron incidence is normal to the specimen surface, the objective lens pole piece must be large enough to accommodate either the light objective or the specimen. Other instrument designs use nonnormal electron incidence and a specimen tilted toward the light microscope and/or x-ray spectrometer. This geometry eases the electron optical requirements but creates new analytical difficulties because the electron distribution below the specimen surface is no longer symmetrical.

The maximum current $i$ attainable in a probe of diameter $d$ is proportional to

$$i \propto E_0 \beta d^{8/3} C_s^{-2/3} \qquad (2)$$

where $E_0$ is the incident electron energy, $\beta$ is the gun brightness, and $C_s$ is the spherical aberration coefficient of the objective lens. Optimum values for conventional EPMA optics are 1 $\mu$A for a 1-$\mu$m probe at 30 keV and 2 nA for 0.1-$\mu$m probe at 30 keV. Currents lower than approximately 1 nA are not practical for electron probe microanalysis, because the corresponding x-ray signals are very weak.

The current $i$ for a given probe size can be increased significantly only by an increase in gun brightness or a decrease in the spherical aberration coefficient. Both of these require radical design changes. Two very promising high-brightness guns are already available for commercial scanning electron microscropes. They are the $LaB_6$ gun developed by Broers[4,5] and the tungsten field emission gun developed by Crewe *et al.*[6] Both guns require better vacuum systems than most present EPMA provide ($10^{-6}$ Torr for $LaB_6$, $10^{-9}$ Torr for tungsten field). The field emission gun gives a significant brightness advantage only for small probe diameters ($<0.2$ $\mu$m).

The spherical aberration coefficient can be decreased only by a very different objective lens design. Iron-free minilenses are a possible means of decreasing both $C_s$ and the physical size of the conventional objective lens. The iron pole pieces and casings of conventional lenses smooth out the magnetic field but also significantly increase the number of ampere-turns required to produce a net field of a given magnitude. A decrease in the physical size of the lens is of considerable advantage, because it increases the space available for spectrometers and other detection systems. Thin "pancake" lenses have been discussed and recommended by Bassett and Mulvey[7]; a flat objective lens is used in a recent commercial instrument.

Magnetic beam deflection systems are built into most EPMA. The scanning coils are generally located between the condenser and objective lenses and make it possible to scan the focused beam over an area on the specimen in synchronization with the sweeping spot on an oscilloscope. Electrostatic deflection systems are not as satisfactory, because they are easily affected by surface contamination of the electrodes.

*2.1.1.2. X-Ray Optics.* The purpose of the x-ray analyzing system is to disperse the polychromatic x-ray signal emitted by the specimen and to measure quantitatively the intensity of the characteristic lines belonging to specific elements. The dispersion of x rays can be accomplished either by wavelength dispersion or by energy dispersion, and the two approaches lead to very different spectrometers. The two techniques are complementary, and both types of x-ray analyzers are available for most modern instruments.

Wavelength dispersion consists of diffraction of the polychromatic x-ray beam from a single crystal according to the Bragg equation

$$n\lambda = 2d \sin \theta \tag{3}$$

The crystal lattice spacing $d$ is known, the angle $\theta$ between the x-ray beam and the crystal surface is varied continuously, and the product of the wavelength $\lambda$ and the order of diffraction $n$ is determined. The electron beam produces a point source of x rays; it is thus of considerable advantage to use a curved crystal focusing spectrometer that can intercept a large solid angle of x-rays from the source. For a fully focusing spectrometer, the x-ray source, the crystal, and the detector slit must lie on a focusing circle of fixed radius $R$. The crystal is curved to a radius $2R$; the concave side is then ground to a radius $R$ to fit the focusing circle. Radiation from the point source will strike the curved and ground crystal surface at the same angle $\theta$ along its entire length on the focusing circle. After diffraction, the radiation converges to a line image at the detector slit.[8]

In the most common spectrometer geometry, the crystal moves along a line through the fixed x-ray source and rotates around its own center. The distance between the source and the crystal center is $2R \sin \theta$ (proportional to $\lambda$) for the diffraction angle $\theta$. The detector path is a portion of a "four-leaved rose" with a source-to-detector distance of $2R \sin 2\theta$. The detector also rotates to keep the slit pointed toward the crystal center. The value of $R$ is generally between 4 and 6 in., and the $\theta$ range is typically from $10°$ through $70°$.

The choice of analyzing crystal determines $2d$ and the maximum $\lambda$ that can be diffracted. A number of good diffracting crystals exist for the wavelength range of 1–10 Å. The choice of a specific crystal is to some degree a question of personal preference and depends on whether spectral resolution or peak intensity is of primary interest and whether or not strong high-order reflections are desired. As electron probe microanalysis has been extended to the low elements, for which $\lambda$ becomes very large (O, $K_\alpha$ = 23.6 Å; C, $K_\alpha$ = 44.7 Å; Be, $K_\alpha$ = 114 Å), increasingly esoteric crystals or pseudocrystals have been developed. Some of these consist of 100 or more individually deposited "soap film" layers such as lead stearate decanoate ($2d = 100$). The performance and stability of the large-$d$ pseudocrystals is relatively poor and is one of the limiting factors in soft x-ray analysis. The usefulness of many pseudocrystals is restricted to three or four elements.

Gas proportional detectors are used almost exclusively in wavelength dispersive spectrometers. For these detectors, the energy of the detector output pulse is proportional to the energy of the x-ray photon producing it. Energy discrimination may be used to separate overlapping spectral lines corresponding to different orders of diffraction (and different energies). For soft x-rays, thin window detectors are required to reduce x-ray absorption by the window material. These windows are not vacuum tight, and the counting gas must be continuously flowed through the detector.

Energy dispersion relies totally on the energy discrimination of the detector. This method has become very popular in recent years because of the development of high resolution lithium-drifted silicon radiation detectors, which operate in the following manner: The silicon crystal is located between two metal electrodes across which a bias voltage of approximately 700 V is applied. X-rays impinging upon the crystal create electron hole pairs by photoelectric absorption and impact ionization; the number of charge carriers is proportional to the x-ray energy. The integrated current charge, collected under the influence of the bias voltage, is fed to a field effect transistor in a high-gain, low-noise preamplifier. Both the silicon detector and the field effect transistor are cooled by liquid nitrogen to reduce thermal noise and prevent the drift of lithium under the applied bias. The ouput voltage pulse from the preamplifier is further amplified and shaped and passed to a multichannel analyzer. In the multichannel analyzer, the voltage pulses are separated by amplitude and stored in the memory channel corresponding to a particular energy increment. The resulting spectrum can be displayed on an oscilloscope or recorded on strip chart or punched tape. Computer deconvolution of spectra from multielement specimens is often necessary because of the relatively poor spectral resolution.

The solid state detector cannot be mounted within the vacuum system of an electron beam instrument because the cold silicon crystal would act as a trap and rapidly become contaminated. The detector is also sensitive to light and backscattered electrons. For these reasons, the detector is mounted in a separate vacuum chamber that is isolated from the beam column by a thin beryllium window. Both the beryllium window (which absorbs backscattered electrons and long-wavelength radiation) and the inherent properties of present detectors limit the usefulness to x-ray energies above 1 keV.

The primary advantages of energy dispersion over wavelength dispersion are speed, low beam current requirements, and mechanical simplicity. The data for a qualitative analysis of the major constituents of an unknown specimen may be obtained in approximately 5 min and 1 hr by energy and wavelength dispersion, respectively. Very low beam currents are of relatively little advantage for EPMA but have made an energy dispersive spectrometer a very desirable accessory for scanning electron microscopes. Mechanical simplicity is an advantage because crystal spectrometers have to be aligned periodically. Wavelength dispersive spectrometers have a higher peak-to-background ratio and much greater spectral resolution. They give more accurate results for quantitative analyses, especially at low concentration levels.[1] All of the data for the examples discussed in this chapter were obtained by wavelength dispersion.

*2.1.1.3. Specimen-Viewing Optics.* A light microscope greatly facilitates the positioning of the sample under the electron beam so that the desired area can be analyzed. In most instruments, the light optics are concentric with the electron optics and enable the operator to observe the sample while it is being analyzed. The resolution and contrast of a probe microscope tend to be somewhat inferior to those of a typical metallurgical microscope because a reflecting curved mirror objective must be used and all concentric light optical components have a hole drilled through the center to allow passage of the electron beam. Nonconcentric light optics with better resolution are possible if the specimen is tilted or has separate viewing and analyzing positions.

With instruments equipped with $x$–$y$ scanning systems, secondary electron, backscattered electron, or specimen current images may also be used for specimen examination. The contrast in these images depends upon topography and/or atomic number and may be quite different from the light microscope contrast, which depends upon reflectivity and color. The magnification of the electronic images may be varied over a large range, but the magnification of the light microscope is fixed at approximately 300× in most instruments.

One important function of the light microscope that is occasionally overlooked is to provide a reproducible guide for the correct specimen height. For a properly aligned instrument, the light microscope image is in focus when the specimen x-ray source is located on the focusing circle(s) of the x-ray spectrometer(s) and the specimen-to-objective lens distance is correct for the optimum focal length of the lens.

*2.1.1.4. Specimen Stage.* The specimen is typically a metallographic mount 1 in. in diameter. For quantitative analyses it is often necessary to mount and polish the specimen and standards separately, and some means must be provided to exchange standard and specimen mounts without breaking the beam column vacuum. The mechanical stage should allow the motions of a good optical microscope—precise $xyz$ motion and rotation within the field of view. A scanning or step-scanning motor may be attached to the $x$ and/or $y$ drive. These motors can be computer interfaced for automatic data accumulation on many specimen areas.

### 2.1.2. Data Accumulation and Readout

After the pulses from the x-ray detector are amplified, they may be displayed or recorded in a variety of ways.[1,2,8] The most applicable method of data handling for a particular sample depends upon the nature of the analysis and the sophistication of the instrumentation available.

There are two basic types of analyses, and an example of each is shown in Figure 2. A spot analysis consists of an analysis for all detectable elements on one spot of a much larger specimen. This analysis may be representative of the entire specimen, or it may be an analysis of an unusual region. A distribution analysis determines the distribution of one or more elements as a function of position on the specimen. A distribution analysis is used to

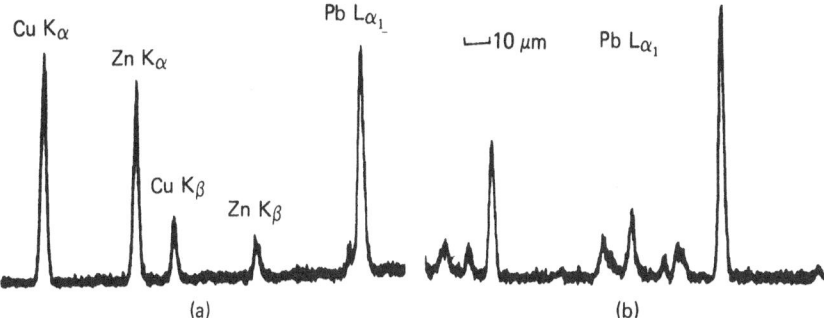

Figure 2. Data readout—strip chart scans of a commercial leaded brass. (a) Analysis of a lead-rich spot, x-ray intensity versus wavelength. (b) Distribution of lead along a line, lead intensity versus distance.

detect compositional gradients on a specimen surface; the average composition of the specimen may be known from a bulk analysis performed by other methods.

A qualitative spot analysis is often required to identify an unknown material. To obtain this analysis, the spectrometers are scanned through the portion of the x-ray spectrum detectable with the instrument and the signal is recorded on strip chart or $x-y$ recorder, yielding a plot of x-ray intensity versus wavelength or diffraction angle. Peaks at specific wavelengths are assigned to emitting elements with the help of tables[9,10] or previously recorded data. Since the intensity of each line is related to the amount of the emitting element present, an experienced operator can read a good qualitative analysis from the strip chart. The spectrum for a qualitative analysis may also be obtained very quickly by energy dispersion, and this is one of its best applications.

A quantitative analysis requires considerably more analysis time and is generally meaningful only if the specimen composition is reasonably uniform or if the phase to be analyzed is no less than 2 $\mu$m in its smallest dimension. Relative x-ray intensities of specimen to standard are measured to cancel out uncertain terms, such as the absolute spectrometer efficiency. The standards may be either pure elements or well-characterized compounds or alloys. The measured x-ray intensities are corrected for background, detector nonlinearity, and possible instrumental drift. The corrected intensity ratio for each element is then adjusted by the corrections for quantitative analysis discussed in Section 2.3 to obtain the specimen weight concentration for that element. The magnitude of these corrections depends upon the composition of the specimen, and calculations are done by successive approximations.

If the quantitative data are collected with an energy dispersive spectrometer, computer processing of the raw data is usually required before the quantitative corrections in Section 2.3 can be applied. Since energy spectra have a peak-to-background ratio that is typically lower by an order of magnitude than that of wavelength spectra, the background determination takes on increased importance. Some overlap of spectral lines is also more frequent in energy spectra and has to be dealt with by computer unfolding. The reader is referred elsewhere[11-14] for methods for dealing with spectrum fitting and background subtraction. The writer feels very strongly that wavelength spectrometers are more reliable for quantitative analyses if there is a choice. The manufacturers of energy dispersive systems have unfortunately oversold the technique and have created the illusion that anyone can obtain an instant accurate analysis by pushing the right buttons.

The second type of analysis involving the distribution of particular elements has become increasingly more important. Many physical properties of materials depend upon homogeneity or upon the size and distribution of a second phase to a greater degree than they do upon average composition. In any determination of homogeneity, resolution is of utmost importance, since few materials are homogeneous on an atomic scale. In practice, it is very difficult to find alloys or compounds that are homogeneous on a micron scale.

To determine the variation in a specific element, a spectrometer is peaked for an x-ray line of that element, and the beam is moved across the specimen while the x-ray intensity versus position is recorded. A one-dimensional line scan may be produced by sweeping the beam across a stationary specimen with the magnetic deflection coils or by driving the specimen stage under a stationary beam. The stage-drive method yields more accurate x-ray intensities for long scans with the curved crystal focusing spectrometers in common use, because it avoids the problem of spectrometer defocusing as the beam is swept away from the focusing circle.

The sensitivity of a line scan depends considerably upon the counting and recording system used. Ideally, the response time of the system should be very small to permit detection of rapid signal changes, and decade selection or some other scale expansion should be available to facilitate the detection of small variations in a large signal.

Two-dimensional scans may be of considerable help in the analysis of samples containing several phases. They are produced by sweeping the beam over an area on the specimen in synchronization with the sweeping spot on an oscilloscope. The intensity of x-ray or current signals is used to modulate the brightness of the oscilloscope spot. The resulting images are qualitative compositional maps.[1,2,8,15-17]

Examples of typical oscilloscope images are shown in Figure 3. The line scan in Figure 3a is similar to Figure 2b, but is in this case produced by sweeping the beam across a stationary specimen with the deflection coils. A scan of this type is often superimposed on a current image (see, for

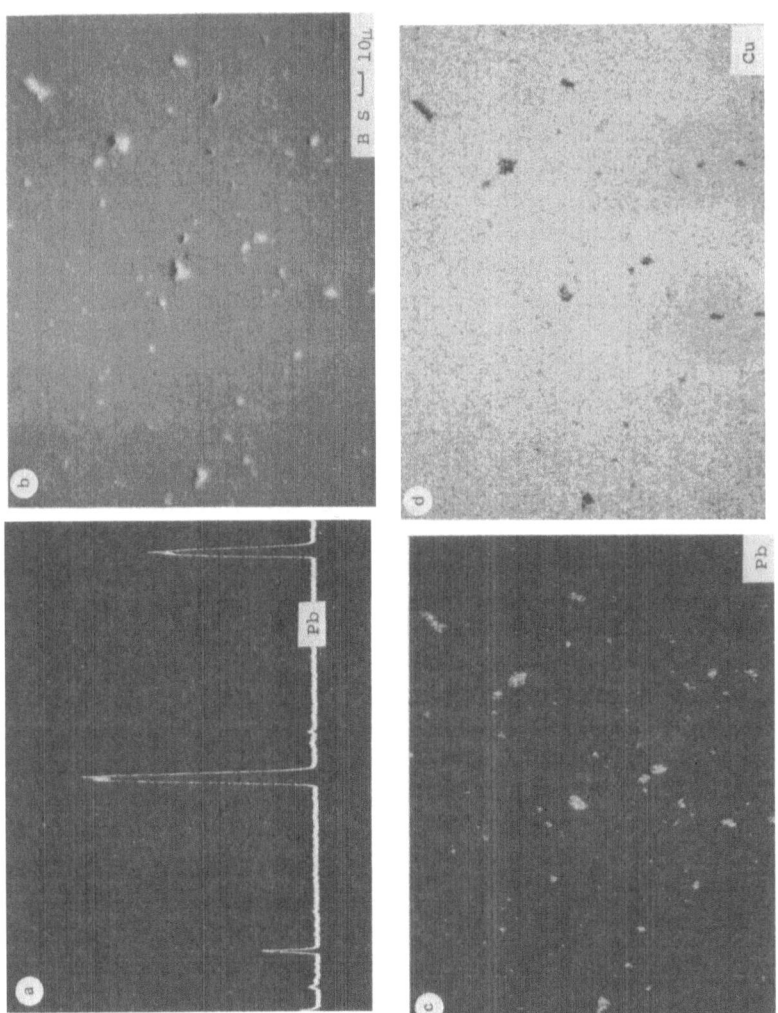

Figure 3. Data readout—oscilloscope images of a commercial leaded brass; (a) distribution of lead along a line; (b) backscattered electron current image; (c) lead x-ray image; (d) copper x-ray image.

example, Figure 15). The backscattered electron current image (Figure 3b) shows atomic number and some topography contrast. The copper x-ray image shows the effect of spectrometer defocusing by a slight darkening at the corners.

Current scans may be obtained by imaging secondary electrons, backscattered electrons, or electrons absorbed by the specimen (specimen current). Secondary electrons have the highest spatial resolution and are normally used at very low beam currents to image specimen topography. They are the standard signal of scanning electron microscopy and can also be used to image voltage gradients during the analysis of active microcircuits. Backscattered electrons give intermediate spatial resolution, and their intensity depends both on atomic number and specimen topography. In some instruments, either atomic number or topography contrast can be emphasized by the addition or subtraction of signals from two symmetrically placed detectors. Specimen current images show primarily atomic number contrast and give the poorest spatial resolution of the current images. Both backscattered electron and specimen current images are usually used in conjunction with x-ray analysis. The current signals are amplified analog signals and are relatively noise-free in modern instruments.

X-ray images inherently have less spatial resolution and are noisier than any of the current images. The resolution is related to the actual x-ray source size, which is larger than the beam diameter because of electron scattering. The scans are noisy, because only a relatively small number of x-ray quanta are measured during the time period during which one frame is displayed or photographed. The x-ray signal is processed as a pulse signal, and each measured x-ray quantum produces a white dot on the oscilloscope. The size and brightness of each individual dot can be adjusted in modern instruments. If a very small dot is chosen to reduce image graininess, a much longer exposure time will be required to adequately fill the image area with white dots. In practice, two-dimensional x-ray scans are best suited for sorting out large compositional variations in complex samples. Small variations in an x-ray signal of high intensity and large compositional variations in a very small area tend to be indistinguishable from noise. In the profile or line scan mode (Figure 3a) of x-ray signal display, an analog signal taken from the rate meter is processed. This method is more sensitive to small variations in the x-ray signal, because the measured x-ray quanta are distributed along a line instead of over an area.

There is a continued trend toward more and more automation in both data collection and data processing. Multispectrometer two-dimensional step scanning instruments with on-line data reduction are powerful tools in materials research and development. They can combine the accuracy of quantitative determinations with distribution analysis for complex multiphase materials and relieve much operator drudgery associated with point-by-point data collection. These superinstruments are very impressive, yet

*Table 1. Typical Detectability Limits with Wavelength Dispersion*

| Type of analysis | Spectral scan | Quantitative (100 sec) |
|---|---|---|
| **Bulk** | | |
| Best cases of hard K radiation in low-$Z$ matrix | 0.05 wt.% | 0.005 wt.% $<10^{-15}$ g/$\mu$m$^3$ |
| Poorest cases of soft L or M radiation in high-$Z$ matrix | 0.5 wt.% | 0.05 wt.% |
| Surface layer, 20-keV beam | $1 \times 10^{-6}$ g/cm$^2$ | $5 \times 10^{-8}$ g/cm$^2$ $<$one atomic layer |

they still depend upon people to operate and program them and to prepare specimens. Under unfavorable circumstances, automated microprobes can turn out erroneous data at an amazing speed.

## 2.1.3. Detectability Limits

Modern electron probes can detect x-ray lines of all elements in the periodic table, from beryllium or boron through the transuranium elements. The low atomic number elements ($Z = 4$ or 5 to 10) are generally detectable only when present at a concentration greater than several tenths of a weight percent. Typical detectability limits with wavelength spectrometers for the other elements are given in Table 1. Values are listed for spectral scans and for quantitative determinations with a 100-sec counting time. The detectability limit in a bulk specimen depends upon the intensity of the x-ray line measured, the attenuation by other elements present in the specimen, and the intensity of the continuous radiation background produced by the specimen. Thus, the detectability limit of two elements may differ by as much as an order of magnitude. The sensitivity for thin surface layers is good. An example is given in Figure 4, which shows a spectral scan of a very thin layer of chemically deposited copper and nickel on aluminum foil.

## 2.1.4. Measurement Precision

For a quantitative analysis, the intensity ratio $k$ for an x-ray line in the specimen and standard must be determined. The typical procedure is to measure the counts accumulated over a time interval $t$ for peak and background spectrometer positions on both specimen and standard. Each peak and background measurement is corrected for the dead time of the detector and associated circuitry, and $k$ is calculated from

$$k = (P - B)/(P_0 - B_0) \tag{4}$$

where $P$ and $B$ are the accumulated counts on the specimen after the dead time correction; the corresponding standard measurements are indicated by

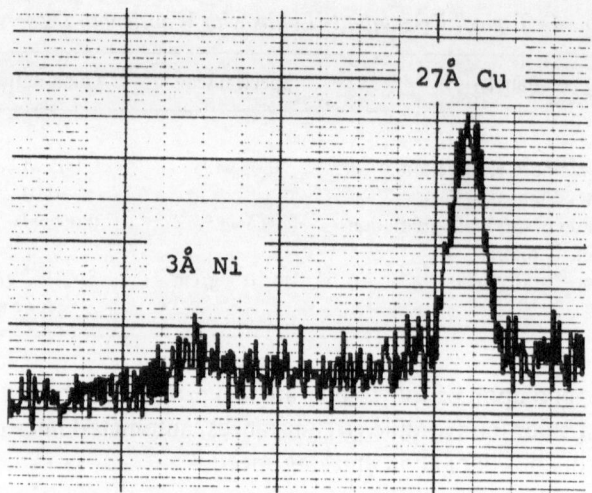

Figure 4. Sensitivity for thin surface layers. Spectral scan over aluminum foil with $0.26 \, \mu g/cm^2$ nickel and $1.97 \, \mu g/cm^2$ copper deposited on it.

a zero subscript. Several individual determinations of $k$ are usually made and averaged.

*2.1.4.1. Dead Time Correction.* The dead time correction corrects for the limited paired pulse resolution of the x-ray detector and associated electronics. At high counting rates, some x-ray photons occurring very close together are not counted separately. The effective dead time of the system may be determined by several methods[18,19] and is typically in the range of 1–3 $\mu$sec. The true counting rate $N$ in counts per second can be calculated from the observed rate $N'$ by the equation

$$N = N'/(1 - \tau N') \qquad (5)$$

where $\tau$ is the experimentally determined dead time. The Ruark–Brammer expression given in equation (5) is an approximation that is valid only for $\tau N' < 0.10$. For higher counting rates, the valid expression depends upon the nature of the electronic circuits following the x-ray detector.[20]

The measurement precision is determined by the uncertainty in the measured value of $k$. The error in $k$ is a combination of random errors that can be treated statistically and systematic errors that are in some instances very difficult to evaluate. Values of $k$ with a relative accuracy of 0.5–1% can be measured routinely by an experienced analyst. However, it is necessary to investigate or at least be aware of the possible measurement pitfalls.[1,8,21,22]

*2.1.4.2. Random Errors.* Random errors consist of statistical counting errors, random instrumental fluctuations, and random variations due to specimen preparation. They can be evaluated easily by the comparison of a number of individual measurements on the same uniform specimen and standard. The random errors determine the precision to which two areas on the same specimen can be compared and the concentration level at which segregation or inhomogeneity can be detected. The experimental fractional standard deviation of $k$ is given by

$$\frac{\varepsilon_{k_{ex}}}{k} = \frac{1}{\bar{k}} \left[ \frac{\sum_{i=1}^{n} (\bar{k} - k_i)^2}{n-1} \right]^{0.5}$$ (6)

where $\bar{k}$ is the average experimental value and $n$ is the number of measurements.

The standard deviation due to the random nature of x-ray emission can be calculated from Poisson statistics.[23] The standard deviation of $P$ counts is $P^{0.5}$, and fractional standard deviation of $k$ due to counting statistics is

$$\frac{\varepsilon_{k_c}}{k} = \left[ \frac{P+B}{(P-B)^2} + \frac{P_0+B_0}{(P_0-B_0)^2} \right]^{0.5}$$ (7)

The value of $\varepsilon_{k_c}$ decreases with increasing counting time and increasing concentration of the measured element within the specimen.

The experimental standard deviation determined by equation (6) will always be larger than the value of equation (7); the difference is due to instrumental and specimen fluctuations and tends to approach a limiting value ($\sim 0.002$–$0.003$) for long counting times. Likely random instrument errors are: (1) voltage and current fluctuations in lenses, power supplies, detectors, etc., and (2) slight variations in x-ray spectrometer resetting and light optics focus. Random specimen errors are (1) deviation from flatness and (2) density variations or microporosity.

*2.1.4.3. Systematic Errors.* Systematic errors are more difficult to detect and evaluate, because they cannot be characterized by repetitive measurements of the same specimen. Some error sources produce similar effects, and the problem may be exasperatingly difficult to isolate. In general, the systematic errors do not affect the precision of relative measurements on the same specimen in applications such as homogeneity testing. They do, however, cause an uncertainty in an absolute quantitative analysis, and the magnitude of this uncertainty may be difficult to estimate. Systematic errors may be caused by the instrument or the specimen; a discussion of error sources follows.

1. *Instrumental parameters.* The incident electron energy $E_0$ and the x-ray emergence angle $\psi$ are input parameters in the equations to correct for interelement effects (see Section 2.3). If the actual values of $E_0$ and $\psi$ are different from the nominal ones, the error is propagated in the quantitative corrections.

2. *Instrumental drift.* Random and systematic drift may be caused by a number of electronic, mechanical, and temperature-dependent variations. The best procedure to minimize error due to systematic drift (which is not necessarily linear) is to measure standards frequently throughout the data collection period.

3. *Dead time error.* The experimentally determined value of $\tau$ has an uncertainty of $\sim 10\%$. This uncertainty in $\tau$ will produce a systematic dead time correction uncertainty that is equal to $\sim 10\%$ of the deviation from unity.

4. *Pulse amplitude shifts.* At high counting rates, the average amplitude of individual gas proportional counter pulses decreases. If the counting rate on sample and standard is very different and if energy discrimination is used to suppress other orders of diffraction, some pulses of the measured x-ray line may fall outside of the preset energy interval. This problem can be eliminated through the use of a wide energy window and a clean detector anode wire[1,18,24-26] Other systematic errors in the detection circuitry are possible and may be related to mismatching of individual circuits in the detector train, poor ground or interconnections, and numerous interference or overload effects. The presence of these malfunctions can often be detected from precise measurements of counting rates or intensity ratios as a function of beam current.

5. *Background.* The best background measurement procedure is to determine the intensity at spectrometer positions above and below the spectral line. This is not always possible or accurate if interfering lines or absorption edges are present near the measured line. The background correction becomes increasingly important for minor and trace constituents and may cause a systematic error.

6. *Sample preparation.* Systematic uncertainties introduced during sample preparation include embedding of abrasives, relief, preferential polishing, or etching and are discussed in Section 3.1.2.

7. *Electron-beam-induced concentration changes.* The movement of sodium and potassium ions in glasses during electron bombardment has been observed by several investigators.[27-29] The writer has recently noted the redistribution of cadmium in complex glasses and mixed oxides. The motion of these ions may be toward or away from the bombarded volume, depending on the specimen and the experimental conditions. Organic or biological specimens tend to decompose (e.g., paper becomes charred) and elements of interest may be volatilized or redistributed. These instabilities can generally be detected by observing the emitted x-ray intensity as a function of time.

8. *Wavelength shifts due to chemical bonding.* The wavelength, line shape, and intensity from an element in a compound often depend upon the chemical bonding of the element (see Section 2.2.2.3 and Figure 25). If the

state of bonding changes from specimen to standard, serious errors may result, especially for very soft x rays. The effect may be minimized by the use of a standard of the same chemical state. In some instances, it is helpful to decrease the wavelength resolution of the spectrometer by opening the detector slit.

## 2.1.5. Limitations

Every probe analyst has experienced some dismal failures; these are generally carefully omitted in the published literature on electron probe microanalysis. As materials and problems become more complex, it occurs more and more frequently that the problem that prompted the analysis is never satisfactorily solved. These analytical failures are typically due to one or more of the following problems.

1. *Specimen instability.* The specimen must be stable in vacuum and during electron bombardment. Nonconductors should be coated with an evaporated conducting film to prevent electrical charging of the specimen surface and resulting beam instability. Most instability problems are due to local heating and/or charging, which cause decomposition or melting of the specimen. The problem can be severe with glasses, organic materials, and easily decomposed compounds. Local heating of poor thermal conductors can be minimized by placing a relatively thin layer of the specimen on a metal backing and coating the surface with a conducting film. A defocused electron beam will lower the current density on the specimen and may improve stability.

2. *Sensitivity.* Under normal conditions, the detectability limit will not be lower than approximately 50 ppm for a quantitative analysis. This means, for example, that typical semiconductor doping levels are not detectable. The poorest sensitivity is for very soft x rays in a specimen that is highly absorbing for the measured radiation.

There are some cases of spectral interference, although they are relatively rare. The most severe interference is for the L spectra of the elements near silver in the periodic table. Several lines of adjacent elements in this region are superimposed, and a small quantity of element $Z$ is difficult to detect in the presence of a large quantity of $Z - 1$. Other interferences involve the elements below atomic number 18, which have only one strong x-ray line. If the $K_\alpha$ line is masked by a strong line of another element, the sensitivity may be poor.

3. *Localized trace impurities.* The usefulness of electron probe analysis for the detection and identification of trace impurities that cover a small fraction of the analyzed surface area is generally overrated by scientists only slightly familiar with the technique. If the impurity sites are not visible and the identity of the impurity element is not known, the statistical odds for a

successful analysis are very unfavorable. If the problem areas can be "decorated" to make them visible or can be "tagged" with a known element by electrochemical means, the probability of a successful analysis is good. It is therefore advantageous to spend considerable effort on the development of a technique to localize the areas to be analyzed on specimens of this type.

4. *Specimen geometry limitations.* In some samples, the material of interest may be part of a complex geometry. An example is an impurity in a deep hole or crevice. The x rays generated within the deep hole will be very highly attenuated by the surrounding material before they can be detected by the spectrometer. If the specimen surface is polished, some of the material of interest may be removed or the projecting material may be smeared over it.

In some multilayer specimens, problems occur primarily at interfaces. The success of these analyses depends upon the possibility of evenly removing the upper layers chemically or mechanically to expose the material to be analyzed.

5. *Problem unrelated to composition.* Some specimen problems are unrelated to chemical composition but are rather due to allotropic transformations, line or point defect structure or density, and other crystallographic variations. The probe has no sensitivity for long-range order or structural variations in its normal mode of operation.

## 2.2. Interaction of Electron Beam and Specimen

The electron beam energy is dissipated by the generation of heat, the continuous x-ray spectrum, and a number of potentially useful signals, including characteristic x rays, secondary electrons, Auger electrons, and cathodoluminescence. The propagation of high-energy electrons within the specimen and the generation of the various useful signals are complex statistical processes that can be treated easily for one electron or one photon but that are difficult to sum or integrate over the entire bombarded specimen volume.

### 2.2.1. Electron Scattering

As the high-energy electrons penetrate into the specimen, they are scattered elastically by the screened atomic nuclei and inelastically by the atomic electrons. The proportion of elastic and inelastic collisions depends upon the atomic number of the specimen. *Elastic scattering is more pronounced in a high-atomic-number element; energy loss is more rapid in a low-atomic-number element.* The following approximate numbers for gold and carbon will give an indication of the degree of variation with atomic number. For an electron energy of 30 keV, approximately 70% of all scattering events in gold are elastic, and the average scattering angle per event is near 2°. In carbon, less than 20% of all scattering events are elastic, and the average scattering angle per event is approximately 0.35°. The energy loss per unit mass distance $\Delta \rho s$ along the path of a 30-keV electron is at least three times as great in carbon as in gold.

The scattering and energy loss of electrons in a specimen containing a mixture of gold and carbon is intermediate and considerably different from both pure elements. This difference in electron propagation affects the production of characteristic x rays and the shape and size of the excited x-ray volume. A thorough understanding of electron scattering is therefore a requirement for quantitative electron probe analysis. Cosslett and Thomas[30] have published a series of outstanding papers correlating existing scattering theory with experimental data obtained by transmitting 5–30 keV electrons through thin films of aluminum, copper, silver, and gold. A detailed discussion of electron scattering applicable to electron probe analysis can also be found in the thesis of Bishop.[31] A brief description of the most important features is given in the following paragraphs.

*2.2.1.1. Energy Losses.* The magnitude of the small discrete energy loss at each inelastic collision is difficult to measure or to calculate. In practice, it is necessary to replace discrete energy losses by a continuous mean loss. In the treatment of Bethe and others, energy loss is described by the stopping power $S$, defined as

$$S = -(1/\rho)\, dE/ds \qquad (8)$$

where $dE$ is the average increment of energy lost by an electron along the path length $ds$ in a material of density $\rho$. A suitable nonrelativistic expression for $S$ is

$$S = (7.85 \times 10^4)\,(Z/A)\,(1/E)\,[0.1534 + \ln(E/J)] \qquad (9)$$

where $Z$ and $A$ are the atomic number and weight, $E$ is the electron energy in electron kilovolts and $J$ is the mean ionization energy of the target atoms in electron kilovolts. Relativistic effects can be introduced by the inclusion of the

$$\text{relativistic multiplier} = (2E + 511)/511 \qquad (10)$$

in equation (9) as suggested by Brown *et al.*[32]; 511 is the electron rest energy. Values of $J$ are approximately proportional to $Z$. Berger and Seltzer[33] have suggested the expression

$$J = 0.00976Z + 0.0588Z^{-0.19} \qquad (11)$$

as the best fit to the experimental data for elements with $Z > 13$. The experimental measurements of $J$ have been made at very high energies with protons or other heavy particles, because the required energy loss straggling and multiple scattering corrections are smaller and easier to calculate. They are therefore not necessarily accurate for EPMA conditions. Some analysts prefer to use the empirically determined $J$ values of Duncumb and Reed.[34] The accuracy of these $J$ values is not established either.

The Bethe stopping power in the form given in equation (9) is widely used in EPMA calculations and is a good continuous energy loss approximation. However, it becomes too large as the energy $E$ approaches $E_0$. A better approximation is given by the Spencer–Fano theory,[35] which shows a reduction in the rate of energy loss due to a reduction of possible loss mechanisms as $E_0 - E$ approaches zero.

*2.2.1.2. Elastic Scattering.* The degree of elastic scattering determines the spatial profile of electrons below the specimen surface and also the number and energy of electrons scattered back out of the surface. The scattering cross section at constant energy varies approximately as $Z^2$. The probability of scattering through a given angle during the traverse through the layer thickness $\Delta\rho z$ varies approximately as $Z^2/E^2$. Thus, the electrons lose their initial direction very quickly in a high atomic number element, and the electron profile within the specimen approaches the shape of a hemisphere. In a low atomic number element, elastic scattering is less pronounced, and the electron profile is somewhat more narrow and deep.

The high-energy electrons that are scattered back out of the specimen while still retaining much of their energy are called "backscattered electrons." Their number and energy distribution as a function of $Z$ can be measured experimentally with a good degree of accuracy. The backscattered electron fraction $\eta$ is given by

$$\eta = (i_b - i_s)/i_b \qquad (12)$$

where $i_b$ is the incident beam current and $i_s$ is the current measured through the specimen. Measurements of $\eta$ by Bishop[36] and Heinrich[37] are shown in Figure 5. More recent measurements[38–40] are in good agreement. Electrons backscattered from a high atomic number element have a greater average energy than those backscattered from a low atomic number element.[36,38,41] Thus, much more of the incident beam energy is unavailable for x-ray production in a high atomic number element; more electrons are backscattered, and the average backscattered electron has a greater energy.

### 2.2.2. X-Ray Emission

The generation of a characteristic x-ray photon requires the ejection of an inner-shell electron from a specimen atom by a high-energy incident beam electron. The second step is the transition of an outer-shell electron to

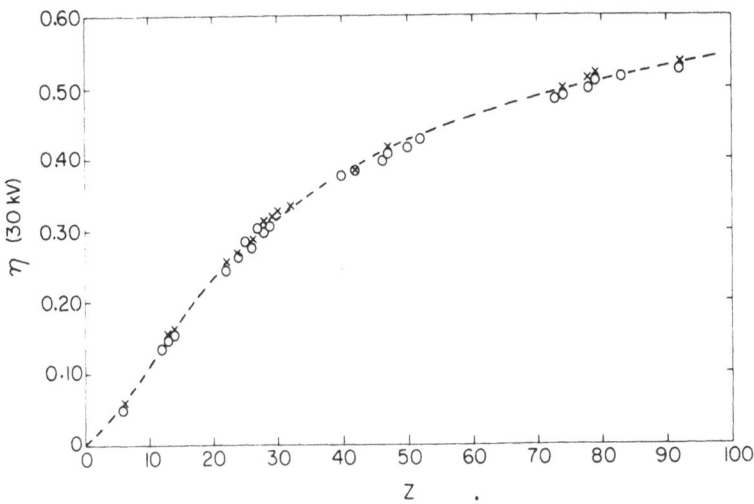

Figure 5. Experimentally determined backscattered electron fraction $\eta$ as a function of atomic number. Data from Bishop[36] ($\times$) and Heinrich[37] (○).

the vacant site. The final step is the emission of a photon with an energy equal to the energy difference between the excited and final atomic states. The energy (or wavelength) of the x-ray photon is thus characteristic of the emitting atom. The production of an x-ray photon is a relatively rare event.[42] A 30-keV electron has 3.3 times the energy required for copper K shell ionization, yet its probability of generating a $CuK_\alpha$ photon before it comes to rest is only approximately $3 \times 10^{-3}$.

*2.2.2.1. Ionization Cross Section.* The probability of c subshell ionization is given by the ionization cross section $Q_c(E)$. (The letter c will be used as a general subshell designation; c = K, $L_1$, $L_2$, etc.) The ionization cross section is defined as the number of c subshell ionizations per unit length along the path of an electron of energy $E$ divided by the number of atoms per unit volume. The magnitude of $Q_K$ is in the range from $10^{-22}$ to $10^{-20}$ for typical electron probe beam energies. Absolute values of $Q$ are extremely difficult to measure.[43]

The energy dependence of $Q$ is most easily expressed as a function of the overvoltage $U = E/E_c$, the ratio of the electron energy $E$ to the ionization energy $E_c$ required to remove an electron from the c subshell. The precise form of the energy dependence is open to debate and depends upon the boundary conditions chosen in the derivation of $Q$. Figure 6 gives relative values of $Q$ as a function of $U$ obtained experimentally and calculated from several energy dependencies. The writer favors an expression of the form

$$Q = \text{const} \, (1/E_c^2) \, (\ln U) \, (1/U^m) \tag{13}$$

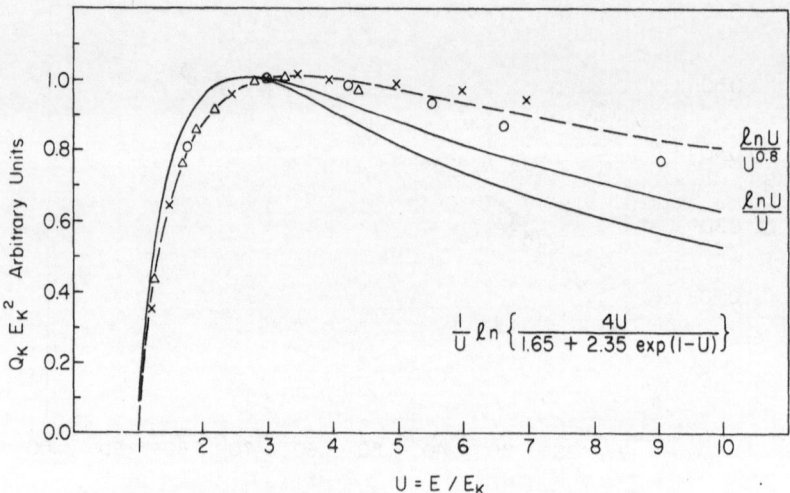

Figure 6. Relative values of the ionization cross section as a function of $U$. All curves and experimental data have been fitted to the same value for $U = 3$. Experimental data are from Webster *et al.*[44] ($\times$), Pockman *et al.*[45] ($\bigcirc$), and Green[46] ($\triangle$). The curves show three different energy dependencies.[47]

with $m$ between 0.7 and 1. More complex equations offer no apparent advantage. For calculations in Section 2.3, $m = 1$ is used.

*2.2.2.2. Fluorescence Yield.* The probability of an electron vacancy in the c subshell resulting in the emission of an x-ray photon is given by the atomic fluorescence yield $\omega_c$. Values of the fluorescence yield vary from $< 0.01$ to $0.60$ for x-ray lines used in electron probe analysis. The fluorescence yield for any particular subshell increases with $E_c$. For low energy values, a competing process, Auger electron emission, is much more efficient than characteristic x-ray emission. Auger electron emission also requires the creation of an inner-shell vacancy and the transition of an outer-shell electron to the vacant site. However, the difference in binding energy is transferred to another bound electron, which is ejected from the atom. Thus, the atomic fluorescence yield is really the probability that the difference in binding energy is released as an x-ray photon rather than as an Auger electron. Extensive theoretical calculations and experimental measurements of atomic fluorescence yields have been made and are reviewed in comprehensive papers by Fink *et al.*[48] and Bambynek *et al.*[49]

*2.2.2.3. Characteristic X-Ray Spectra.* The characteristic x-ray spectra in the wavelength range most easily analyzed with the microprobe (1–10 Å) are quite simple for low atomic number elements and become moderately complex for high atomic number elements. The radiation of highest energy for every element corresponds to transitions from the K level and consists primarily of the $K_\alpha$ and $K_\beta$ doublets. Up to atomic number 33, the K lines are the only lines emitted of wavelengths less than 10 Å; and the $K_\alpha$ line is normally used for analysis.

The L series is much more complex, since it consists of transitions from three L levels to all outer shells. There may be considerable L spectrum interference between adjacent elements. Both L and M lines are easily measured for elements above the rare earths in the periodic table. The higher-energy L lines tend to be better for quantitative bulk analysis, because the absorption correction is smaller. The M lines tend to be more sensitive for thin surface layers and can be excited at much lower beam voltages.

The strong K, L, and M spectral lines normally used for analysis are referred to as diagram lines, and their wavelengths can be accurately predicted from an energy level diagram of the emitting element. The intensity of the lines is closely related to the densities of the electronic states involved, and the transitions obey specific quantum mechanical selection rules. A great number of weak nondiagram lines can also be detected in the typical spectrum of many elements. The largest group of nondiagram lines are the satellite lines occurring on the high-energy side of many diagram lines.[9,50] They are apparently due to transitions in atoms from which two or more electrons have been ejected. Certain other lines or bands appear or disappear with chemical combination, and the origin of these lines is not totally understood.

The satellite lines become increasingly stronger and more clearly separated from the parent lines with decreasing atomic number of the emitting element. For elements below argon, the relative intensity, wavelength, and shape of the satellite lines and some of the diagram lines are noticeably dependent upon the chemical state and short-range order of the emitting atoms. This dependence can be used to differentiate between compounds of the same element and is of considerable theoretical interest. A general discussion of changes in x-ray emission spectra due to bonding and alloying is given in review papers by Baun[51,52]; a practical example from the author's laboratory is given in Section 3.3.5.4. The very low-energy x-ray band spectra required for the detailed study of valence bands[53] are beyond the range of capabilities of most electron probes.

*2.2.2.4. Continuous Spectrum.* The continuous spectrum (Bremsstrahlung) results from the deceleration of the incident electrons through interaction with the strong fields surrounding the atomic nuclei. It gives an undesirable background signal that limits the ultimate detectability. The continuous spectrum may also cause fluorescent excitation of characteristic lines. The intensity of the continuous spectrum increases with $E_0$ and $Z$. The number of continuous spectrum photons with energies between $E$ and $E + dE$ generated per primary electron is often approximated by the

Kramer relation

$$dN(E) = KZ \left[ (E_0 - E)/E \right] dE \qquad (14)$$

The absolute value and possible $Z$ dependence of the "constant" $K$ is under dispute.

## 2.3. Interelement Effects—Corrections for Quantitative Analysis

### 2.3.1. Qualitative Discussion of Interelement Effects

The measurement of $CuK_\alpha$ in a 40Cu–60Au specimen and a pure copper standard is used as an example through part of Section 2.4. The use of an example simplifies the discussion; this particular example was chosen because all four types of interelement effects are significant.

As a first approximation, the measured intensity ratio $k_{Cu}$ of $CuK_\alpha$ in the alloy and in pure copper is equal to the weight fraction $C_{Cu}$ of Cu in the alloy.

*First Approximation:*

$$k_{Cu} = C_{Cu} \qquad (15)$$

Inherent in this approximation are the assumptions that the number of $CuK_\alpha$ photons generated depends only on the density of copper atoms in the analyzed volume and that the generated photons are attenuated similarly during their propagation to the surface of the specimen and the standard. The first approximation is not very accurate. (In other systems, the error may vary from a few percent to a factor of 3.) To obtain a more accurate analysis, the effect of the presence of the gold on the measurement of copper in the specimen must be calculated. The accuracy of this calculation plus the precision of the x-ray intensity measurement determine the absolute accuracy of the analysis.

The interelement effects can be separated into the following four categories:

1. The atomic number effect is due to the different electron scattering and energy loss in sample and standard. In the Cu–Au alloy, the presence of gold increases the elastic scattering and reduces the stopping power. The net effect is an increase in the alloy of the number of $CuK_\alpha$ photons generated per incident electron per copper atom.

2. The x-ray absorption effect is due to the difference in the degree of absorption of generated characteristic x-ray photons in the specimen and standard. Absorption of generated $CuK_\alpha$ photons is greater in the Cu–Au alloy than in pure copper. This causes a net decrease in the $CuK_\alpha$ signal measured in the alloy.

3. The characteristic line fluorescence effect is due to the production of characteristic x rays by the characteristic x-ray lines of other elements present in the specimen. In the Cu–Au alloy, some $CuK_\alpha$ photons are produced as a result of the absorption of photons from gold L lines. This causes a net enhancement of the $CuK_\alpha$ signal measured in the alloy.

4. The continuous-spectrum fluorescence effect is due to the production of characteristic x rays as a result of the absorption of photons of the continuous x-ray spectrum. In pure copper some $CuK_\alpha$ photons are produced by the copper continuous spectrum. Gold has a stronger continuous spectrum [see equation (14)], and therefore, more $CuK_\alpha$ photons are produced by the continuous spectrum in the alloy.

### 2.3.2. Methods of Calculation of the Emitted X-Ray Intensity

It would be ideal to be able to calculate accurately the absolute number of x-ray photons emitted by a specimen under certain analytical conditions. Inaccuracies are reduced when the calculation is limited to relative numbers of x-ray photons or relative x-ray intensity, because the constants in the equations for $Q$ and $S$ need not be known and $\omega_c$ and some systematic errors cancel out of much of the calculation. This is one reason why relative x-ray intensities are measured in quantitative electron probe microanalysis. The basic problem is the calculation of the distribution below the specimen surface of the generated radiation as a function of specimen composition. In order to calculate the depth distribution of x-ray production, it is necessary to know where the electrons are traveling and what their energies are.

In this section, four computational approaches are considered. The first two methods (Monte Carlo techniques and electron transport equation) are fundamental, and their accuracy is limited primarily by input parameters rather than by the basic structure of the computation. The calculations are at present not used for day-to-day analyses because of their mathematical complexity and requirements for considerable computer time. Their main value lies in the illumination of trends and the substantiation of occasional experimental measurements. They can also help to shed some light on basic problems, such as how to treat specimens with concentration gradients or sharp discontinuities. Brown[54] has compared the strengths and weaknesses of the two computational techniques in a very readable review paper.

The third technique is an empirical procedure that has been used successfully in a number of metallurgical and mineralogical systems. It is based on coefficients determined experimentally from known binary standards.

The fourth correction scheme consists of a separate correction for each of the four interelement effects listed in Section 2.3.1. It is by far the most popular correction scheme and combines relative computational ease with some fundamental basis.

*2.3.2.1. Monte Carlo Calculations.* The Monte Carlo technique makes a random sampling of calculated paths of individual electrons and the x-ray photons produced by them. The electron trajectories can be simulated by a random number process that calculates the direction and energy of the electron after each scattering event from known scattering distributions and energy loss relations. The accuracy of this calculation is limited primarily by uncertainties in the elastic scattering cross sections and in the energy loss for inelastic collisions. Several thousand electron trajectories must be considered for a statistically significant sampling; the detailed calculation may therefore be prohibitively expensive in computer time. In practice, the detailed description of particle histories is often abandoned, and analytical solutions are used for certain aspects of the multiple scattering problem. In the treatment of Bishop,[55] for example, each electron trajectory is divided into 25 equal steps of length $\Delta s$. During each step, the electron is assumed to scatter through an angle selected from the scattering distribution given by the Goudsmit–Saunderson multiple scattering theory.[56] The reader is referred elsewhere[31,55,57–60] for more details on this type of computation.

Until now, Monte Carlo calculations have been performed primarily for pure element targets, and the results have been compared with experimentally determined energy distributions of backscattered electrons and the depth distribution of characteristic x-ray photons excited within the specimen. Results have been in qualitative agreement. The significance of differences between calculation and experiment for pure elements is obscured by the fact that the crucial experimental measurements are very difficult and may contain systematic errors. An important contribution is a test of averaging procedures by the direct comparison of Monte Carlo calculations of pure elements and alloys of systems that have a large spread in atomic number.[61,62] Monte Carlo calculations of the x-ray intensity generated in thin films of a low-$Z$ element on a high-$Z$ substrate (and vice versa) as a function of film thickness are of considerable interest. These samples have an electron scattering discontinuity at the film–substrate interface. They are ideally suited for the Monte Carlo approach but difficult to treat accurately by other methods. Several laboratories have worked on this calculation.

*2.3.2.2. Electron Transport Equation.* In this approach, electrons are treated not individually but in terms of a distribution function $f(z, \theta, s)$. This distribution function is the probability that an electron which has traveled a distance between $s$ and $s + ds$ will be located at a distance between $z$ and $z + dz$ below the specimen surface and will be traveling in a direction at an angle between $\theta$ and $\theta + d\theta$ to the specimen surface normal. If it can be assumed that (1) scattering is predominantly through small angles, (2) the specimen is homogeneous, and (3) only the variation of electron density in the $z$ direction is of interest, the transport equation may be written in the form

$$\frac{\partial}{\partial s}[\sin \theta f(z, \theta, s)] = -\cos \theta \frac{\partial}{\partial z}[\sin \theta f(z, \theta, s)]$$

$$+ \frac{1}{\lambda(s)}\frac{\partial}{\partial \theta}\left[\sin \theta \frac{\partial}{\partial \theta} f(z, \theta, s)\right] \tag{16}$$

The parameter $\lambda$ is the transport mean free path and contains the average cross section for electron scattering. Equation (16) can be solved numerically, provided $\lambda$ can be expressed as a function of $s$. Since scattering is a function of energy rather than path length, it is necessary to introduce analytical expressions for $\lambda$ as a function of $E$ and $E$ as a function of $s$. Uncertainties in the expressions for scattering cross sections and energy loss are, therefore, a probable source of error also in the transport equation calculations.

Brown and others[32,54,63] have detailed the structure of the computation and the choice of input parameters. In a numerical solution, it is possible to use different approaches at different stages in the scattering problem. Thus, the electron distribution within the electron probe specimen may be calculated in three stages by using (1) a single scattering equation for very short path lengths, (2) a transport equation for intermediate path lengths, and (3) a diffusion equation for long path lengths. The single scattering treatment minimizes the effect of the small-angle approximation inherent in the transport equation as given in equation (16); the diffusion equation for long path lengths reduces computation time. Brown et al.[32] have used this approach to calculate x-ray intensity ratios and x-ray depth profiles that are accurate within experimental measurement error.

2.3.2.3. *Empirical Method.* In the empirical approach suggested by Ziebold and Ogilvie,[64] the weight fraction and the measured intensity ratio for element X are related by the expression

$$C_X/k_X = \alpha_X + (1 - \alpha_X)C_X \tag{17}$$

For a specimen containing elements XYZ, $\alpha_X$ is given by

$$\alpha_X = (\alpha_{XY}C_Y + \alpha_{XZ}C_Z)/(1 - C_X) \tag{18}$$

Expressions for $\alpha_Y$ and $\alpha_Z$ are analogous. For a ternary alloy, there will be three equations of the form of equation (17). These can be solved simultaneously for the $C$, $k$, or $\alpha$ values if the other two sets of variables are known.

The empirical method has generally been successful when experimentally determined $\alpha$ coefficients have been used. It has been applied to ternary alloys[65,66] and to complex geological samples[67,68] The limitation is the availability of known standards for the $\alpha$ determinations. The $\alpha$ coefficients may be calculated by the $\bar{Z}AF$ scheme to be discussed subsequently but will then contain the error introduced by the calculation. Equation (17) implies that, for a binary system, $C/k$ versus $C$ is linear with a slope of $1 - \alpha$ and an intercept of $\alpha$. This linearity of $C/k$ versus $C$ is not observed for all systems. Some physical assumptions inherent in equation (18) may also lead to errors.[65]

*2.3.2.4. The $\bar{Z}AF$ Method of Individual Corrections.* By far the most widely used method for the calculation of interelement effects is the combination of several individual corrections for average atomic number, absorption, and fluorescence. This method has been nicknamed the $\bar{Z}AF$ method; the letters stand for the individual corrections. The division into separate corrections is somewhat artificial but none-the-less very useful. The accuracy of the $\bar{Z}AF$ scheme is good with "normal" specimens for which the corrections are not very large.

The corrections are expressed as multiplying factors, and their order is not important. Thus, the weight concentration of copper in the Cu–Au alloy is given by

$$C_{Cu} = k_{Cu} \begin{bmatrix} \text{atomic number} \\ \text{correction} \\ \text{factor} \end{bmatrix} \begin{bmatrix} \text{absorption} \\ \text{correction} \\ \text{factor} \end{bmatrix} \begin{bmatrix} \text{fluorescence} \\ \text{correction} \\ \text{factor} \end{bmatrix} \quad (19)$$

with each of the factors above or below unity. The value of the correction factors depends upon the composition of the specimen and standard. Thus, several rounds of iteration through all applicable corrections are required for a specimen of unknown composition.

Very simple models are used for the calculation of the correction factors in equation (19). Constants have been adjusted to fit experimental data or have been determined empirically. The models and computations are described in detail in Sections 2.3.3–2.3.6. The current trend is to feed all quantitative data into a time-shared computer for calculation; a number of good programs have been written for this purpose.[69,70]

## 2.3.3. Atomic Number Correction

The variation of electron scattering with the atomic number of the specimen has the following three separate consequences: (1) a decrease in stopping power for high $Z$, (2) an increase in the backscattered electron fraction for high $Z$, and (3) a more shallow distribution of generated x-ray photons within the specimen for high $Z$. The first two items directly affect the *number* of x-ray photons produced and are calculated as the two parts of the atomic number correction. The third item only affects the *distribution* of the generated x-ray photons within the specimen and is treated as part of the absorption correction.

For the 40Cu–60Au alloy, a decrease in the stopping power $S$ due to the presence of gold means that the average beam electron will travel a greater mass distance $\rho s$ in the alloy than in pure copper. Along the longer path, the average electron has a greater probability of producing a $CuK_\alpha$ photon in the alloy. (The average length $s$ along the electron path is considerably longer for high $Z$, but the perpendicular mean depth $z$ of x-ray production is slightly smaller for high $Z$; $s$ and $z$ are very different because of elastic scattering.) If backscattering of electrons and fluorescence effects are neglected temporarily, the ratio of the number of $CuK_\alpha$ photons generated in the alloy and in pure copper can be expressed as

$$C_{Cu} \frac{\int_{E_{CuK}}^{E_0} (Q_{CuK}/S_{CuAu})\, dE}{\int_{E_{CuK}}^{E_0} (Q_{CuK}/S_{Cu})\, dE} \tag{20}$$

$E_0$ is the incident electron energy and $E_{CuK}$ is the ionization energy below which $CuK_\alpha$ cannot be excited. Constants and the atomic fluorescence yield common to sample and standard have been dropped. After substitutions of equations (9) and (13) in equation (20), the integrals are of the form

$$\int_1^{U_0} \frac{\ln U\, dU}{\sum_i C_i (Z_i/A_i)[0.1534 + \ln(E/J_i)]} \tag{21}$$

where $U$ is the overvoltage $E/E_c$. The value of $S$ depends upon specimen composition, and the denominator of equation (21) must be summed by weight fraction over the $i$ elements present in the specimen.

Springer[71] and Philibert and Tixier[72] have proposed the formal integration of equation (21) in terms of the logarithmic integral (li) function:

$$\int_1^{U_0} \frac{\ln U\, dU}{\sum_i C_i (Z_i/A_i)[0.1534 + \ln(E/J_i)]}$$

$$= \frac{1}{\sum_i C_i (Z_i/A_i)}\left( U_0 - 1 - \frac{\ln G}{G}[\text{li}(U_0 G) - \text{li}(G)]\right)$$

where

$$G = \exp\left(\frac{\sum_i C_i(Z_i/A_i)[0.1534 + \ln(E_c/J_i)]}{\sum_i C_i(Z_i/A_i)}\right)$$

$$\text{li}(x) = \text{const} + \ln|\ln x| + \sum_{m=1}^{\infty} \frac{(\ln x)^m}{m \cdot m!} \tag{22}$$

$$\text{const} = \text{Euler's constant} = 0.5772\ldots$$

The integration can be added to existing $\bar{Z}$AF programs and is the preferred computation method. For hand calculation, Thomas[73] and Duncumb and Reed[34] have suggested an approximation based on the fact that the ratio of the stopping power in standard and specimen is a slowly varying function of $E$. $S_{Cu}$ and $S_{CuAu}$ in equation (20) may be evaluated at a mean energy $\bar{E} = \frac{1}{2}(E_0 + E_c)$ and taken out of the integrals. The $Q$ integrals then cancel. This reduces equation (20) to $C_{Cu}(\bar{S}_{Cu}/\bar{S}_{CuAu})$. The $\bar{S}$ approximation introduces an error of less than 1% in most cases, but this is significant if an absolute accuracy of 1–2% is sought.

The largest probable error in the evaluation of equation (20) is due to the uncertainty in the values of $J$. Duncumb et al.[34,74] have proposed $J$ values determined empirically from the analysis of known standards. Their values do not agree with experimental $J$ measurements made with high-energy protons and fitted by equation (11). The writer has found that the Duncumb $J$ values tend to overcorrect when used in conjunction with a correction for fluorescence by the continuous spectrum. The Duncumb $J$ values may contain a partial continuous spectrum fluorescence correction, since this effect was not taken into account during the empirical $J$ determination. The $J$ uncertainty is most serious for low atomic number elements and for low values of $U_0$.

The second part of the atomic number correction is the calculation of the x-ray intensity lost because some of the beam electrons are backscattered while they still retain much of their initial energy. More energy us lost due to backscattering in the Cu–Au alloy than in copper, and this causes a decrease in the number of CuK$\alpha$ photons generated in the alloy. The backscatter effect is always in the opposite direction as the stopping power effect. The stopping power correction is generally dominant and causes an enhancement of a low-$Z$ element in a multielement sample.

The correction is made by the introduction of a coefficient $R$, defined as the ratio of the intensity actually generated to that which would be generated if all of the incident electrons remained within the specimen. $R$ is always less than unity, and $1 - R$ is the intensity lost due to backscattering. If all backscattered electrons retained their original energy $E_0$, then $1 - R$ would be equal to $\eta$. Since they possess a range of energies from $E_0$ down, $1 - R$ is less than $\eta$ by an amount that depends upon the energy distribution of the backscattered electrons.

Both the values of $\eta$ and the shape of the energy distribution curves are very nearly independent of $E_0$ in the 10–30 keV range. It is thus most convenient to plot the energy distribution as a function of $W = E/E_0$ rather than $E$. The $d\eta/dW$ curve for a particular element can then be used for all values of $E_0$. From experimental data for $\eta$ and $d\eta/dW$ as a function of $Z$, it is possible to calculate $1 - R$ by numerical integration of the equation

$$1 - R = \frac{\int_{W_c}^{1} (d\eta/dW) \int_{E_c}^{WE_0} (Q/S) \, dE \, dW}{\int_{E_c}^{E_0} (Q/S) \, dE} \tag{23}$$

This computation has been carried out by Duncumb and Reed, and they have published a table of $R$ versus $Z$ and $U_0 = E_0/E_c$ that may be interpolated to calculate all values of $R$. This table[34] is used almost exclusively by electron probe analysts for the calculation of $R$.

For the 40Cu–60Au example, the decrease in the CuK$\alpha$ signal due to backscattering is given by $R_{CuAu}/R_{Cu}$. The atomic number correction factor in equation (19) is given by

$$\text{Atomic number correction factor} = \frac{R_{Cu}}{R_{CuAu}} \frac{\int_{1}^{U_0} (Q/S_{Cu}) \, dU}{\int_{1}^{U_0} (Q/S_{CuAu}) \, dU} \tag{24}$$

### 2.3.4. Absorption Correction

Some x-ray photons generated below the specimen surface will be absorbed by overlying layers of the specimen. The degree of absorption varies as an exponential function of $\mu \rho z \csc \psi$. The parameter $\mu$ is the mass absorption coefficient in square centimeters per gram for a particular wavelength in a particular absorbing element; values of $\mu$ have been tabulated by Heinrich[75] and others. The absorption path length is $\rho z \csc \psi$, where $\rho z$ is the perpendicular mass depth and $\psi$ is the instrumental x-ray emergence angle defined in Section 2.1.1. If the absorber is a compound or an alloy, a weight fraction average of $\mu$ is used. The absorption parameter $\chi = \mu \csc \psi$ is substituted for ease of notation in much of the probe literature and in the remainder of this discussion.

The absorption function $f(\chi)$ is defined as the ratio of the intensity actually measured to the intensity that would be measured in the absence of absorption. The evaluation of $f(\chi)$ requires integration over the x-ray ionization function $\phi(\rho z)$, which gives the distribution of excited x-ray intensity as a function of mass thickness $\rho z$.

$$f(\chi) = \frac{\int_{0}^{\infty} \phi(\rho z) \exp(-\chi \rho z) \, d(\rho z)}{\int_{0}^{\infty} \phi(\rho z) \, d(\rho z)} \tag{25}$$

If $\phi(\rho z)$ is known from experimental measurements, $f(\chi)$ can be evaluated by numerical integration of equation (25). The function $\phi(\rho z)$ varies rapidly with $E_0$ and more slowly with $Z$. It has been determined experimentally by the following trace layer method: A thin trace layer of element X is evaporated onto a polished solid piece of element Y. The trace layer is then covered by different thicknesses of evaporated layers of Y. The x-ray intensity from the trace layer of X buried under different thicknesses of Y is measured and corrected for absorption by the overlying layers of Y. The data are normalized by the intensity measurement of trace layer X without a substrate (i.e., supported on a fine grid); this intensity measurement is given the value of $\phi = 1$.

The original trace layer data of Castaing and Descamps[76,77] at 29 keV are replotted on a linear scale in Figure 7. More recent measurements of lead with a bismuth tracer[78] and transport equation calculations[32] indicate that the gold curve may be consistently too low in Figure 7. The error is probably due to excessive thickness of the trace layer. The shape of the curves is easily explained in the following qualitative terms: Each curve has a maximum at some depth below the surface because the incident electrons lose their original direction by elastic scattering while stilll retaining much of their energy. Thus, at depth $\rho z$, the layer $\Delta \rho z$ is traversed at various angles by the

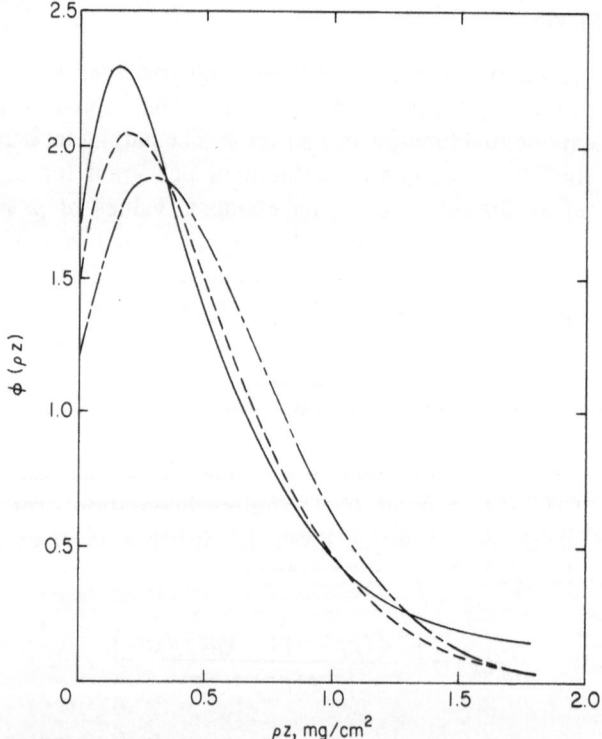

Figure 7. X-ray ionization functions $\phi(\rho z)$ experimentally determined by Castaing and Descamps for a beam energy of 29 keV. The curves are for gold with a bismuth trace layer (——), copper with a zinc trace layer (– – –), and aluminum with a copper trace layer (— – —).

scattered electrons and is of greater effective thickness than $\Delta\rho z$ at the surface. The maximum for gold occurs at a smaller mass depth than that for aluminum, because elastic scattering is more pronounced in gold. The value of $\phi$ at $\rho z = 0$ (often referred to as $\phi_0$) is greater than unity due to the presence of the bulk substrate. Some of the incident electrons are backscattered and may excite x-rays during both traverses through the trace layer on the surface.

Any $\phi(\rho z)$ function determined experimentally by the trace layer method is strictly valid only for the particular electron energy, trace layer, and primary element chosen. The three variables are only loosely interconnected, and their effects can be separated to a large degree by varying them one at a time. Other $\phi(\rho z)$ functions that have been experimentally determined at several voltages are $MgK_\alpha$ in aluminum,[79] $VK_\alpha$ in titanium,[78] $ZnK_\alpha$ in copper,[80] and $BiL_\alpha$ in lead.[78]

The first $f(\chi)$ curves were calculated by numerical integration over the $\phi(\rho z)$ curves shown in Figure 7. Philibert[81] derived an analytical expression for $f(\chi)$ from a simplified scattering model and fitted it to the available experimental data. The Philibert model was improved by Duncumb and Shields[82] through the introduction of a dependence on $E_c$, and new constants were suggested by Heinrich.[74,83] The most commonly used expression, which is fairly accurate for $f(\chi) > 0.5$, is

$$f(\chi) = \frac{1+h}{(1 + \chi/\sigma_c)[1 + h(1 + \chi/\sigma_c)]} \tag{26}$$

where $h = 1.2 \sum_i C_i A_i / Z_i^2$ and $\sigma_c = (4.5 \times 10^5)/E_0^{1.65} - E_c^{1.65})$. The relatively small $Z$ dependence is contained in the $h$ term. The primary dependence of $f(\chi)$ is on $\chi/\sigma_c$. Since $\chi = \mu \csc \psi$ is generally fixed by the specimen and the available instrument, the only means of minimizing absorption is to use a relatively low beam energy $E_0$.

Recently Love and others[84,85] have used extensive experimental data on binary systems to optimize the expressions for $h$ and $\sigma_c$ in equation (26). They found that $f(\chi)$ is not very sensitive to the expression for $h$ and to the averaging method used to determine $h$ for alloys. Bishop[86] has suggested that the absorption correction be separated into a mean depth function and a shape function. The shape function predominates in cases of high absorption and is a function of the overvoltage ratio and the mean atomic number of the specimen.

The most significant change has been suggested by Heinrich and Yakowitz.[87] They show that the effect of atomic number in the absorption correction is smaller than the statistical spread of the experimental data and that equation (26) contains a systematic overcorrection. Their argument is that data reduction procedures should not be complicated needlessly by an atomic number dependence that has not been demonstrated experimentally, even though such a dependence is expected theoretically. The expression proposed by Heinrich and Yakowitz simplifies the computation with an apparent increase in accuracy:

$$1/f(\chi) = (1 - a\gamma\chi)^2 \qquad (27)$$

where $a = 1.2 \times 10^{-6}$ and $\gamma = (E_0^{1.65} - E_c^{1.65})$.

For the 40Cu–60Au example, the absorption correction factor in equation (19) is given by

$$\text{Absorption correction factor} = \frac{f(\chi)_{Cu}}{f(\chi)_{CuAu}} \qquad (28)$$

### 2.3.5. Characteristic Line Fluorescence Correction

Fluorescence by characteristic x-ray lines is possible only if spectral lines of higher energy than the line being measured are emitted by another element present in the specimen. The effect is strongest when the exciting lines have an energy between approximately 1.1 and 1.5 times the energy of the line that is being enhanced. This situation exists for the analysis of copper in Cu–Au, and all of the AuL lines can excite $CuK_\alpha$.

To correct for fluorescence by characteristic lines, it is necessary to calculate the ratio

$$\gamma_f = I_f'/I_p' \qquad (29)$$

where $I_f'$ is the measured intensity due to excitation by characteristic line fluroescence and $I_p'$ is the measured intensity due to primary excitation by electrons. $I'$ is the measured intensity after absorption; $I$ will be used to designate generated intensity; $I'/I = f(\chi)$. If the total measured intensity is $I_f' + I_p'$, the correction factor in equation (19) is

$$\text{Characteristic fluorescence correction factor} = \frac{1}{1 + \gamma_f} \qquad (30)$$

In the Cu–Au alloy, several AuL lines can excite $CuK_\alpha$ photons by fluorescence. For the highest accuracy, the contributions of the individual lines should be computed separately and added. Expressions for the following factors need to be known in order to complete this calculation:

1. The intensity (in photons per primary electron) of every AuL line excited in the specimen.
2. The probabilities that photons from the specific AuL lines will produce $CuK_\alpha$ photons in the specimen.
3. The intensity of $CuK_\alpha$ excited by electrons in the specimen.
4. Absorption of primary and fluorescent radiation in the specimen.

Items 1 and 3 may be calculated from the equation giving the intensity excited by electrons of the d line from the c subshell of element X.

$$I_{pX} = 0.5\omega_c p_{cd}\left(\sum_i C_i R_i\right)\frac{C_X}{A_X} bn_c\left[\int_1^{U_0}\frac{\ln U\, dU}{\sum_i C_i(Z_i/A_i)\,[0.1534 + \ln(E/J_i)]}\right] \quad (31)$$

$R$ and $S$ have to be summed over the elements present in the specimen; the evaluation of the integral is given in equation (22). For each computation, the $U_0$ corresponding to the $E_c$ of the specific line should be used. The parameter $p_{cd}$ is the "weight" of the d line (i.e., the intensity of the line d relative to the total intensity of all lines emitted by the ionized c subshell), $\omega_c$ is the fluorescence yield for the c subshell, $b$ is a constant in Bethe's equation for $Q$ ($b \sim 0.76$), and $n_c$ is the number of electrons in the excited c subshell.

The probability of a $CuK_\alpha$ photon being produced by a $AuL_j$ photon generated within the specimen is

$$\frac{I_f(CuK_\alpha)}{I_p(AuL_j)} = 0.5 C_{Cu}\omega_{CuK}p_{CuK_\alpha}\left(\frac{\mu(AuL_j \text{ in Cu})}{\mu(AuL_j \text{ in CuAu})}\right)\left(\frac{r_{CuK} - 1}{r_{CuK}}\right) \quad (32)$$

The absorption edge jump ratio $r_{CuK}$ is the ratio of the absorption constant of copper for radiation on each side of the CuK absorption edge. For the correction procedure, the ratio of measured fluorescent and primary $CuK_\alpha$ intensity is of interest; therefore, the primary $CuK_\alpha$ intensity and absorption terms have to be added. The full equation for $\gamma_f$ is a variation of that given by Hénoc *et al.*[88]

$$\frac{I_f'(CuK_\alpha)}{I_p'(CuK_\alpha)} = \sum_j \frac{I_p(AuL_j)}{I_p(CuK_\alpha)} 0.5 C_{Cu}\omega_{CuK} p_{CuK_\alpha}\frac{\mu(AuL_j \text{ in Cu})}{\mu(AuL_j \text{ in CuAu})}\frac{r_{CuK} - 1}{r_{CuK}}$$

$$\times \left\{\frac{f[\sigma_{AuL_j}, \chi(CuK_\alpha \text{ in CuAu})]}{f[\sigma_{CuK_\alpha}, \chi(CuK_\alpha \text{ in CuAu})]}\right\}\left[\frac{\ln(1 + u)}{u} + \frac{\ln(1 + v)}{v}\right] \quad (33)$$

where

$$u = \frac{\chi(CuK_\alpha \text{ in CuAu})}{\mu(AuL_j \text{ in CuAu})} \qquad v = \frac{\sigma_{AuL_j}(1 + h)}{\mu(AuL_j \text{ in CuAu})(1 + 2h)}$$

The expression for $v$ in equation (33) is that given by Heinrich.[89] The primary intensities for $CuK_\alpha$ and the $j$ AuL lines are calculated from equation (31), which also contains $\omega$ and $p$. For the analyzed line (in this case $CuK_\alpha$), $\omega$ and $p$ cancel. However, $\gamma_f$ is directly proportional to $\omega$ and $p$ for the exciting (AuL) lines and thus strongly affected by any uncertainty in them. Values of $\omega_c$ are fairly well known for K lines but are uncertain for L lines. Experimental measurements generally determine a mean value $\bar{\omega}_L$. The calculation of the three subshell yields from $\bar{\omega}_L$ is complicated by the radiationless Coster–Kronig transitions between subshells.[48,49]

If the excited line is an L line, the factor containing $r$ has to be modified to take into account the multiplicity of L absorption edges. Thus, for $L_\alpha$ radiation, which is emitted from the $L_3$ subshell

$$\frac{r-1}{r} \to \frac{r_{L_3}-1}{r_{L_1}r_{L_2}r_{L_3}} \tag{34}$$

The use of the full equation (33) is most rigorous, but obviously it requires computer calculation. Castaing[90] and Reed[91] have previously derived simpler expressions for fluorescence of K lines by K lines, and these expressions are more amenable to hand calculation. Both of the simpler equations make the following approximation:

1. The exciting radiation is assumed to be monochromatic. This eliminates the summation over $j$ lines. This simplification is not very serious for K lines, for which more than 90% of the intensity is contained in the $K_\alpha$ line. The "weight" $p$ of the excited line is dropped, and all calculations are made using absorption constants for $K_\alpha$.

2. The $f$ ratio in curly brackets in equation (33) is dropped. This factor is always slightly larger than unity for elements of interest.

3. An analytical expression is used for the primary intensity ratio of the two elements. Castaing uses the expression

$$\frac{I_p(Y)}{I_p(X)} = \frac{C_Y \omega_Y A_Y E_c(X)}{C_X \omega_X A_X E_c(Y)} \tag{35}$$

Reed uses

$$\frac{I_p(Y)}{I_p(X)} = \frac{C_Y \omega_Y A_Y}{C_X \omega_X A_X}\left(\frac{U_Y-1}{U_X-1}\right)^{1.67} \tag{36}$$

Both also use a simpler form of $v$, leaving out the terms containing $h$.

Reed[91] suggests the use of his expression also for LK or KL fluorescence after the addition of the empirical adjustment factor $P_{ij}$. He then breaks up the equation into several functions that are given in graphical or tabular form and need only be multiplied together. Hand calculation by the Reed method is very easy, but the errors introduced by the approximations have not been systematically evaluated. The assumption of monoenergetic radiation is much more serious for the L than for the K spectra.

### 2.3.6. Continuous-Spectrum Fluorescence Correction

The correction for fluorescence by the continuous spectrum is small in many cases,[92] and the correction is frequently omitted. It differs from the characteristic line fluorescence correction, because continuous-spectrum fluorescence also occurs in a pure element standard. The correction can only be calculated by computer and typically doubles the computation time. The correction tends to be largest for short-wavelength radiation and a large spread in atomic number. If both the characteristic and continuous-spectrum fluorescence corrections are required, it is easiest to combine them into one correction factor in equation (19).

The measured intensity ratio of $CuK_\alpha$ in Cu–Au and copper may be expressed as

$$k_{Cu} = \frac{I'_p(CuAu) + \sum I'_f(CuAu) + I'_{cont}(CuAu)}{I'_p(Cu) + I'_{cont}(Cu)} \tag{37}$$

If $I'_f / I'_p = \gamma_f$ and $I'_{cont} / I'_p = \gamma_{cont}$

$$k_{Cu} = \frac{I'_p(CuAu)}{I'_p(Cu)} \left[ \frac{1 + \gamma_f(CuAu) + \gamma_{cont}(CuAu)}{1 + \gamma_{cont}(Cu)} \right] \tag{38}$$

The atomic number and absorption corrections operate only on the $I'_p$ ratio and change it into a weight fraction ratio equal to $C_{Cu}$. The inverse of the factor in square brackets is the fluorescence correction factor in equation (19) for $CuK_\alpha$ excitation by gold L lines and the gold continuous spectrum.

The $\gamma_{cont}$ values are calculated in a manner similar to that for the $\gamma_f$ values for both sample and standard. The intensity of $CuK_\alpha$ in CuAu excited by continuous-spectrum fluorescence is given by

$$I'_{cont}(CuK_\alpha) = 0.5 C_{Cu} \omega_{CuK} p_{CuK_\alpha} \left( \frac{r_{CuK} - 1}{r_{CuK}} \right) \left[ K \frac{\sum_i C_i Z_i}{\chi(CuK_\alpha \text{ in Cu Au})} \right]$$

$$\times \left\{ \left[ \int_{\lambda_0}^{\lambda_c} \mu(\lambda \text{ in Cu}) \ln \left[ 1 + \frac{\chi(CuK_\alpha \text{ in CuAu})}{\mu(\gamma \text{ in Cu Au})} \right] \left( \frac{1}{\lambda_0} - \frac{1}{\lambda} \right) \frac{d\lambda}{\lambda} \right\} \tag{39}$$

The wavelengths $\lambda_c$ and $\lambda_0$ correspond to $E_c$ and $E_0$; $K$ is Kramer's constant that has been postulated to be independent of specimen composition. The parameters $\omega$ and $p$ cancel in the $\gamma$ ratio, since they are also contained in $I'_p$. The primary dependency in equation (39) is on $\sum C_i Z_i$ and on the absorption constants. The numerical integration of the $\lambda$ integral is time-consuming for multielement specimens, since a number of absorption edges are likely to be located between $\lambda_0$ and $\lambda_c$. Hénoc[93] has developed a method for formal integration between absorption edges which is based on the assumption that $\mu$ varies as $\lambda^3$.

All corrections for fluorescence by the continuous spectrum in current use are based upon Kramer's equation [see equation (14)] and may have to be revised as more data on the intensity of the continuous spectrum are accumulated.[42,94] Also inherent in equation (39) are the assumptions that the continuous-spectrum radiation is isotropic and originates from a point source on the specimen surface. The error caused by these assumptions has not been investigated. Some empirical constants or expressions for the other corrections (e.g., $h$, $\sigma_c$, $J$) were established without the continuous-spectrum fluorescence correction. They may, therefore, contain a partial compensation for continuous-spectrum fluorescence, which leads to overcorrection when a separate continuous-spectrum correction is used.

### 2.3.7. Soft X Rays

The original interest in long-wavelength x-ray analysis with the electron probe was for the analysis of low-$Z$ elements, which emit only soft x-ray lines. This is still the primary use of soft x-ray analysis, but in addition, soft L or M lines of higher-$Z$ elements are measured in some samples in order to improve the spatial resolution, improve the absolute detection level (the number of atoms that can be detected), or study the effects of chemical combination on the soft x-ray spectra. Long-wavelength lines can be excited at low beam energies, and this results in a much lower analytical volume. Low-beam-energy excitation is particularly important to minimize the spreading of the electron beam in organic or biological specimens, which typically have a density of less than 1 g/cm$^3$.

Surprisingly good analyses of oxygen, carbon, and boron in compounds with elements up to $Z = 26$ have been reported by Shiraiwa *et al.*[95,96] On the other hand, a considerable lack of success has been reported by Openshaw *et al.*[97] for the carbon analysis of $UC_2$. The problems inherent in soft x-ray analysis were documented by Andersen[98] and by Duncumb and Melford[99] soon after the instrumentation came into general use. More recent investigations of the problems have been done.[85,86,100–103] A qualitative discussion of the limitations follows.

*2.3.7.1. Chemical Bonding Effects.* Soft x-ray emission involves outer-shell electrons that are highly affected by or directly involved in chemical bonding. Since the wavelength and intensity of a specific x-ray line will depend upon the chemical state of the emitting element, both standard and sample should be in the same chemical state. The bonding effect is especially pronounced for the L spectra of iron and neighboring elements for which the intensity ratio of the $L_{\alpha_1}$ and $L_{\beta_1}$ lines can change by as much as a factor of 5 with bonding.[98]

*2.3.7.2. Mass Absorption Coefficients.* For soft x-ray lines, the mass absorption coefficients are large, inaccurate, and often totally unknown. The coefficients are nearly impossible to measure directly by the transmission of soft x rays through foils. They may be approximated indirectly by a technique that relates $\chi$ to the maximum of an x-ray intensity versus $E_0$ curve[98,104] or by a thin film model.[100]

*2.3.7.3. Absorption Correction.* The absorption correction based on the Philibert model and given in equation (26) is not valid for very large $\mu$ and very small $E_c$. The derivation, based on a simplified exponential scattering model,[81] contains the approximation that $Q = \text{const}$, which is not true for the very large values of $U$ typically encountered for soft x-rays. (At a beam energy of 20 KeV, $(1/U)\ln U$ for $CK_\alpha$ with $E_K = 0.284$ keV is only 16% of its maximum value at 0.76 keV.) The simplest form of the Philibert correction utilizes a $\phi(\rho z)$ function with $\phi_0 = 0$ and a compensating sharp increase and high peak for small $\rho z$. This unrealistic shape introduces very little error for low $\mu$ but is not appropriate for high $\mu$ and $U$.

Figure 8 shows the low $\rho z$ portion of three $\phi(\rho z)$ curves for $CK_\alpha$ in $Fe_3C$ at a beam energy of 20 keV. The absorption curve for $\psi = 20°$ is also shown. For the experimental conditions, all measured radiation originated from a layer less than 0.15 $mg/cm^2$ thick. Thus, the $\phi(\rho z)$ curves must be accurate at low $\rho z$ to obtain an accurate value for $f(\chi)$. The dashed Philibert curve with $\phi_0$ equal to the calculated value 1.9 is better than the solid curve with $\phi_0 = 0$ but still suffers from the $Q = \text{const}$ approximation. The upper curve, calculated by Monte Carlo methods, shows the effect of the greater $Q$ for backscattered electrons and incident electrons that have lost some of their energy.

*2.3.7.4. Atomic Number Correction.* The atomic number correction fails for low beam energies, which would reduce the effect of absorption. The Bethe stopping power [equation (9)] is valid only for $E \gg J$. For soft x rays in a high-$Z$ specimen, $E_c$ may be considerably less than $J$, and this would make $S$ calculated from equation (9) zero or negative. This is due entirely to the mathematical form of the equation and has no physical basis. Any $E/J$ less than approximately 10 may introduce error into the $S$ calculation; the magnitude of this error is not known.

### 2.3.8. Analytical Accuracy to Be Expected in Practice

The analytical accuracy that can be achieved on a "typical" sample depends on both the measurement precision and the accuracy of the correction for interelement effects. Measurement errors have been discussed in Section 2.2.4 and will be approximately $\pm 0.5\%$ for a major constituent in a carefully prepared specimen if no systematic errors are present. Errors in the interelement corrections are due to uncertainties in the instrumental and fundamental input parameters (values of $\psi$, $E_0$, $J$, $\mu$, $\omega_c$, etc.) and to errors in the basic models or equations.

Studies of error propagation due to input parameter uncertainties have been pursued rigorously by Heinrich and Yakowitz.[105-108] Their curves showing the effect of probable errors in the individual input parameters are rather sobering. The only general conclusion that can be drawn is that

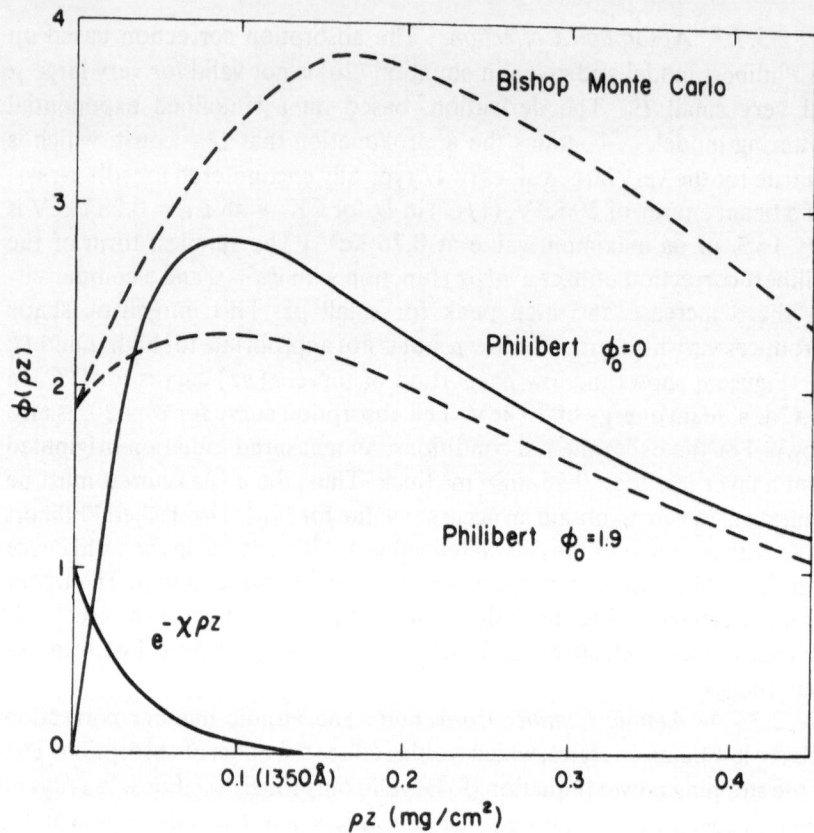

Figure 8. X-ray ionization functions and the x-ray absorption curve for $CK_\alpha$ in $Fe_3C$ at 20 keV. Data from Bishop[31] and Duncumb and Melford.[99]

relative errors of 1–2%, due to input parameter uncertainties only, are quite probable in specimens with a large atomic number spread and/or moderately high mass absorption coefficients. For any particular analysis, the uncertainty due to input parameters can be estimated by variation of each parameter within its probable error range and recalculation of the interelement corrections. To minimize errors, the individual correction factors should be kept as close to unity as possible. To achieve this, different elements in the same specimen must often be measured at different beam energies. A rough calculation before any quantitative data are taken can be helpful in making the optimum choice.

Basic errors in the models are obscured by the uncertainties in the input parameters. Better values of $J$, $\mu$, and $\omega_c$ are required before the reexamination of the x-ray generation equations is likely to lead to significant improvement. The short cuts for hand calculation—the $\bar{S}$ approximation for atomic number, the Reed fluorescence correction, and the omission of the continuous-spectrum fluorescence correction—can cause significant error for some systems. The more detailed calculations are limited by scientific knowledge rather than ease of computation. They give consistent results, and deviations between experiment and calculation can generally be attributed to input parameter uncertainty. The most detailed program available at this time is the COR program developed at the National Bureau of Standards.[69,89,109] It is limited, however, to instruments with beam incidence normal to the specimen surface.

The ultimate test of analytical accuracy is the analysis of standards of accurately known composition. Therefore, many electron probe analysts continue to test their instruments, measurement techniques, and inter-element correction schemes by the analysis of known standards. The National Bureau of Standards has issued a series of standards (Fe–Si alloy,[110] Cu–Au alloy series, and Ag–Au alloy series[111]) that are of high reliability with regard to absolute composition and homogeneity. Standards from other sources may in some cases be suspect on both counts. Homogeneity can be determined with EPMA itself; the absolute composition has to be confirmed by more than one independent technique and/or laboratory. "Stoichiometric" compounds often deviate from stoichiometry by 1% or more. The requirement for reliable absolute standards is by no means trivial.

Analyses of reliable standards by various laboratories are generally within ±2% of the nominal value. This is a relative accuracy that can generally be obtained with sufficient care. If the corrections are small, an absolute accuracy of ±1% of the amount present is possible. With the shortcut calculations, the error may be as large as ±3–5%. These estimates of analytical accuracy do not apply for elements with $Z < 11$, for x-ray lines of long wavelength, or for specimens with very high x-ray absorption for the measured line.

## 3. Applications

### 3.1. Bulk Characterization

#### 3.1.1. Introduction

The types of analyses discussed in this section comprise a large fraction of the published applications of electron probe microanalysis. The specimens are large compared to the volume analyzed by the electron probe, but the small beam is required to look for compositional variation within the sample. The object of the analysis is to characterize the bulk material either qualitatively or quantitatively.

No attempt will be made by this writer to review the large number of general and specific application papers that have been published in the fields of metallurgy and mineralogy. Only a few examples are given in the following sections to indicate the possibilities. The following papers are recommended for general reading: a review of metallurgical applications by Goldstein[112]; reviews of mineralogical applications by Keil[113,114] and by Sweatman and Long[115]; suggestions for biological specimens by Andersen,[116] by Hall,[117] and by Robinson[118]; applications to glass technology by Kane[119]; and applications of bonding studies by White.[120] Papers on more specific applications can generally be found in the journals appropriate for the study or specimen. Some papers on application are also given in the proceedings of the international conferences on x-ray optics and microanalysis,[121-124] the extended abstracts of the yearly MAS conferences,[125] the Denver x-ray conference series,[126] and other collections of probe papers.[127-130]

### 3.1.2. Specimen Preparation

Since electron probe microanalysis is an analysis of a relatively thin surface layer, great care must be taken to make certain that the analyzed surface is representative of the bulk. The specimen will normally have to be sectioned and polished. If there is considerable preferred orientation, a number of sections in different planes may be necessary to characterize the volume sufficiently by a two-dimensional analysis. A number of specimens and/or standards can often be mounted and prepared together. The writer has found a commercially available copper–lucite mixture to be a very convenient mounting medium.

The polishing abrasive may become imbedded in soft samples, such as high-purity metals, or in hard but porous samples, such as sintered oxides. In these cases, an abrasive that does not contain any of the elements to be measured in the specimen should be used. Most polishing compounds are nonconducting and thus may cause beam instability. Also, any amount of imbedded abrasive will lower all other x-ray signals.

If the phases present are of very different hardness, a layer of the softer material may be smeared over the harder material. This effect can usually be minimized by using a very hard abrasive, such as diamond, and a polishing cloth with little or no nap. It may be desirable to remove the smeared layer by light etching of the specimen. However, etching or other chemical treatment has its own pitfalls, since it can leach out an element preferentially or deposit another element on the surface. Several cycles of alternate light etching and light repolishing are possibly the best compromise for the preparation of problem specimens. Height differences between phases are a frequent result of polishing or etching, but they are not nearly as detrimental to the precision of a quantitative analysis as chemically altered surface layers.[131]

Electropolishing is not recommended as a general method for specimen preparation. Preferential removal of one phase or element in multiphase alloys during electropolishing is a common occurrence. The writer has also observed marked preferential removal of one element in binary alloys consisting of a single phase.[132]

Nonmetallic specimens normally require somewhat different polishing procedures. Sintered oxides, ceramic materials, and some geological samples tend to be brittle, and entire grains may be pulled out of the sample during grinding. This can be disastrous if a particular phase is pulled out preferentially. Such samples may have to be polished for as long as 3 hr under very gentle conditions to give a representative surface; automatic polishing equipment is practically mandatory for successful sample preparation. Glasses and plastics will flow and smear readily and should be ground and polished with very little pressure and large volumes of cooling water.

Nonconducting specimens have to be coated with a thin conducting layer to dissipate charge and heat. Thin evaporated films of carbon or aluminum are satisfactory for this purpose; the resistivity of the evaporated film should be no greater than 1 M$\Omega$ per square.

### 3.1.3. Phase Equilibria

Many of the early probe studies dealt with the quantitative analysis of secondary phases in metallurgical samples and with the establishment or confirmation of phase diagrams. Both are still important applications.

For some binary systems, all phases present at a given temperature can be determined by the analysis of one diffusion couple of the pure elements annealed at that temperature. In the cross-sectioned diffusion couple, intermediate phases will appear as bands between the pure elements. The concentration-versus-distance profile in the diffusion direction will be continuous through solid solution regions and will exhibit sharp discontinuities at phase boundaries.[112]

All intermediate phases may not form in a pure element couple. This may be due to extremely slow diffusion or to very limited nucleation of the missing phase, so that it is consumed more rapidly in forming the next phase than it is nucleated. As an example, consider the metals X and Y, which form the intermetallic phases $X_2Y$ and $XY$. If $XY$ nucleates much more readily than $X_2Y$, only $XY$ may appear as an intermediate phase in a pure element couple. A diffusion couple of X and $XY$ would be required to establish the presence of $X_2Y$; and a diffusion couple of X and $X_2Y$ would be required to confirm the absence of intermetallics more X-rich than $X_2Y$. The kinetics of phase transformation applicable to this problem have been treated by Kidson.[133] An example of missing phases with a pure element diffusion couple for the U–Ni system has been described by Adda et al.[134] The diffusion couple technique is also useful for the exploration of ternary systems, but the number of couples and annealing runs required for a complete determination of all phases may become very large.

The primary advantage of EPMA for phase identification over other chemical techniques and x-ray diffraction is the small volume of material required. Any secondary phase region approximately $2 \mu m$ wide can be analyzed quantitatively. Also, a sample containing several different secondary phases can be dealt with; this is not generally true for the other techniques. It has been shown by Goldstein and Ogilvie[135,136] that EPMA can be used to determine phase equilibria in multicomponent systems even if the various phases analyzed are not homogeneous. This is true because the electron probe may be used to measure the interface compositions of two coexisting phases, and equilibrium conditions must exist at phase interfaces.

The accuracy of phase diagram determination is limited by uncertainties in the interelement corrections and by the relative magnitude of concentration gradients and the x-ray source diameter. Several analysts have cautioned that fluorescent excitation of adjacent material may occur. For example, pure nickel adjacent to iron may appear to contain several weight percent of iron even though no beam electrons excite iron. This is because the $NiK_\alpha$ radiation has a much greater range than the beam electrons and may excite $FeK_\alpha$ by fluorescence.

In typical industrial materials conditions are nonideal for the compositional determination of a second phase. The primary limitation is normally the small size and dendritic nature of the second phase. An example is given in Figure 9, which shows a finely divided chromium-rich second phase in a Ni–Cr primary phase. Even if the analyzed secondary phase region is larger on the specimen surface than the area excited by the beam, it may be too shallow in the depth direction, which the analyst cannot see. Thus, some of the primary phase may be included in the analyzed volume.

Intermixing of phases is even more prevalent in many oxide systems. An example is given in Figure 10, which shows a transmission electron micrograph of a shadowed surface replica of a polished and etched ceramic capacitor material. The sample consists of stoichiometric barium titanate to which a small amount of titanium dioxide has been added. The $BaTiO_3$ and $TiO_2$ were milled together in powder form, pressed into disks and sintered. The $TiO_2$ serves as a grain growth inhibitor during sintering, but if mixing is poor, it may also cause the growth of a secondary phase with a higher titanium concentration than the matrix. The second phase grows in widely separated snowflakelike structures containing many fine branches and corresponds to the light and smooth areas in Figure 10. The line scan of a

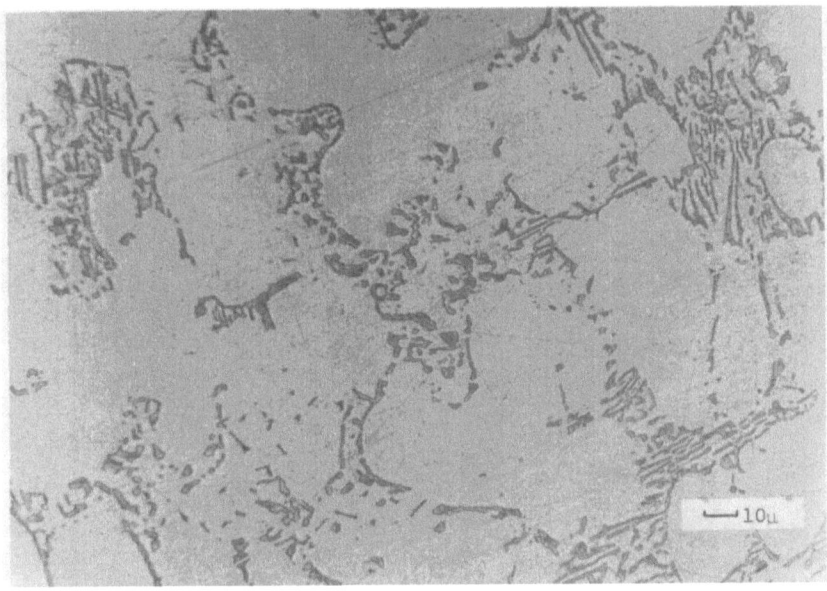

Figure 9. Dendritic chromium-rich second phase in Ni–Cr alloy.

polished section in Figure 11 shows that the second phase is titanium-rich; a similar scan for barium shows that the second phase is barium-poor. It is also evident that the primary and secondary phases are intermixed on a smaller scale than can be completely resolved. As a result, the composition of the second phase cannot be determined accurately, since some matrix is always measured with it. The highest titanium readings in the line scan closely approach the titanium concentration of $BaTi_2O_5$, and this may be the actual composition of the second phase. However, the phase $Ba_2Ti_5O_{12}$ can also form under the processing conditions and cannot be completely ruled out by the analysis.

### 3.1.4. Diffusion

Diffusion constants may be determined with EPMA.[112] The concentration gradient can be measured directly on a cross-sectioned annealed diffusion couple. For binary systems, the primary advantage of EPMA over older techniques is the faster and more convenient measurement of the concentration gradient. Annealing cycles may also be shorter, since steeper gradients can be measured. The accuracy depends on the uncertainties in the corrections for quantitative analysis and, in the case of very steep gradients,

Figure 10. Transmission electron micrograph of surface replica that shows finely divided titanium-rich secondary phase in $BaTiO_3$-based ceramic material.

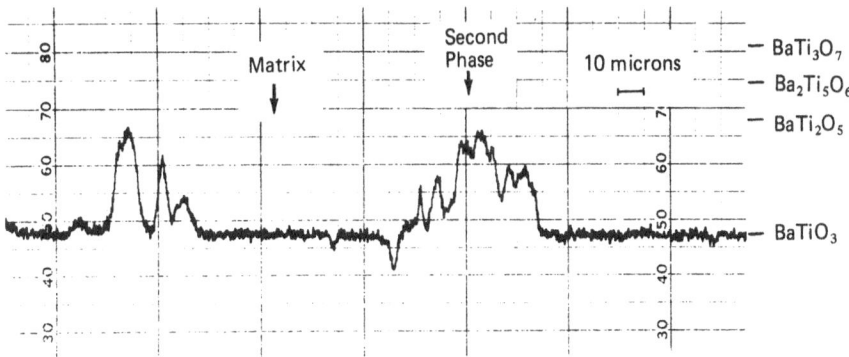

Figure 11. Line scan for $TiK_\alpha$ on $TiO_2$-rich $BaTiO_3$. The composition of the second phase cannot be determined accurately because it is mixed with the primary phase.

on the relative length of the gradient and the diameter of the excited volume for one measurement.

It is possible to perform diffusion studies with EPMA that would not have been possible with conventional methods. The effect of grain boundary diffusion can be very effectively demonstrated by electron proble diffusion studies in bicrystals.[137,138] The small analyzed volume makes it possible to determine isoconcentration lines across the grain boundary and to compare these profiles with the prediction of theoretical models.[139,140] In general, the degree of grain boundary diffusion has been found to vary with the degree of misalignment across the bicrystal boundary.

Ternary interdiffusion has been studied in the Cu–Ag–Au system by Ziebold and Ogilvie[66] and in the Fe–Co–Ni system by Vignes and Sabatier.[141] Ternary diffusion is much more complex than binary diffusion, since there are two independent composition variables. Concentration profiles generally do not vary monotonically through solid solution ranges in ternary systems, and four interdiffusion coefficients are required to describe the transport process.

### 3.1.5. Microsegregation, Inhomogeneity, Trace Analysis

Studies of microsegregation and inhomogeneity are a real forte of EPMA, since these analyses cannot be done directly by any other technique. The primary advantage is the small size of the analyzed volume. Qualitative data are often sufficient for the characterization of impurity segregation or inhomogeneity. This is fortunate, since the concentration gradients are often very difficult to characterize in a quantitative fashion; they may be very steep and vary in depth as well as laterally. A few examples follow to illustrate the potential and limitations of the method.

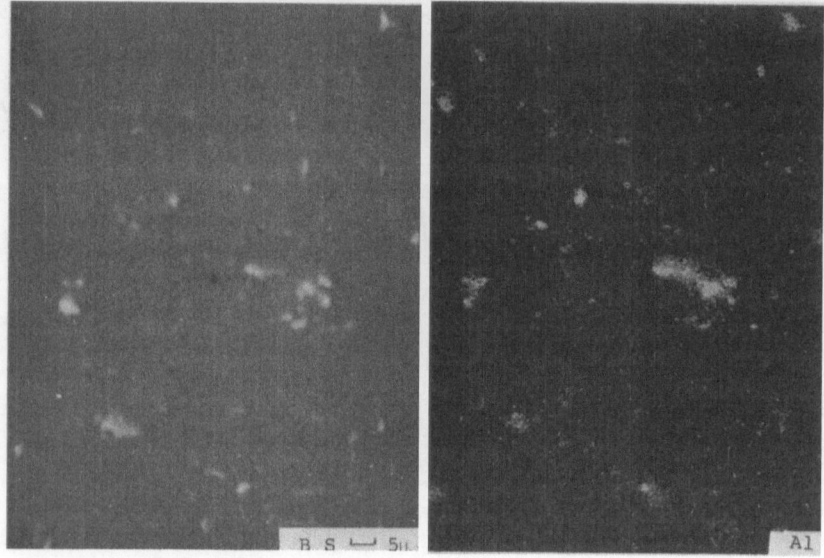

Figure 12. Aluminum segregation in beryllium. Backscattered electron current (*left*) and aluminum x-ray images (*right*).

In many materials impurities are segregated at defect sites, grain boundaries, or in a second phase. An example of such a case is shown in Figure 12. The sample was beryllium of high nominal purity; it was polished with diamond slurry to eliminate any possibility of contamination with polishing alumina. Concentrations of over 1 wt.% aluminum were measured in some areas, but aluminum could not be detected in other areas. The aluminum was a bulk impurity and could have been detected by any suitable bulk analytical technique. The additional information obtained from electron probe analysis was that the aluminum was very segregated and apparently present in an intergranular second phase.

Inclusions in various industrial materials often differ considerably from the matrix composition and may contain major concentrations of impurity elements. Such inclusions can often be characterized by x-ray images from two-dimensional scans. An example is the study of inclusions in sheet steels by Tenenbaum.[142]

Some studies of grain boundary segregation have also resisted analysis by EPMA because the volume analyzed with the probe is still too large for the problem. Efforts to correlate the stress corrosion susceptibility of many Al–Zn–Mg alloys to solute depletion near grain boundaries have been largely unsuccessful.[143,144] The problem here is one of measuring relatively small differences of the major or minor constituents in submicron-wide regions adjacent to grain boundaries. Other unsuccessful studies involve grain boundary pinning by a few atomic layers of impurity atoms. In this case, the weight fraction of impurity atoms within the analyzed volume is so small that the element is not detectable. Both types of studies also involve the problem of locating the grain boundary. If the specimen is chemically etched to make the grain boundary visible, some of the material of interest may be removed. Thermal grain boundary grooving produces a trench and possibly a ridge on each side of it that will affect the measured intensity because of nonflatness of the specimen.

Many alloys or compounds are inhomogeneous even though they consist of a single phase. The degree of inhomogeneity may greatly affect the mechanical, electrical, or chemical behavior of the material, yet it is very difficult to determine by any other method than electron probe micro-analysis. An example is shown in Figure 13. The sample is an alloy with the nominal composition of 42 wt.% chromium and 58 wt.% nickel. This alloy is in a single-phase region of the phase diagram above 1000°C and in a two-phase region below 1000°C. The alloy was to be used as a sputtering source for making thin-film Ni–Cr resistor films of the same composition.[145] Since nickel and chromium sputter at different rates, it was important that the alloy consist of a single phase and be as uniform as possible.

The left side of Figure 13 shows a $CrK_\alpha$ line scan of the alloy after solidification and quenching to room temperature. The instrumentation was set to record the second digit of the signal full scale; the actual relative intensity is 200 plus the value shown on the strip chart. The highest and lowest intensity values on a long scan were 283 and 247, respectively. These values correspond to a deviation of ±6.8% from the average chromium counting rate. The right side of Figure 13 shows a similar $CrK_\alpha$ scan of the

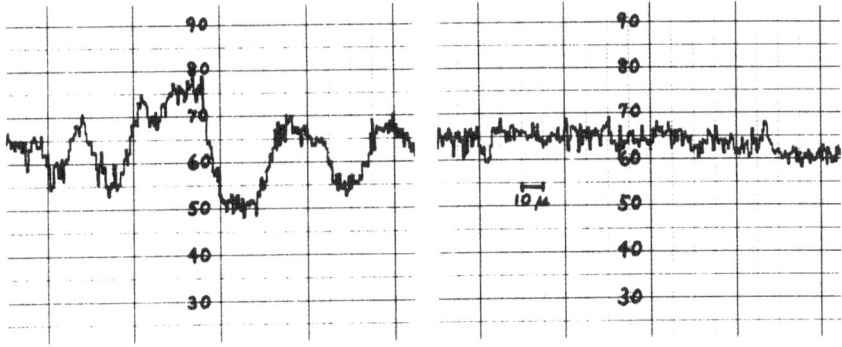

Figure 13. Line scan for $CrK_\alpha$ in 42Cr–58Ni alloy before and after long anneal for homo-genization. The actual intensity is twice full scale plus what is shown.

same specimen after a long anneal above 1000°C. Any remaining inhomogeneity is now difficult to distinguish from statistical fluctuation.

The type of scan shown in Figure 13 can be completed quite rapidly
(approximately $\frac{1}{2}$ hr per element) and can give very valuable information to
material scientists. Annealing, sintering, or reaction cycles can be optimized
by this technique, and abnormal behavior can often be explained. Solid
solution inhomogeneity is not readily detected by techniques other than
electron probe microanalysis; no difference was found in the Ni–Cr alloy
before and after annealing by standard metallographic and x-ray diffraction
techniques.

In analyses for minor or trace constituents, the measurement errors
may be more severely limiting than the interelement correction errors. If the
peak-to-background ratio is very low, the primary error will be due to
counting statistics. Under these circumstances, it is often advisable to
increase the specimen current by a factor of 10 or 20 from the value used for
the analysis of major constituents in order to obtain a high counting rate.
Pure element standards can then no longer be used, because the extremely
high peak-counting rates would result in a very large and inaccurate dead
time correction.

The most suitable standards for high specimen current analyses are
intermediate standards with an intensity of less than 5% of the pure element
standards. The intermediate standards may be thin films, alloys, or
compounds. Their only requirements are that they be homogeneous and
contain no elements with interfering spectral lines. The composition of an
intermediate standard need not be known, but the intensity of the element of
interest relative to an absolute standard or the pure element must be
accurately measured at a low specimen current. This calibration measurement need be done only once for any combination of high voltage and
emergence angle. The use of an intermediate standard in essence replaces
one very large correction (the large dead time correction) by two much
smaller corrections (a small dead time correction and the error in the relative
intensity calibration). With this procedure, detectability limits of 10 ppm
have been achieved by Schneider[146] for strong $K_\alpha$ lines in geological
specimens.

### 3.1.6. Oxides and Glasses

Quantitative analyses of nonconducting specimens, such as geological
samples, refractory materials, ceramics, and glasses, are generally more
difficult than quantitative analyses of metallurgical samples. Some of the
samples may be porous and thus tend to become filled with polishing slurry.
Others are such poor heat conductors that volatile elements tend to be lost
from the surface during analysis. In addition, surface features may be

difficult to distinguish after the conducting coating has been evaporated onto the surface.

The evaporated conducting layer enables a stable beam at the surface, but it cannot prevent a charge build-up deeper within the specimen. The effect is greatest at high beam energies and with refractory materials of a high breakdown voltage. The resulting electric field within the specimen may result in the redistribution of mobile ions.[27-29] It will also tend to produce a more shallow electron profile (and therefore a more shallow distribution of generated x-ray photons) in the nonconductor compared to a metallic specimen. The effect on quantitative analyses is that the absorption and fluorescence corrections tend to overcorrect. The calculated concentrations for nonconducting specimens measured relative to metal standards will be too great. There is also a probable error in the stopping power factor of the atomic number correction, which is difficult to evaluate. These errors are partially compensated for if nonconducting standards are used. If the composition of samples and standards is not very different, the accuracy may be very good. For example, the writer has obtained very satisfactory results in the analysis of $Pb_x Ba_{1-x} TiO_3$ crystals using $PbTiO_3$ and $BaTiO_3$ standards.[147] However, an analysis of $PbTiO_3$ relative to metallic lead and titanium was very poor; both the lead and titanium concentrations calculated by the procedures in Section 2.3 were several percent too high. Known oxides or minerals are commonly used as standards for geological specimens.

In spite of the preparative and quantitative difficulties, EPMA has become a common and desirable tool in geological studies. The great deal of effort that has gone into the characterization of lunar specimens with the electron probe may be cited as an example. Recent studies of new minerals generally include electron probe analyses and some of these are listed by Mead.[148] Probe characterization is particularly useful for multiphase or inhomogeneous specimens; even qualitative studies can be very helpful. The writer has studied several complex glass systems used for electrical passivation. Susceptibility to cracking under thermal or electrical stress could in each case be related to large compositional gradients within the glass. The concentration gradients could be detected with simple line scans similar to that shown in Figure 13.

### 3.1.7. Biological Specimens

The study of biological materials has been limited by the difficulties arising from the delicate nature of typical biological specimens. Many of the problems in preparation and quantitation are not unsurmountable, however, and some excellent work has been completed in the biological field.

Mineralized materials, such as bones and teeth, have been studied by methods suitable for mineralogical specimens. The major constituents (calcium, phosphorus) have been measured to study the mineralization process. Minor and trace constituents (fluorine, zinc, tin, and others) have been measured in teeth to study the effects of topical fluoride treatments, the reaction with dental amalgam, and the effects of diet.

The analysis of soft (noncalcified) tissues is more difficult because of their low resistance to heat, their highly variable mass density, and the probable low concentration level of the elements to be measured.[149] The very low density of dried tissues leads to a large excited volume at beam energies of ~20 keV, since scattering varies as $\rho z$. Soft x-ray lines and low beam energies are therefore mandatory for thick specimens. An alternate method for preserving spatial resolution is to study thin tissue sections at beam energies of 30 keV and up.

The procedures normally used to prepare thin tissue sections for transmission electron microscopy involve a danger of the displacement or loss of the elements to be studied. The specimen is taken through a series of liquid baths in the course of fixation, embedding, staining, and sectioning. The preferred procedure is to freeze a block of tissue, section it frozen, place the frozen section on a cold support, and dry it by sublimation. Displacement of ions can also occur during analysis as a result of electric fields produced by charge accumulation in the nonconducting tissue.

In typical biological tissues, the dry mass per unit volume varies considerably from one area to another. Thin sections of constant linear thickness may therefore have a very inhomogeneous distribution of mass per unit area. This can result in very abrupt changes of the measured x-ray intensity without any change in the concentration of the measured element. Hall[117] has developed a method for quantitation in thin specimens based on the use of the continuous-spectrum intensity as a measure of mass per unit area. The ratio of the intensities of the characteristic and the continuous-spectrum radiation is a relative measure of the weight fraction of the element in a thin section. The method has been extended to provide absolute values of the weight fractions by the introduction of the physics of x-ray production. Testing of the method with known thin metal films gave an error of approximately ±10%.[150]

The following types of studies have been completed on soft biological tissues: (1) the effect of implant materials on surrounding tissue; (2) the identification of foreign particulate matter in the lungs—asbestos particles, coal dust, paint pigments—and their effect on surrounding tissues; (3) the effect of toxic materials such as the accumulation of lead in rat kidney due to lead poisoning; (4) localized concentrations of specific elements due to metabolic defects—such as high localized copper concentrations in Wilson's disease; (5) the study of body fluid residues; (6) normal and abnormal plant pathology; (7) the testing of the effectiveness of histochemical stains by testing of the spatial correspondence of an element in the stain and an element in the tissue component the stain is intended to display.

## 3.2. Surface Characterization

### 3.2.1. Introduction

The specimens analyzed with the electron probe may be separated into two basic groups. The first type of sample has been cut from a larger specimen, and the surface has been carefully polished. The aim of the analysis is a characterization of the bulk material, even though the analysis is of a 1-$\mu$m-thick surface layer. These analyses have been discussed in Section 3.1. The second type of sample is the as-reacted or as-deposited surface obtained after the specimen has undergone a specific chemical or mechanical process. In this case, the surface is known (or suspected) to be chemically different from the bulk material, and the aim of the analysis is a qualitative or quantitative characterization of the surface. The most common samples in this class involve corrosion, an electrochemical reaction, or industrial processing contamination. Surface roughness may be a problem, and very sharp concentration gradients are often present. The samples are typically very poor candidates for quantitative analysis, and special techniques are required to extract the maximum amount of information from them. These analyses are the subject of this section.

### 3.2.2. Surface Impurities and Failure Analysis

The electron probe is a very good tool for the identification of surface impurities. These impurities are normally a very small fraction of the sample volume and are therefore very difficult to identify by any bulk analytical technique. More often than not, the distribution of the surface impurity is spotty and the sample may have a mottled appearance. The regions free of impurity may then be used for standard or background readings. If the impurity sites are widely spaced, it is helpful if they are visible with the probe microscope or if the sample can be decorated in some way to make them visible. Surface films can normally be seen most readily on oxide films having strong interference colors (such as the $SiO_2$ on silicon of microcircuits) or on highly reflecting metal surfaces.

The surface of high-purity tantalum after an anodic oxidation at 20 V is shown in Figure 14. Most of the sample is covered by an oxide approximately 300 Å thick; this gives the sample a deep purple interference color. The rod-shaped areas in the photomicrograph are bare or covered by a much

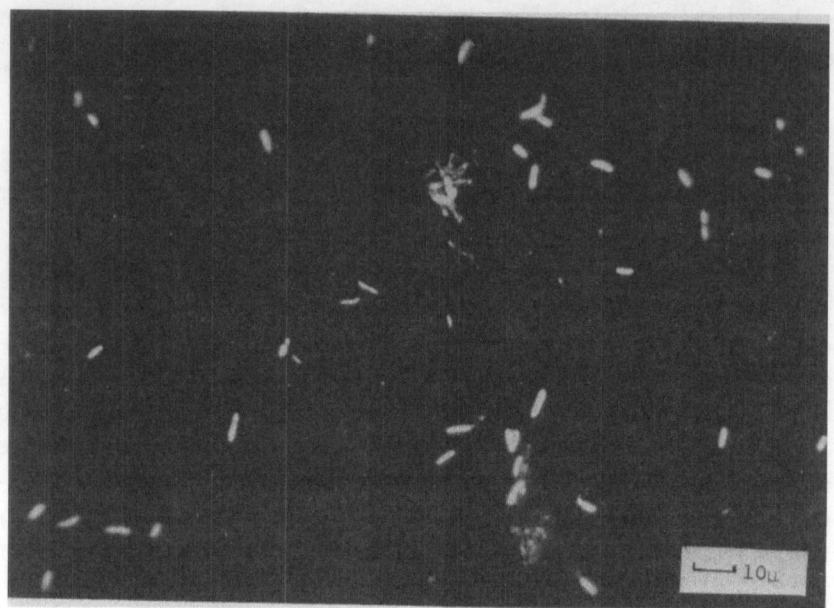

Figure 14. Photomicrograph of tantalum foil surface after anodic oxidation at 20 V. The white rod-shaped areas showed a high surface concentration of iron.

thinner oxide and are lighter in color. The thin spots were all found to contain iron; for this reason, the oxide has not grown properly. In a capacitor, these regions would contribute to a very high leakage current. The iron concentration measured in the rod-shaped spots with a beam energy of 20 keV corresponds to a bulk concentration of 500–1000 ppm. Analyses with a 14-keV beam gave considerably higher iron levels and indicate that the iron impurity is concentrated in a shallow surface layer. The average bulk iron content of this specimen, as measured by emission spectroscopy, is 8 ppm; this concentration is not detectable with the electron probe.

It should be emphasized that the analysis on the tantalum specimen was successful primarily because the spots could be seen on the anodized tantalum. The spots could be checked for a number of likely impurity elements, and iron was found to be the culprit. Without the decoration technique, random searching for several elements would probably have been unsuccessful.

An example of a failure analysis is shown in Figure 15. The sample is a silicon-based microcircuit that has failed electrically. The backscattered electron image reveals that the aluminum interconnect strips are apparently full of holes. Strange raised spheres are also visible. Note how in this case the backscatter image was used to emphasize topography; the aluminum thickness is only 1 $\mu m$; the oxide steps are much thinner. The aluminum x-ray

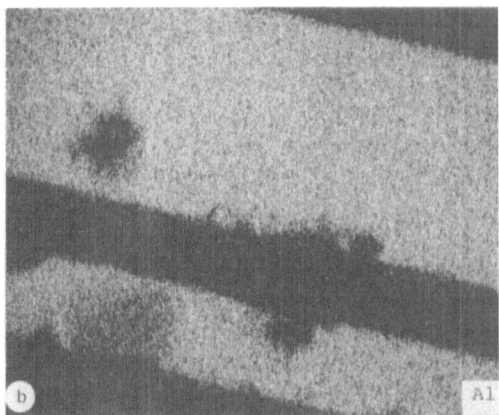

Figure 15. Failure analysis of microcircuit: (a) backscattered electron current image; (b) aluminum x-ray image; (c) backscattered electron current image with chlorine x-ray line scan superimposed.

image confirms the apparent holes in the interconnect; the raised spheres do contain aluminum. Spectral scans on the raised spheres showed aluminum, oxygen, and a trace of chlorine. The last photograph shows that the chlorine is confined to the raised spheres. The interconnect was partially destroyed during a long period of electrical operation by hydrolysis due to the presence of moisture and a trace of chlorine. The spheres are aluminum hydroxide with a trace of chlorine.

A different type of failure is shown in Figure 16. The highly branched tree growth is silver from the end metallization of a multilayer ceramic capacitor. The silver has migrated from the anode (bottom of photomicrograph) toward the cathode over a long period of time at 150°C under the influence of high humidity and an applied potential of 400 V. Eventually, this surface migration caused intermittent shorting of the capacitor. Electron probe microanalysis was required to identify the thin metal deposit as silver. The silver metallization was covered by a tin–lead solder, but no evidence of tin or lead was found in the tree growth. This type of failure was later totally eliminated by a thin moisture-resistant silicone coating on the ceramic.

### 3.2.3. Effect of Topography on Quantitative Surface Analysis

Can a quantitative analysis be performed on a surface of irregular topography? In general, the answer is yes, although the analytical accuracy may be much poorer than for a carefully prepared flat specimen. Many specimens with irregular surfaces also have an inhomogeneous surface composition. The compositional variation may or may not be related to the topography (i.e., hills may be X-rich, valleys X-poor). The problem then becomes one of separating the topographical and compositional variations in the measured signal.

First, a discussion of what is meant by a "flat" or by an "irregular" surface is in order, as these are obviously relative terms. A carefully polished surface will appear nearly featureless when viewed at $1000 \times$ with a metallograph. However, the only requirement for this "featureless" appearance is that the imperfections be smaller than a quarter wavelength of the light used for illumination (i.e., less than 1000 Å). A replica of the same surface may show considerable structure on a 500-Å scale when viewed in a transmission electron microscope. Fortunately, defects less than 1000 Å in diameter have little effect on the x-ray source size that is typically of the order of 1 $\mu$m (10,000 Å) or more in diameter. The incident electron beam may be much smaller, however, and in high-resolution electron probes and scanning electron microscopes small defects will be visible in backscattered electron images. The discussion here is limited to x-ray analysis, and "flat" or "smooth" will henceforth mean "featureless when viewed at high magnification with a good optical microscope."

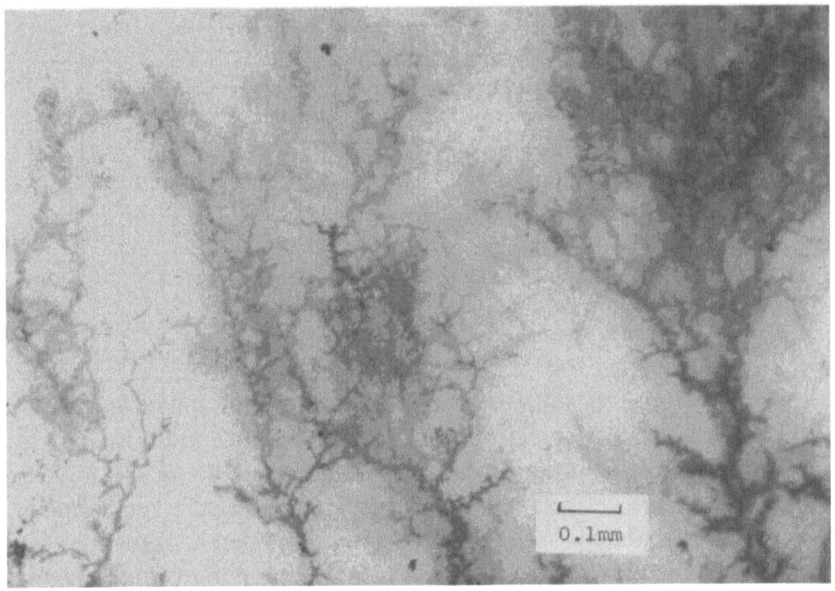

Figure 16. Silver migration under the influence of applied voltage and humidity from the anode (*bottom*) to the cathode of a ceramic capacitor.

As an irregular surface is traversed, a number of negative and positive tilt angles relative to the horizontal plane of a flat specimen are encountered. This affects the x-ray measurement in two ways: (1) the electron beam incidence is no longer normal and (2) the absorption path length becomes longer or shorter because of a change in the effective value of $\psi$. (For nonnormal beam incidence and/or tilted-specimen instruments, the tilt becomes greater or smaller and similar relationships hold.)

The effect of the nonnormal beam incidence is to tilt the electron distribution slightly toward the surface. This increases electron backscattering (thus decreasing $R$ of the atomic number correction) and also gives an unsymmetrical depth distribution of x rays (thus reducing x-ray absorption in the downhill direction and increasing it in the uphill direction). For the approximately $\pm 10°$ range of tilt angles for which this discussion applies rigorously, the change in electron backscattering may be neglected. The absorption change due to nonnormal beam incidence is small and may be compensated for by a slightly larger change in the effective $\psi$.

The larger or smaller effective $\psi$ primarily affeccts the absorption correction. The sample $f(\chi)$ must now be calculated for $\chi = \mu \cos \psi$ above and below the nominal instrumental value. The magnitude of the absorption change depends strongly upon the nominal value of $\psi$ and upon the value of $\mu$ for the specimen, as can be seen from the typical values listed in Table 2.

*Table 2. Typical Changes in f(χ) Due to a Change in ψ of
+5° and −5°*

| | Change[a] in $f(\chi)$ | |
|---|---|---|
| $\mu$ | Nominal $\psi = 15.5°$ | Nominal $\psi = 52.5°$ |
| 100 | +2.6% | +0.2% |
| | −4.7% | −0.3% |
| 2000 | +24.0% | +2.9% |
| | −29.5% | −3.5% |

[a] Evaluated for 20 keV, $\sigma_c = 3500$, $h = 0.1$.

How, then, does one obtain accurate concentrations if the value of $\psi$ is not known? The first step is to collect accurate relative intensity data for each element present in the specimen at a number of points. At any one point, *all* elements must be measured with the *same* linear spectrometer. Different spectrometers "see" the specimen in different directions and would thus have different values of $\psi$ for any one point on an irregular surface. This would make a unique solution impossible.

If the specimen contains $i$ elements, there are $i + 1$ unknown quantities to be determined for each point, namely, $C_1, C_2, \ldots, C_i$ and $\psi$. There are also $i$ equations of the form

$$k_i = \bar{Z}_i A_i F_i C_i \tag{40}$$

and the equation

$$C_1 + C_2, \ldots, C_i = 1.00 \tag{41}$$

Thus, there are $i + 1$ unknowns and $i + 1$ equations allowing a unique solution. The solution is complicated by the fact that the interelement corrections $\bar{Z}_i A_i F_i$ (corresponding to corrections for atomic number, absorption, and fluorescence) also depend on the unknowns.

The situation is not very different from the reiterative matrix calculations for flat samples when $\psi$ is known. The same computer programs can be used, but each point must now be calculated for a number of $\psi$ values (e.g., ±10° nominal $\psi$ in 1° steps). Some of the calculations with the wrong choice of $\psi$ may not converge; thus, a convergence test to stop reiteration may be inadequate. It is most convenient to have the computer always calculate a fixed number of iterations (four is sufficient) and then print out the unnormalized $C_i$ for the last and next-to-last iterations. The double printout serves primarily to indicate nonconvergence. The different format can be achieved with relatively little doctoring of existing programs. For each point on the sample, one will then have a printout of two iterations for

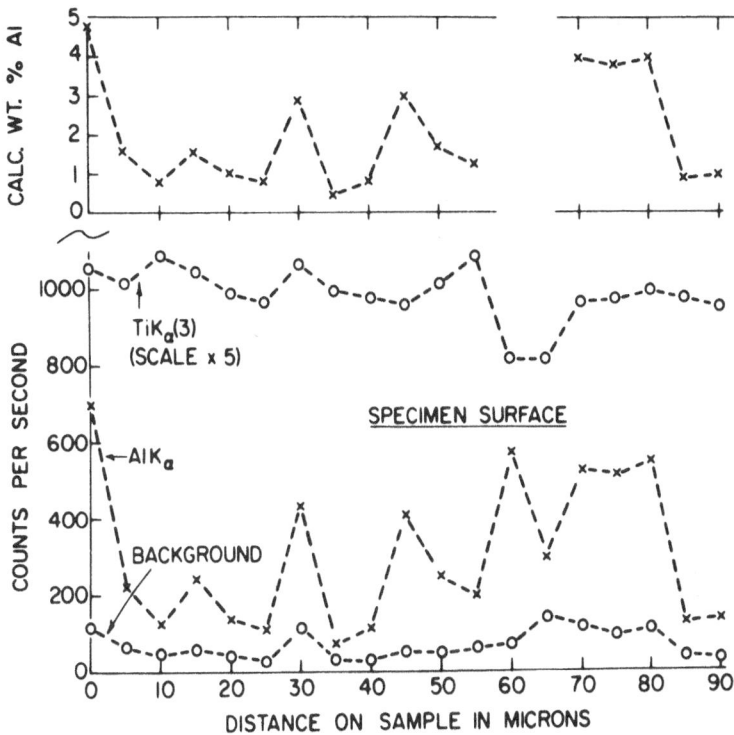

Figure 17. Determination of aluminum concentration on the surface of a rolled sheet of a Ti–Al alloy.

approximately 20 $\psi$ values at 1° intervals. The most nearly correct value of $\psi$ and the corresponding $C_i$ may be selected with the help of the requirement that the sum of the $C_i$ be equal to one.

To determine whether or not the surface composition is homogeneous, a number of points have to be measured and calculated individually by the above procedure. Thus, a 20-point analysis would require 400 separate calculations and 800 printouts; this might severely tax a time-shared computer. An experienced analyst will be able to make an "educated guess" of the approximate value of $\psi$ at each point from inspection of the raw data, because the value of $\psi$ affects primarily the absorption correction. With the educated guesses, the number of necessary calculations for any one point can be reduced from twenty to approximately five unless there are large variations in composition.

The previous discussion has been limited to surface angles of ±10°. This is certainly the limit for low-$\psi$ instruments and for samples containing an element requiring a large absorption correction. Because of the necessity of using the sum of the calculated $C_i$ values to choose the proper $\psi$, a large error in one element will be propagated to the whole analysis. For high-$\psi$

instruments, deviations of as much as $\pm 20°$ from the nominal $\psi$ may be possible with some samples. However, the effect of the tilt on $R$ and on the fluorescence correction can then no longer be neglected. Very rough surfaces, such as intergranular fractures, can very rarely be totally analyzed by this method, but a few suitable areas may be found. In any analysis of this type, the most accurate $C_i$ values will be for effective $\psi$ values close to the nominal instrumental value.

An example of a quantitative analysis on the surface of a rolled sheet is shown in Figure 17. The nominal composition of the alloy was 1Al–99Ti, but it showed an unusual surface enrichment of aluminum that was to be characterized by the analysis. By treating $\psi$ as a variable, the contribution to the $AlK_\alpha$ intensity of topography and aluminum concentration could be separated. Quantitative data were taken with a focused beam at 5-$\mu$m intervals in a direction perpendicular to the rolling marks. The spectrometer used had a nominal $\psi$ of 20° and was fairly sensitive to scattered electrons. The background measurement was therefore an indication of $\psi$. The upper portion of Figure 17 shows that the surface aluminum concentration was enhanced in most measured spots. The aluminum concentration appears to be related to topography in some areas (see 15–35 $\mu$m readings), but not in other areas (70–90 $\mu$m readings). No consistent composition could be calculated for the 60-$\mu$m and 65-$\mu$m readings. These spots apparently correspond to high porosity or a very irregular topography.

### 3.2.4. Shallow Depth Gradients

In the last two sections, the lateral distribution of impurities and major or minor constituents has been considered. The distribution in depth may also be important, and it may be necessary to differentiate between surface and bulk impurities in industrial materials. Surface impurities are typically introduced during manufacture or handling of the part, while bulk impurities may have been present in the starting material. A long-range depth gradient is treated by cross sectioning the specimen and treating it like a lateral gradient. Shallow depth gradients may be characterized by surface analyses at several beam voltages or by analysis of an angle-lapped specimen.

*3.2.4.1. Analysis at Several Beam Voltages.* This method is based on the fact that the depth of the analyzed volume is a function of the electron beam energy. At a low beam voltage, a surface layer will be a much greater fraction of the total analyzed volume than at a high beam voltage. The primary advantage of the method is that it allows the characterization of the depth distribution of an element present on a very small surface area, such as the iron on the tantalum foil shown in Figure 14. Lapping or sectioning of such a sample may be extremely difficult. There are also applications in forensic studies or failure analysis in which the evidence may not be destroyed or distorted by lapping.

If the element of interest has been measured on the sample and on a homogeneous standard at several beam voltages, the depth distribution can be characterized as follows: (1) For a uniform bulk distribution, the calculated concentration (after quantitative corrections) should be constant with increasing voltage. (2) A surface layer should show a rapid decrease in calculated bulk concentration with increasing voltage, but a thin-film calculation should give a constant film thickness. (3) An intermediate case, showing a gradual decrease of the calculated concentration with increasing accelerating voltage, indicates a compositional gradient close to the surface. Examples of such gradients are a bulk constituent concentrated at the surface or a surface impurity extending several microns into the bulk at reduced concentration. A semiquantitative concentration profile with depth can be calculated for such cases from measurements at four or more beam energies. The practical mass depth limit of this method is $2 \times 10^{-3}$ g/cm$^2$.

*3.2.4.2. Analysis of Angle-Lapped Specimens.* Compositional gradients that are several microns deep are easily characterized by the analysis of an angle-lapped sample. The technique is most effective for samples having a depth gradient but little or no lateral gradient. The purpose of the angle lap is to magnify the depth direction; on a 3° lap, approximately 20 $\mu$m laterally corresponds to 1 $\mu$m in depth.

Thin, flat samples, such as crystal platelets or semiconductor chips, are most easily angle-lapped after they are mounted face up on a lapping fixture that determines the angle and minimizes angular distortion. To obtain the correct angle, it is necessary that the two faces of the flat sample be parallel. More irregular samples may be mounted face down by normal metallographic methods. The desired angle is obtained by propping up one side of the specimen with shim stock. For quantitative measurements, a cross section of the angle-lapped specimen may be required to determine or confirm the lap angle. Some samples have a built-in calibration, such as a known oxide thickness or the spacing between internal electrodes. The useful range of lap angles is from 15° to 3°, corresponding to a magnification of 4× to 20× relative to a perpendicular cross section.

The depth resolution of an analysis on an angle-lapped sample is limited by the thickness of the layer in which x-rays are excited. It is desirable to make this effective range as small as possible by using low accelerating voltages and soft x rays for the analysis. The effect of the finite depth of x-ray excitation on intensity profiles obtained from angle-lapped specimens is shown schematically in Figure 18. The upper sketch shows a specimen of material C covered by layers of A, B, and again A. The effective range of x-ray excitation is shown by a dashed line. Note that in this example the effective range is greater than the thickness of B and the lower A layer. The lower portion of Figure 18 gives the A, B, and C relative intensity profiles that would be obtained from the hypothetical specimen. Each lower layer is detected as soon as it is within the effective range, and the x-ray intensity for a thin layer peaks before the layer is visible on the surface. The curve for C gives the broadening of a sharp boundary due to the effective range. This broadening corresponds to the effective range times the magnification factor $1/\sin \alpha$.

For quantitative analyses on angle-lapped specimens, it is necessary to take into account a possible change in the effective x-ray take-off angle, which may increase or decrease from the nominal value depending on the relative orientation of the lap and the spectrometer crystal. A change in $\psi$ will primarily affect the absorption correction. The effect of the slight change in x-ray emission due to nonnormal electron beam incidence is negligible for the lap angles ordinarily used.

The lead concentration on a 3° lap of a flux-grown $Pb_xBa_{1-x}TiO_3$ crystal is shown in Figure 19. The effective depth of $PbM_\alpha$ excitation was $0.3\mu m$; thus, a discontinuous boundary would be broadened to approximately 6 $\mu m$ on the scan. At the lap edge, there is a very slight peak in the lead signal due to a 3° increase in the effective $\psi$. After approximately 12 $\mu m$ ($0.6 \mu m$ in depth), the lead content drops rapidly from 22 to 10 mole %. The drop is very sharp, and the total breadth on the scan can be attributed to effective range broadening. After a gradual decrease to 8 mole %, the lead

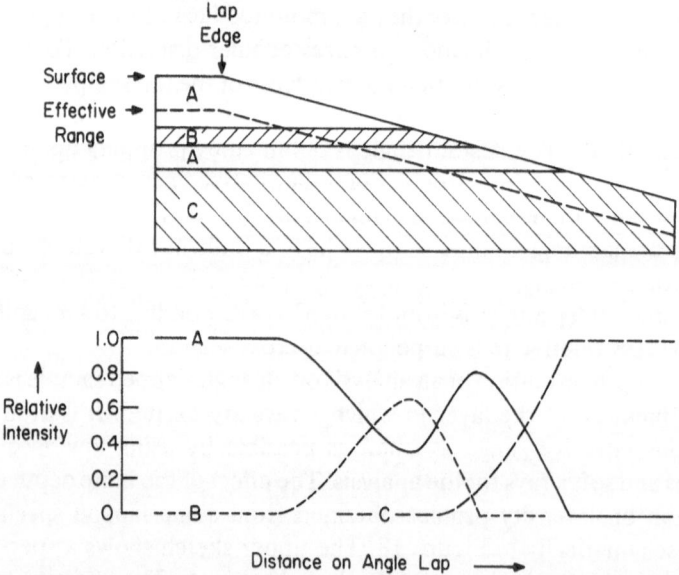

Figure 18. Relative x-ray intensities expected for hypothetical angle-lapped specimen.

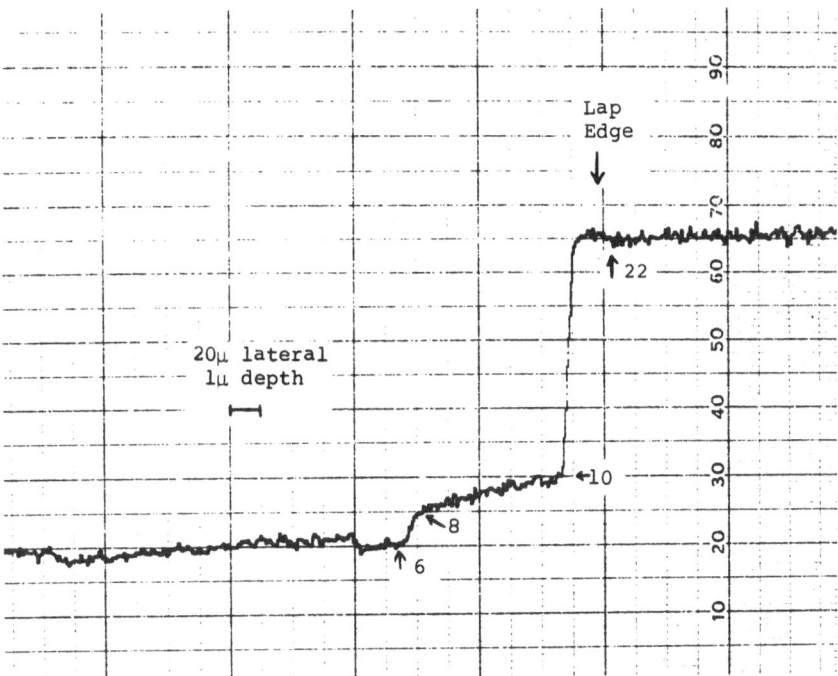

Figure 19. Line scan for lead on a 3° angle lap of a $Pb_x Ba_{1-x} TiO_3$ crystal. The right side of the scan corresponds to the crystal surface. The numbers give the value of $x$ in mole percent lead at the indicated positions in the scan.

concentration again drops suddenly to 6 mole %. The difference between the sharp and gradual drops is rather striking, and the 20× magnification achieved by angle lapping makes quantitative determinations possible. The regions bounded by the sharp concentration gradients correspond to different stages of crystal growth.[147]

### 3.2.5. Analysis of Particles, Powders, and Porous Specimens

Occasionally, a sample to be analyzed is in the form of individual particles or a fine powder. This type of sample may present a mounting problem, as well as an irregular surface. Individual particles can often be mounted by dispersing them on a glass slide and pressing a polished piece of soft metal down on them. The particles will be partially embedded in the metal, and the whole piece can then be handled conveniently and covered with an evaporated conducting coating if necessary. If the particles have some relatively flat sides or facets, these will tend to be the sides originally touching the glass slide and will be up on the mounted particles. An alternate method of mounting individual particles is to drop or shake them onto a substrate coated with a very thin layer of an adhesive. The primary difficulty with this method is that the adhesive tends to flow over the top of the particles, and some trial and error may be necessary. Eastman 910 Adhesive and nitrocellulose in amyl acetate have been used successfully by the writer. All adhesive samples have to be coated with a conducting layer. Large particles (>30 $\mu$m in diameter) can be mounted and sectioned with standard metallographic equipment and a very gentle touch. A small tube or washer surrounding the particle(s) or some other marker is useful for locating the area of interest in a large mount. Fine powers can sometimes be compacted and/or sintered into disks to facilitate handling. The possibility of contamination or loss of volatile constituents should be kept in mind. Some materials cannot be compacted without the addition of an organic binder; the binder can normally be volatilized at a relatively low temperature. Sintered disks tend to be porous but can often be mounted and polished to give relatively flat surfaces.

If a powder specimen can be neither compacted nor sintered without possible compositional change, the following preparation procedure may be used. A small amount of power is mixed with a few drops of a 0.1% solution of nitrocellulose in amyl acetate to make a thin slurry. A drop of this slurry is placed on a substrate and allowed to dry. Successive drops are added, if necessary, to build up a layer of suitable thickness. A sample prepared in this manner will generally contain much less than 1 wt.% nitrocellulose; this is barely enough to tack the powder particles together. The sample will have little physical strength and must be coated with an evaporated conducting layer to prevent charging. It is suggested that the nitrocellulose solutions sold by electron microscopy suppliers for replication be used for mounting; these solutions are of high spectral purity and contain no lumps. The replication solutions should be diluted with reagent grade amyl acetate.

Some of the worst samples from an analytical standpoint that the writer has encountered are glass powders or frits. A photomicrograph of a glass frit sample mounted by the technique described in the last paragraph is shown in Figure 20. The largest particle (upper center) is approximately 40 $\mu$m across; most of the other particles are considerably smaller. The height differences are obvious in the photomicrograph, since many particles are out of focus. Most particles have rough fractured surfaces and there are deep crevices between particles. The extent of the analytical difficulty is shown in Figure 21, which is a line scan for a major constituent of the glass shown in Figure 20. The scan is for $CaK_\alpha$ at 20 keV with $\psi = 18°$; the intensity variations are due largely to topography. The low-intensity portions of the scan correspond to crevices. The radiation emitted at the bottom of a crevice will have to travel through considerable material in the direction of the analyzing crystal, and it is greatly attenuated before it can be measured. Because of the large number of holes and crevices, the "average" intensity reading will be too low. The problem will be less severe for a high-$\psi$

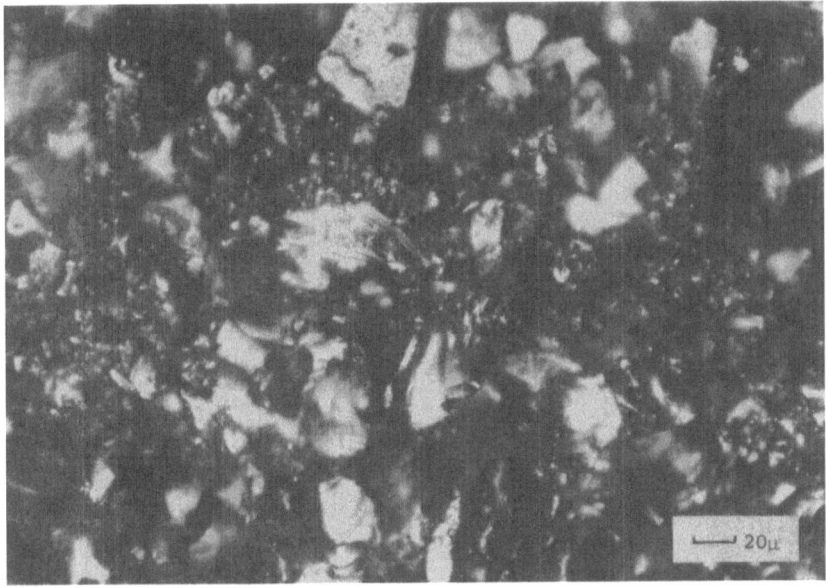

Figure 20. Photomicrograph of glass frit mounted with nitrocellulose.

instrument, but it cannot be eliminated. The only reasonable analytical approach for a sample of this type is to analyze the relatively flat areas by the variable $\psi$ method and hope that these results are representative of the entire specimen.

The holes or pores in porous samples will also give low-intensity readings and should be avoided if possible. The pores in some samples may not be visible with a microscope; such samples can be problematic, because the reason for the low-intensity readings is not apparent. If suspected microporosity is confirmed by another technique, an empirical correction is required to correct the intensity data. This may simply be a normalization to 100 wt.% if the mass absorption constants and atomic numbers of the elements in the specimen are not widely spread.

## 3.3. Thin-Film Characterization

### 3.3.1. Introduction

The electron probe has very good sensitivity for the thickness measurement and compositional analysis of thin films. The primary advantages are that very little material is required for an analysis and that the film need not be removed from the substrate or chemically altered in any way.

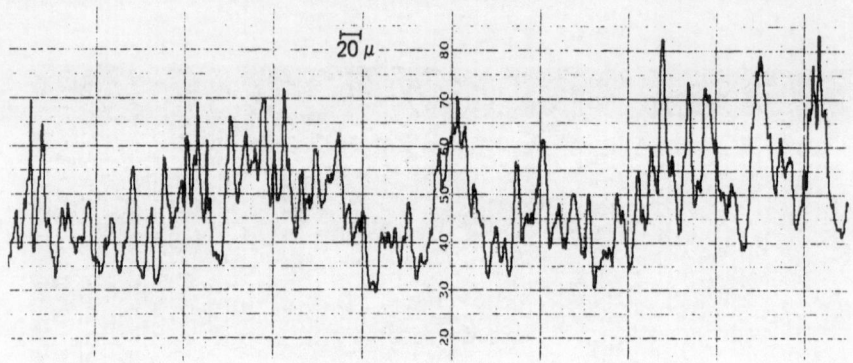

Figure 21. Line scan for CaK$_\alpha$ on specimen shown in Figure 20. Most of the variation is due to topography.

An example of the sensitivity was shown in Figure 4. The possibility of using EPMA as a sensitive method for the determination of film thickness was reported as early as 1960 by Sweeney *et al.*,[151] who constructed empirical calibration curves of measured x-ray intensity versus film thickness. Cockett and Davis[152] suggested that film thickness could be determined from either film or substrate x-ray intensity measurements by numerical or graphical integration of measured $\phi(\rho z)$ curves). Hutchins[153] called attention to the effect of the substrate $Z$ on the x-ray intensity measured for thin films. Several fundamental and semiempirical methods of quantitation have been proposed in recent years,[150,153–161] but additional work is urgently needed. This section details the present state of the art and gives some examples of successful analyses.

The measurement precision of a quantitative thin-film analysis is typically the larger of $\pm 0.05$ $\mu g/cm^2$ or $\pm 1\%$ of the film thickness. The absolute accuracy in mass thickness or weight concentration is not as good because of errors in the conversion factors. The absolute error may be larger than 10% and consists of a small random measurement error and a larger systematic normalization error. The comparison of several films or of several areas on one film can therefore be done more accurately than an absolute determination.

Since electron scattering and energy loss vary as $\rho z$ and not as $z$, mass thickness $\rho z$ in grams per square centimeter is the basic thickness unit for EPMA. The primary dependence of the ionization function $\phi(\rho z)$ (see Figure 7) is on $\rho z$; the atomic number $(Z)$ dependence is only secondary. Because of the great popularity of interferometry, many readers probably have a much better "feel" for linear thickness in angstroms. For this reason, $z$ values are occasionally included in this text; the bulk density $\rho$ is used in the conversion.

## 3.3.2. Substrate Effects

The favorite substrates of the writer are high-purity silicon, pyrolytic graphite, and beryllium. All have a weak continuous spectrum and few spectral lines, can be sliced and polished readily, are electrical and thermal conductors, and are reasonably stable under most deposition and analysis conditions. Microscope slides are poor substrates because they often contain many poorly characterized impurity elements.

A thin-film analysis can be affected by the substrate in several ways. First of all, some of the elements present in the substrate may emit x-ray lines that will interfere with the measurement of x-ray lines emitted by the film. An extreme example is a specimen in which some of the same elements are present in the film and in the substrate; this makes the analysis of the film for these elements nearly impossible. The substrate spectra may also cause excitation of the film elements through secondary fluorescence by spectral lines or the continuum under the conditions detailed in Section 2.3. Both of these effects can normally be eliminated or minimized by careful choice of the substrate and will not be discussed further.

The more fundamental effect of the substrate is that it contributes significantly to the scattering of the incident electron beam. For very thin films, the substrate is a much larger portion of the analyzed sample volume than the film. The substrate, therefore, largely determines the electron penetration and scattering within the specimen. The backscattering of high-energy electrons is of particular importance for thin-film analysis, because each backscattered electron passes through the film a second time in the opposite direction. The following electron paths are possible:

1. The incident electron is transmitted through the film and it eventually comes to rest in the substrate. (Electron passes through film once.)

2. The incident electron is transmitted through the film and backscattered by the substrate. (Electron passes through film twice.)

3. The incident electron is backscattered within the film. (Electron passes through part of film once, part of film twice.)

4. The incident electron is transmitted through the film, backscattered by the substrate, and backscattered again within the film. (Electron passes through part of film once, part of film three times.)

In the limiting case of an extremely thin film, only paths 1 and 2 are possible. Paths 3 and 4 or variations thereof increase in probability with increasing film thickness.

In the remainder of this section, films will be classified somewhat arbitrarily as thick, thin, or very thin films. These designations are not absolute but depend on the relative thickness of the film and the depth of x-ray excitation. The thick, thin, and very thin film classifications are most easily understood with reference to the x-ray ionization functions $\phi(\rho z)$ shown in Figure 7. For 29-keV electrons, any film with a thickness greater than 1500 $\mu$g/cm$^2$ is considered thick and is treated as a bulk sample. (The tails of the ionization curves contribute very little to the total x-ray intensity and can be corrected for by a normalization multiplier.) The thick-film conditions are only useful for the determination of composition. The very-thin-film model applies for thicknesses less than approximately 150 $\mu$g/cm$^2$ for 29-keV electrons. The cutoff for the very-thin-film model is a $\rho z$ value somewhat before the peak of the $\phi(\rho z)$ curve. Both thickness and composition may be determined by the very-thin-film model, which will be discussed in detail. The analysis of films between 150 $\mu$g/cm$^2$ and 1500 $\mu$g/cm$^2$ thick at 29 keV requires a general thin-film model that is mathematically complex and cannot be easily adapted for the determination of composition.

For an extremely thin film ($<100$ Å thick at 20–30 keV), the electron scattering within the film is minimal. The enhancement of the film intensity due to backscattering from the substrate is given by $\phi_0$, the value of the x-ray ionization function at $\rho z = 0$. The value of $\phi_0$ is always greater than unity and is a function of the atomic number of the substrate and of the over-voltage ratio $U_0 = E_0/E_c$ for the thin film x-ray line measured. The substrate $Z$ dependence actually consists of two separate variables—the backscattered electron fraction $\eta$ and the energy distribution of backscattered electrons $d\eta/dW$. The angular distribution of backscattered electrons as they pass through the surface film may also depend on $Z$. The dependence on $U_0$ is due to the form of the ionization cross section of the measured x-ray line

$$Q = \text{const}\,\frac{\ln U}{U^m} = \text{const}\,\frac{\ln WU_0}{(WU_0)^m} \qquad (42)$$

where the exponent $m$ is between 0.7 and 1.0 and $W = E/E_0$.

If an effective mean emergence angle $\alpha$ of the backscattered electrons can be chosen, the value of $\phi$ may be calculated by numerical integration of the following equation:

$$\phi_0 = 1 + \sec\alpha \int_{W_c}^{1} \frac{d\eta}{dW}\,\frac{Q(WU_0)}{Q(U_0)}\,dW \qquad (43)$$

The value of the integral depends on the value of $U_0$ and on the energy distribution of backscattered electrons. It will increase from approximately $0.5\eta$ to $1.3\eta$ as $U_0$ increases from 1.5 to 100. The angle $\alpha$ is commonly taken to be 60° from the surface normal, and $\sec\alpha = 2$. Thus, the effective average film thickness seen by a backscattered electron is twice the actual perpendicular thickness. Numerical integrations of equation (43) for various values of $U_0$ and $Z$ have been completed by Duncumb and Melford.[162]

They used the $d\eta/dW$ data of Bishop,[31,36] $Q \propto 1/U \ln U$, and $\sec \alpha = 2$. The family of calculated curves is shown in Figure 22.

Experimental measurements of $\phi_0$ may be made by two different techniques. The first method is the one originally used by Castaing and Descamps and consists of measuring the x-ray intensity of film F on substrate S and dividing the result by the x-ray intensity of an identical film F supported on a fine grid. Although simple in principle, these measurements are difficult to perform accurately, because the grid-supported films tend to buckle and tear. A second method, originally suggested by Hutchins,[153] avoids the difficulties inherent in the measurement of grid-supported films. A film is evaporated onto a number of pure-element substrates that have been mounted and polished together. The film x-ray intensity is measured on each substrate, plotted as a function of substrate $Z$ and extrapolated to $Z = 0$. All measured intensities are then divided by this extrapolated value to give $\phi_0$ directly. The variable substrate method has two large sources of error that are illustrated in Figure 23: (1) An accurate extrapolation to $Z = 0$ is difficult. For high $U_0$, the curve of $\phi_0$ versus $Z$ is very steep at the

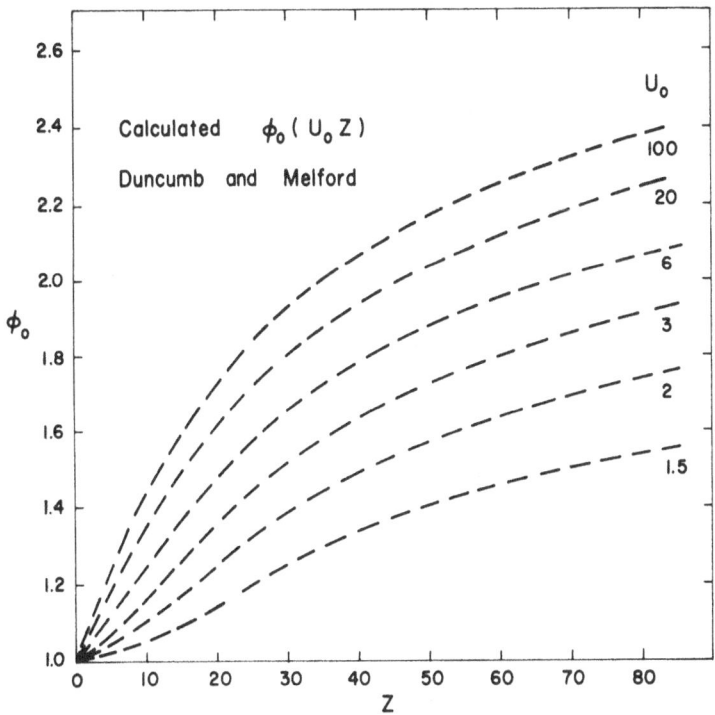

Figure 22. Values of $\phi_0$ calculated by Duncumb and Melford by numerical integration of equation (43).

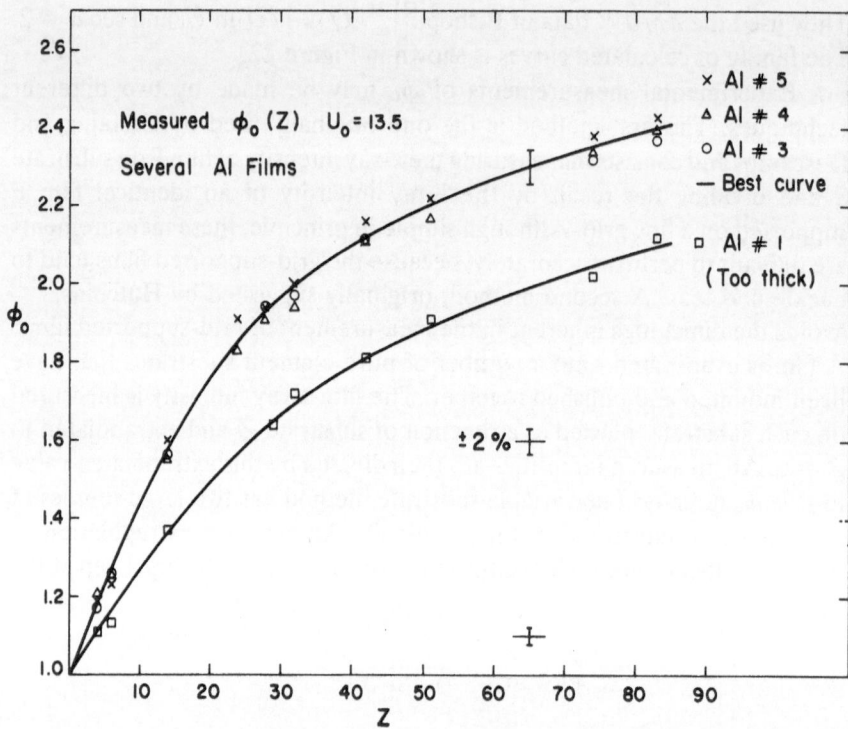

Figure 23. Values of $\phi_0$ obtained for aluminum films by the multiple substrate technique. The upper curve is an average of measurements on three very thin aluminum films; the points for any one film tend to lie above or below the curve due to the error in extrapolating to $Z = 0$. The data for film no. 1 were treated in the same manner and show the effect of scattering by the thick film. Data from Hutchins and Wantman.[163]

low $Z$ end; for low $U_0$, there is an inflection point in the curve below $Z = 10$. (2) The films must be very thin to make scattering by the film negligible. The effect of film thickness is shown in the lower curve of Figure 23. The substrate effect has been diminished and the apparent $\phi_0$ are too low. The correct $\phi_0$ for the thick film would be obtained by raising the lower curve so that it coincides with the upper curve for $Z = 13$.

Reuter[157,158] has measured $\phi_0$ by the multiple substrate technique and has suggested the expression

$$\phi_0 = 1 + 2.8(1 - 0.9/U_0)\eta \qquad (44)$$

as a good fit to his experimental data. The value of $\eta$ versus $Z$ may be read from Figure 5 or may be calculated from Reuter's polynomial fit to Heinrich's 20-keV data,

$$\eta = -0.0254 + 0.016Z - 0.000186Z^2 + (8.3 \times 10^{-7})Z^3 \qquad (45)$$

The $\phi_0$ values calculated from Reuter's equation (44) are always higher than the Duncumb and Melford calculations shown in Figure 22. For $U_0 = 20$, the difference between the curves is within experimental error ($<2\%$), but for $U = 1.5$, the difference is as large as 10%. The discrepancy exists because equation (44) contains no separate dependency upon the energy distribution $d\eta/dW$. This is a valid approximation for $U_0 = 4$ but introduces large errors for $U_0 < 2$. There is an additional problem with equation (44) for low $U_0$, because the $U_0$ dependence becomes too strong and $\phi_0$ approaches $1 + 0.28\eta$ instead of unity as $U_0 \to 1$.

The writer suggests that $\phi_0$ be calculated from equation (44) for all $Z$ values when $U_0 > 4$. For $Z > 30$, equation (44) may be used down to $U_0 = 2$. The Duncumb and Melford curves are likely to be more accurate for $U_0 < 2$ and the combination of $U_0 < 4$ and $Z < 30$. The curves and equation (44) supersede earlier expressions of the form

$$\phi_0 = 1 + (\text{const})\eta \tag{46}$$

that contain no $U_0$ dependence.[153,157]

The value of $\phi_0^*$ for a moderately thick film (e.g., aluminum film no. 1 in Figure 23) should be calculated for an effective value $\eta^*$ that is between the value of $\eta$ for the substrate and the film element. The value of $\eta^*$ can be determined easily if consecutive sample current measurements of the sample and a standard of known $\eta$ are made at the same time the x-ray data are taken. From the definition of $\eta$ in equation (12) and the fact that the beam current $i_b$ is the same for the two measurements, it follows that

$$i_s^*/i_s = (1 - \eta^*)/(1 - \eta) \tag{47}$$

With reasonably stable equipment, $\eta^*$ can be measured routinely to a precision of $\pm 0.01$.

### 3.3.3. Determination of Thickness

The x-ray intensity measured for a film F of thickness $\rho z$ on a substrate S and normalized to the emitted intensity from a bulk standard of the film element can be expressed as

$$k_F = \frac{I_{FS}}{I_{Bulk}} = \frac{\int_0^{\rho z} \phi^*(\rho z)\exp(-\chi \rho z)\,d(\rho z)}{\int_0^\infty \phi(\rho z)\exp(-\chi \rho z)\,d(\rho z)} \tag{48}$$

The asterisk is used to indicate the required modification of the ionization function due to the difference in the backscatter properties of the substrate and the film material. The denominator of equation (48) is a function of $E_c$, $E_0$, $Z$, and $\chi$ and will be constant for a particular element measured under specific conditions. The value of the integral in the numerator will increase systematically with $\rho z$. The intensity ratio $k_F$ can typically be measured to a precision of 1%. What is needed is an accurate conversion from $k_F$ to $\rho z$. From the form of equation (48), it is apparent that an accurate conversion is not to be achieved easily. Several approaches will be considered in this section.

*3.3.3.1. Empirical Methods.* If only a few film elements are to be measured routinely on one or two substrates, it may be easiest to construct empirical calibration curves with films of known thickness. The conversion accuracy will then depend upon the accuracy with which the absolute thickness of the standard films can be determined by other techniques. For this standard thickness measurement, it is desirable to determine $\rho z$ by chemical analysis on a separate control sample prepared simultaneously with the microprobe sample. Marshall and Hall claim a 2% accuracy for colorimetric chemical weighing of thin films.[150] Interferometric measurements are not as reliable, since they only determine $z$, and thin films are not necessarily of bulk density.

After the standard films have been prepared and chemically weighed, $k_F$ may be measured with the electron probe for a number of different characteristic lines and beam energies. Calibration curves can then be drawn for the appropriate analytical conditions. These curves will be nearly linear for small $\rho z$, with the slope dependent upon $E_0$, $E_c$, and $\chi$. For somewhat thicker films, the curves will show a positive or negative deviation from linearity.

Films of known thickness are also required to test the accuracy and validity of the more complex calculations to be discussed subsequently. Thus, the accuracy to which the absolute thickness of standard films can be determined by other methods is a limitation to the ultimate accuracy obtainable by electron probe analysis even with the more complex models.

*3.3.3.2. Very-Thin-Film Model.* This model is valid for $\rho z$ somewhat less than the peak of the $\phi(\rho z)$ curve or approximately one-tenth of the effective electron range. The model requires two approximations, and is an updated version of a previous method by Hutchins.[153]

The first approximation is that the substrate and film dependence can be separated:

$$I_{FS} = \phi_0^* I_F \qquad (49)$$

The value of $\phi_0^*$ is calculated from equation (44) and the measured $\eta^*$ as detailed in the previous section. The intensity $I_F$ corresponds to the intensity measured on an *unsupported* thin film of thickness $\rho z$, and

$$I_F \propto \int_0^{\rho z} \phi_F(\rho z) \exp(-\chi \rho z)\, d(\rho z) \qquad (50)$$

The value of $\phi_F \to 1$ as $\rho z \to 0$ by definition.

Table 3. Values of the Electron Scattering
Parameter B

|            | B     |        |
|------------|-------|--------|
| Beam energy | Low $Z$ | High $Z$ |
| 10 keV | 9000 | 12000 |
| 20 keV | 4000 | 5000 |
| 30 keV | 2000 | 3000 |

The second approximation is that $\phi_F$ can be given by

$$\phi_F(\rho z) = 1 + B\rho z, \qquad \rho z \ll \text{total range} \tag{51}$$

and that $B$ can be treated as a constant for any specific combination of $E_0$, $E_c$, and film $Z$. The parameter $B$ is a measure of electron scattering within the film material. Its value increases rapidly with decreasing $E_0$ and more slowly with increasing film $Z$. Typical values for $B$ are given in Table 3. These values may be computed from the initial slope of $\phi(\rho z)$ curves after a substrate correction or from experimental electron scattering data for thin films. A $\pm 10\%$ accuracy in $B$ is sufficient, since $B\rho z < 1$.

After substitution of equation (51) in equation (50), the integral may be evaluated directly. Since the value of $\chi\rho z$ is less than 1 in the valid thickness range for the model, the exponential factor may be replaced by the first few terms of a rapidly converging series in $\rho z$. This gives

$$
\begin{aligned}
I_F \propto \rho z &+ (\rho z)^2 \left( \frac{B}{2} - \frac{\chi}{2} \right) - (\rho z)^3 \left( \frac{B\chi}{3} - \frac{\chi^2}{6} \right) \\
&+ \cdots + (-1)^n (\rho z)^n \left[ \frac{B\chi^{n-2}}{(n-1)!} - \frac{B\chi^{n-2}}{n!} - \frac{\chi^{n-1}}{n!} \right]
\end{aligned} \tag{52}
$$

Equation (52) predicts a positive or negative deviation from linearity of $I_F$ with $\rho z$ as the film thickness increases. If $B \approx \chi$, $I_F$ will increase nearly linearly with $\rho z$. For $B \neq \chi$, the $(\rho z)^2$ term may be as much as 20% of $\rho z$ and should always be included. The $(\rho z)^3$ term will be less than 1% of $\rho z$ under all conditions for which the approximations are justified; higher-order terms are never required.

In order to make absolute measurements, the bulk normalization has to be included by substitution of equations (52) and (49) in equation (48):

$$
k_F = \frac{I_{FS}}{I_{bulk}} = \frac{\phi_0^* \left[ \rho z + (\rho z)^2 \left( \frac{B}{2} - \frac{\chi}{2} \right) - (\rho z)^3 \left( \frac{B\chi}{3} - \frac{\chi}{6} \right) \right]}{\int_0^\infty \phi(\rho z) \exp(-\chi\rho z) \, d(\rho z)} \tag{53}
$$

The factor in the denominator can be evaluated most accurately from standard films of known $\rho z$. The error in the bulk normalization will then depend primarily on the $\rho z$ error of the standard film.

Since the integral

$$\int_0^\infty \phi(\rho z) \exp(-\chi \rho z) \, d(\rho z) = f(\chi) \int_0^\infty \phi(\rho z) \, d(\rho z) \tag{54}$$

by definition, the bulk normalization may also be determined by calculation of $f(\chi)$ from equation (26) or (27) and graphical or numerical integration of experimental $\phi(\rho z)$ curves. This approach requires no standard films but generally results in larger errors. Voltage series of $\phi(\rho z)$ curves have been determined experimentally for $MgK_\alpha$ in aluminum, $VK_\alpha$ in titanium, $ZnK_\alpha$ in copper, and $BiL_\alpha$ in lead. These curves may be interpolated for other beam voltages and may also be used for adjacent elements. However, there are large gaps in film elements and x-ray lines that cannot be interpolated. Examples are all M spectra and the soft L spectra in the atomic number range 38 through 53. The bulk normalization for these elements can only be done with standard films. Another problem is that $\phi(\rho z)$ curves determined for one instrument geometry are not necessarily valid for other instrument geometries. Brown and Parobek[164] have recently measured the same trace layer samples with several instruments and have demonstrated a dependence on the angle of electron beam incidence.

The form of equation (53) used for a thickness determination is

$$\rho z = \frac{k_F \int_0^\infty \phi(\rho z) \exp(-\chi \rho z) \, d(\rho z)}{\phi_0^* \left[ 1 + \rho z \left( \frac{B}{2} - \frac{\chi}{2} \right) - (\rho z)^2 \left( \frac{B\chi}{3} - \frac{\chi^2}{6} \right) \right]} \tag{55}$$

Successive approximations are used to calculate the unknown $\rho z$; three iterations are generally sufficient. The modification of equation (55) for multielement films is discussed in Section 3.3.4.

The primary limitation of the very-thin-film model is the thickness restriction. The practical upper limit is $150 \ \mu g/cm^2$ (i.e., $1500 \ \text{Å}$ for $\rho = 10$) for 30 keV and considerably less for lower beam energies. Equation (55) is most accurate for extremely thin films ($< 100 \ \text{Å}$), and for these films the errors are due primarily to the bulk normalization. For thicker films, uncertainties in $\phi_0^*$ and $B$ also contribute to the absolute error.

*3.3.3.3. General Thin-Film Models.* Let us take another look at equation (48). If it were possible to develop an expression for $\phi$ and $\phi^*$, $k_F$ versus $\rho z$ could be calculated by numerical integration. This approach would be completely general and would not be restricted to very thin films.

An expression for $\phi(\rho z)$ was derived by Philibert in connection with his treatment of the absorption correction.[81] The Philibert model has two basic flaws: (1) the assumption that $Q = \text{const}$ and (2) an exponential electron transmission model. In the intervening years, considerable experimental data have been generated concerning the propagation of relatively slow 5–30 keV electrons through solid films. These data have given a much better insight into the scattering of electrons in the electron probe energy range. They have also proven many of the expressions extrapolated from high-energy electron measurements to be incorrect. For example, it has been determined that the electron fraction transmitted through a thin film decreases linearly (rather than exponentially) with film thickness for a considerable fraction of the total electron range. At least half of the backscattered electrons are produced by single scattering events, and the backscattered electron fraction increases linearly with film thickness until it reaches approximately half of the bulk material value. Reuter[158] has utilized some of the new data in an expression for $\phi(\rho z)$. He has also applied numerical integration of $\phi(\rho z)$ to the thickness measurement of thin films and has tested this approach with a number of known standard films. The overall absolute accuracy of approximately 10% found by Reuter is somewhat disappointing. However, his measurements included extremes in $E_c$, $E_0$, and film and substrate $Z$ that would not generally be encountered in practice. The program apparently undercorrects for the effects of the substrate; results are likely to be most accurate if $\eta$ for the film and substrate elements are not very different. The Reuter method requires computer calculation and is not easily modified for compositional analysis of films containing elements with a large spread in atomic number.

A graphical method for hand calculation has been suggested by Bishop and Poole.[159] The shape and magnitude of $\phi(\rho z)$ curves varies with $E_0$, $U$, and $Z$; the strongest dependence is on $E_0$. Bishop and Poole suggest that $E_0$ can be eliminated as a separate variable if one exploits the scaling properties of electron scattering distributions and expresses all thicknesses as a fraction of the Bethe range $R_B$. Curves of $\phi(\rho z)$ calculated for $E_0 = 30$ keV and 10 keV for the same $U$ and $Z$ nearly coincide when plotted versus $\rho z / R_B$ rather than $\rho z$. The Bethe range is obtained by integration

$$R_B = \int_0^{E_0} (1/S)\, dE \tag{56}$$

where $S$ is the Bethe stopping power, given by equation (9). It should be noted that the Bethe range gives the length of the electron path $\rho s$ and not the perpendicular thickness $\rho z$; $\rho s$ is considerably longer than $\rho z$, especially for high $Z$.

Bishop and Poole plot families of curves of the normalized integral distribution function $p(t)$ as a function of $1/U$ and $\rho z/R_B$ for a specific $Z$.

$$p(t) = \frac{\int_0^t \phi(\rho z)\, d(\rho z)}{\int_0^\infty \phi(\rho z)\, d(\rho z)} \qquad (57)$$

where $t$ is the mass thickness of the film. If absorption can be neglected, $p(t) = k$, but a simple method is given for an approximate absorption correction. The curve for the nearest $Z$ is used to determine $\rho z/R_B$ from the measured relative intensity $k$ and the known $1/U$. This value of $\rho z/R_B$ is then multiplied by $R_B$ determined from a graph of $R_B$ versus $E_0$ for specific values of $Z$ to give the film mass thickness $t$ in grams per square centimeter. The curves may also be used to determine $t$ by measurement of the substrate radiation and to determine the composition of an alloy film. If the film and substrate $Z$ are very different, the curves for an intermediate $Z$ should be used. The value of the interpolated $Z$ varies with film thickness and depends on whether the film or substrate dominates the electron scattering.

The Bishop and Poole curves were calculated from earlier Monte Carlo calculations by Bishop.[57] They tend to underestimate the film thickness by approximately 15%; this error is possibly due to the simplified scattering cross sections used in the Monte Carlo calculation. Better curves from a more detailed Monte Carlo calculation would be highly desirable, but the present ones are good enough to illustrate the potential of the graphical method.

### 3.3.4. Determination of Composition

The composition of thin films can be calculated with relative ease only for thick-film and very-thin-film conditions. X-ray intensities can be measured equally well for intermediate films, but the conversion to absolute weight concentration becomes very difficult.

*3.3.4.1. Thick Films.* These films are treated as bulk samples and are measured at relatively low electron beam energies (typically 5–10 keV). The film thickness should be $100\ \mu g/cm^2$ or greater for this approach to be successful. Only very soft x-ray lines can be excited efficiently at these low beam energies; the uncertainties in soft x-ray mass absorption coefficients may be a limitation in some cases.

*3.3.4.2. Very Thin Films.* The composition of very thin films can be determined by a procedure that is an extension of the thickness determination. As a first approximation, $(\rho z)_X$ and $(\rho z)_Y$ can be determined separately from equation (55) as detailed in Section 3.3.3. For an XY alloy, the weight concentration of element X is given by

$$C_X = \frac{(\rho z)_X}{(\rho z)_X + (\rho z)_Y} \tag{58}$$

This approximation ignores interelement effects but is a good approximation for very thin films.

For a more exact solution, electron scattering and x-ray absorption in the total film have to be considered. The interelement effects can be treated rigorously without approximations if the quantities of all film elements are expressed in units of $\rho z$. Djurić and Cerović[154] used linear thicknesses for individual elements and a mean density $\bar{\rho}$ for the film in their adaptation of equation (55) for compositional analysis. Their equations are therefore not completely general.

For a one-element film, the integral leading to the series in equation (52) is

$$I_F \propto \int_0^{\rho z} (1 + B\rho z) \exp(-\chi \rho z)\, d(\rho z) \tag{59}$$

The corresponding expression for element X in a two-element film XY is

$$I_X \propto C_X \int_0^{(\rho z)_T} (1 + B_{XY}\rho z) \exp(-\chi_{X \text{ in } XY}\, \rho z)\, d(\rho z) \tag{60}$$

where $(\rho z)_T = (\rho z)_X + (\rho z)_Y$ and $C_X$ is given by equation (58). The expression corresponding to equation (55) for element X in XY then becomes

$$(\rho z)_X = \frac{k_X \int_0^\infty \phi_X(\rho z) \exp(-\chi \rho z)\, d(\rho z)}{\phi_{0X}^* \left[ 1 + (\rho z)_X \left( \dfrac{B_X}{2} - \dfrac{\chi_{X \text{ in } X}}{2} \right) + \dfrac{(\rho z)_Y^2}{(\rho z)_X} \left( \dfrac{B_Y}{2} - \dfrac{\chi_{X \text{ in } Y}}{2} \right) \right]} \tag{61}$$

The third-order terms have been dropped in equation (61). The only new term for the two-element film is the $(\rho z)_Y^2/(\rho z)_X$ term in the denominator. Initial values for $(\rho z)_X$ and $(\rho z)_Y$ are calculated without the mixed terms; final results are obtained by successive approximations for both X and Y. The presence of additional elements in the film leads to additional mixed terms. Both $\phi_0^*$ and the bulk normalization integral are generally different for different elements in the same film. Equation (61) contains an x-ray absorption correction and an atomic number correction (in $\phi_0^*$ and the $B$ terms). No provision has been a made for a fluorescence correction, since this correction is likely to be very small.

*3.3.4.3. Very Thin Unsupported Films.* Two methods have been developed for the compositional determination of very thin films without a substrate such as biological tissue sections or extraction replicas. In both methods, the thickness is not determined and need not be known, but the films must be very thin.

The method developed by Hall[117] and Marshall and Hall[150] primarily for biological thin sections utilizes the x-ray intensity from a specific energy band of the continuous spectrum as a measure of film thickness. The intensity ratio of characteristic to continuous radiation is constant over a considerable thickness range. This constant has to be determined for each $Z$ and $E_0$ of interest by calibration with pure element films or films of known composition. In unknown specimens, the concentration is then given by the measured ratio times the calibration constant. The method is most useful for specimens in which all elements cannot be determined and/or specimens that have large localized variations in mass thickness.

Tixier and Philibert[156] have developed a method for the analysis of unsupported metallic films or extraction replicas. The film intensities are measured relative to bulk pure-element standards to give the intensity ratios $k_X$ and $k_Y$. A correction is made for x-ray absorption and atomic number for the pure element standards, while the film intensities are left uncorrected. The weight ratio of the two elements is given by

$$\frac{C_X}{C_Y} = \frac{k_X}{k_Y} \frac{f(\chi)_X}{f(\chi)_Y} \frac{P_Y}{P_X} \tag{62}$$

where

$$P_X = \frac{\ln U_{0X}}{E_{cX}} \frac{1}{R_X/S'_X}$$

where $1/S'_X$ is the energy integral over $Q/S$. The method can be extended to more than two elements by measuring all film elements relative to one film element.

### 3.3.5. Examples of Thin-Film Analyses

In this section, a few of the thin-film analyses performed in the author's laboratory are summarized to indicate the possibilities. Electron probe analysis is often used in conjunction with other techniques for basic thin-film studies. If the substrate and film contain some of the same elements, the film composition must be determined by another technique (e.g., reflection electron diffraction), unless the films can be grown thick enough for a low-electron-kilovolt bulk analysis. The composition, in turn, must be known to calculate the mixed terms in equation (60) and obtain the highest-thickness accuracy with the electron probe.

*3.3.5.1. Very Thin Nickel–Chromium Films.* Sputtered nickel–chromium films were prepared in a wide compositional range by sputtering from alloy targets and also by sputtering simultaneously from two pure-element targets. The composition of the alloy sputtering targets was determined by bulk electron probe microanalysis. Films sputtered from these targets onto polished silicon slices were found to be of the same composition within experimental error.[145] For films with $\rho z$ approximately 6 $\mu g/cm^2$ (~75 Å), the measurement error in both mass thickness and composition was less than 2% in all cases. The absolute errors were somewhat higher because of errors in the conversion factors.

The composition of nickel–chromium films evaporated from a large volume of melt was studied as a function of melt composition and temperature. The evaporated films all had a much higher chromium concentration than the melts. This chromium enrichment was found to vary systematically with evaporation temperature.

*3.3.5.2. Thick Films of Barium Titanate.* Barium titanate films approximately 5000 Å thick were radiofrequency-sputtered from a stoichiometric target onto polished silicon slices. Some of the deposition variables were the substrate temperature, silicon resistivity, the sputtering gas mixture, and the sputtering power. Film stoichiometry as a function of these variables was measured with the electron probe; the Ti:Ba atomic ratio was found to vary from approximately 0.99 to 1.22. X-ray diffraction, electron microscopy, and reflection electron diffraction were required in addition to electron probe microanalysis to sort out the effects of all the deposition variables. It was eventually concluded that stoichiometric $BaTiO_3$ could be deposited only if the sputtered film was crystalline as deposited. Amorphous films were always titanium-rich, and some dissociated into two phases upon annealing.[165]

*3.3.5.3. Growth Rate of Pd₂Si films.* In one study,[166] the diffusion-controlled growth rate of $Pd_2Si$ films was determined with the electron probe. Palladium films ~900 Å thick were deposited on polished (111) silicon substrates. Samples were annealed in nitrogen at 125, 200, and 250°C for predetermined time periods such that the $Pd_2Si$ phase boundary never reached the palladium surface. The unreacted palladium surface layer was etched off with a $KI + I_2$ solution that does not attack $Pd_2Si$. The thickness of the $Pd_2Si$ formed during the annealing cycle was determined with equation (61) by measurement of $PdL_\alpha$ relative intensity ratios. The composition of $Pd_2Si$ was assumed, since only this phase was detected by reflection electron diffraction. The thickness measurements are given in Figure 24. The growth of the $Pd_2Si$ phase follows a parabolic law:

$$z^2 = k(T)t \tag{63}$$

where $z$ is the linear film thickness, $k(T)$ is the temperature-dependent velocity constant, and $t$ is the annealing time. The dependence of $z$ on $t^{0.5}$ is evident in Figure 24. The velocity constant is

$$k(T) = k_0 \exp(-Q/RT) \tag{64}$$

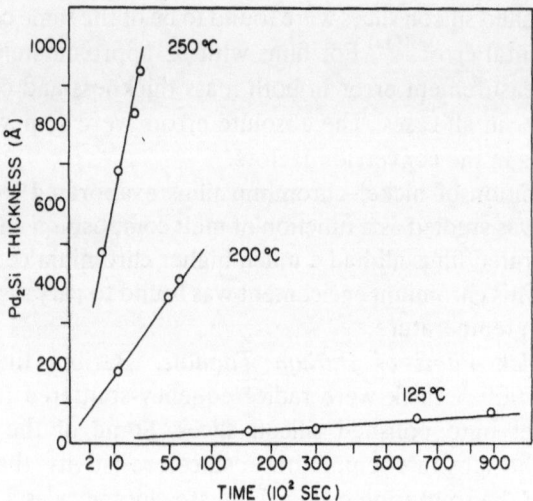

Figure 24. Pd$_2$Si thickness measurements to determine diffusion-controlled growth constant. Time is plotted on a square root scale to show the good fit of equation (63).

where $Q$ is the activation energy. A least-squares fit of the data gave

$$k(T) = (7 \times 10^{-2}) \exp\left(\frac{-29.2 \pm 0.8}{RT}\right) \quad \left(\frac{cm^2}{sec}\right)$$

The activation energy determined by a different method was $29.3 \pm 0.7$ kcal/mole.

*3.3.5.4. Composition of Si$_3$N$_4$ Films.* Silicon nitride films prepared by reactive sputtering from a silicon source[167] or radiofrequency sputtering from a Si$_3$N$_4$ source showed a tendency toward nonstoichiometry. The presence of silicon and/or SiO$_2$ in the film was determined by measuring the intensity-versus-wavelength profile of the SiK$_\beta$ emission band. This x-ray line is broad and shows a marked dependence on chemical combination. Profiles for silicon and Si$_3$N$_4$ measured with a relatively low-resolution spectrometer are shown in Figure 25; the profile for SiO$_2$ is shifted to the right of the Si$_3$N$_4$ profile by about the same amount that the silicon profile is shifted to the left. The shift in the Si$_3$N$_4$ peak due to an excess silicon concentration of 3 wt.% is also shown in Figure 25. This was the lowest concentration of excess silicon that could be reliably detected by the method.

*3.3.5.5. Etch Rate Studies.* Mass thickness measurements may be used to monitor various chemical reaction rates. Very often, it is only necessary to determine the shape of the rate curve (linear, exponential, asymptotic, etc.), and this can be done with relative data. An example is the testing and optimization of a preferential etchant that will remove a surface layer of material X rapidly but will attack the underlying layer Y only at a much slower rate or not at all. To be useful, a preferential etchant has to have an asymptotic rate curve with a relatively long period, during which all X but no Y has been removed. The analysis can be done most easily but cutting the initial specimen into a number of small pieces (the writer typically uses $\frac{1}{4} \times \frac{1}{8}$ in. rectangles). These pieces can then be etched or reacted individually, mounted together, and analyzed successively.

An example of an etching study is shown in Figure 26. The specimens consisted of pyrolytic graphite substrates onto which a layer of $SiO_2$ and a layer of aluminum had been electron-beam evaporated without breaking vacuum between evaporations. Some of the specimens were annealed in nitrogen at various temperatures to study the reaction between $SiO_2$ and aluminum. The untreated specimens corresponded to an Al : Si weight ratio of 0.66. An etchant consisting of a mixture of phosphoric, acetic, and nitric acids was used to measure the degree of reaction of the aluminum with the $SiO_2$. All except a monoatomic layer of aluminum could be removed from a freshly prepared specimen by etching for 4 min. The atomic layer apparently reacted during deposition and could not be removed with an etching time of

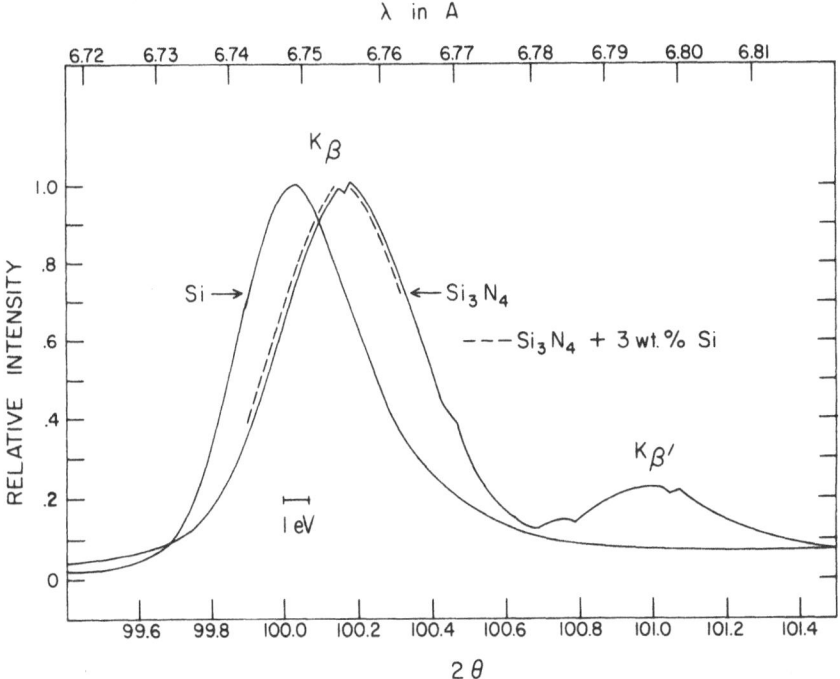

Figure 25. Intensity-versus-wavelength profiles of $SiK_\beta$ band.

4 hr. After a specimen had been stored at room temperature for 2 months, the single atomic layer level of aluminum was reached only after an etching time of 8 min. A specimen heated at 344°C for 170 hr showed a much more gradual etching rate, with an asymptotic decrease to four atomic layers of aluminum. A specimen annealed for 2 hr at 527°C consisted of two phases, which are shown in Figure 27. The dendritic phase was bright blue and had an Al:Si weight ratio of approximately 0.25. The absolute mass thickness of aluminum in these regions was lower than in the matrix, as is shown in the aluminum x-ray image, but the mass thickness of silicon was considerably higher. Thus, the lateral movement of both silicon and aluminum produced this phase. The dendritic phase was not attacked by the etchant. A separate dilute hydrofluoric acid etch series showed that the dendritic phase had the

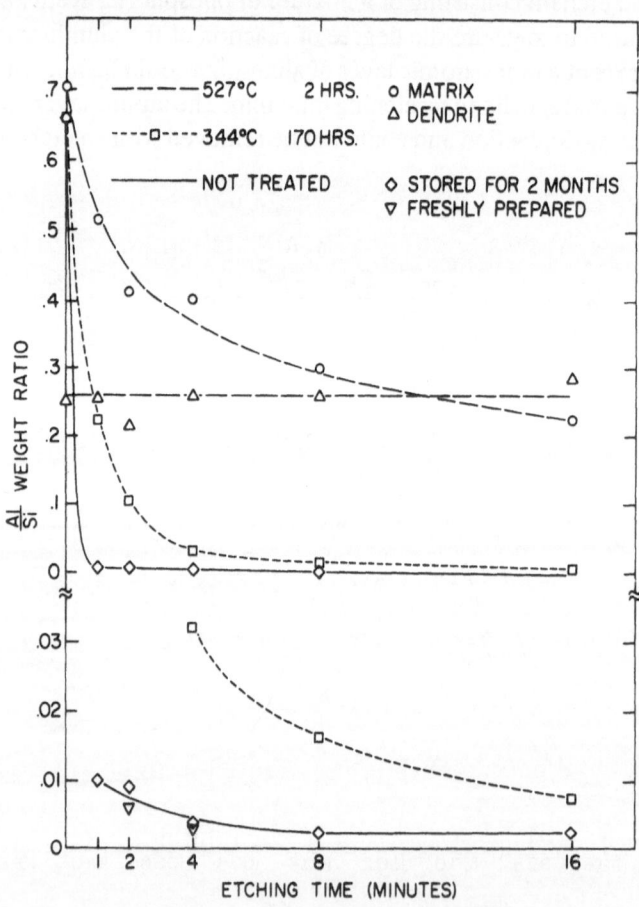

Figure 26. Etching studies on $SiO_2$ + Al films.

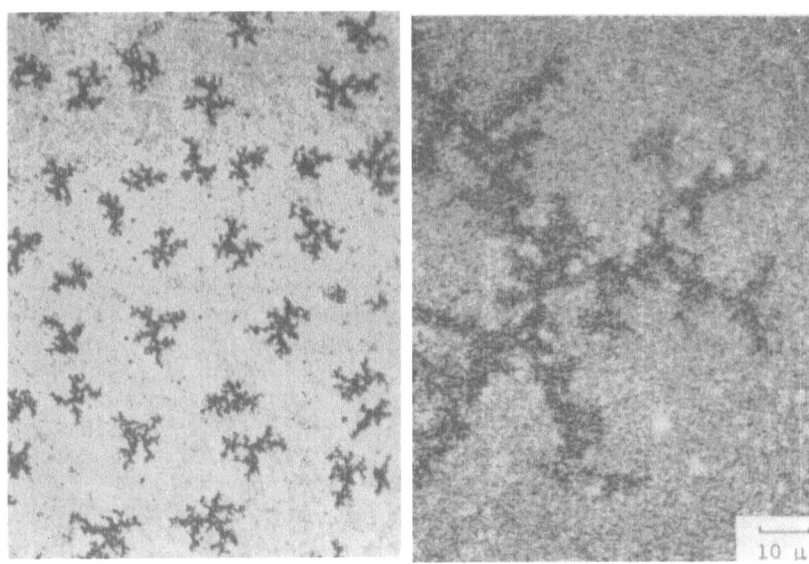

Figure 27. Dendritic second phase in $SiO_2$ + Al film annealed for 2 hr at 527°C; (*left*) light micrograph, (*right*) aluminum x-ray image.

identical Al:Si ratio from surface to substrate. The matrix of the 527°C specimen showed a very gradual decrease to an Al:Si ratio of approximately 0.1.

## *References*

1. D. R. Beaman and J. A. Isasi, *Electron Beam Microanalysis*, STP506, American Society for Testing and Materials, Philadelphia (1972).
2. P. Duncumb, *J. Phys. E* **2**, 553 (1969).
3. C. W. Oatley, *J. Phys. E* **8**, 1037 (1975).
4. A. N. Broers, *J. Appl. Phys.* **38**, 1991 (1969).
5. J. D. Verhoeven and E. D. Gibson, *J. Phys. E* **9**, 65 (1976).
6. A. V. Crewe, D. N. Eggeberger, J. Wall, and L. M. Welter, *Rev. Sci. Instr.* **39**, 576 (1968).
7. R. Bassett and T. Mulvey, in *Fifth Int. Congr. X-Ray Optics and Microanal.* (G. Möllenstedt and K. H. Gaukler, eds.), p. 224, Springer, Berlin (1969).
8. D. M. Poole and P. M. Martin, *Met. Rev.* **14**, 61 (1969).
9. E. W. White and G. G. Johnson, *X-Ray Emission and Absorption Wavelengths and Two-Theta Tables*, ASTM DS 37A, American Society for Testing and Materials, Philadelphia (1970).
10. G. G. Johnson and E. W. White, *X-Ray Emission Wavelengths and keV Tables for Nondiffractive Analysis*, ASTM DS 46, American Society for Testing and Materials, Philadelphia (1970).
11. N. G. Ware and S. J. B. Reed, *J. Phys. E* **6**, 286 (1973).

12. C. E. Fiori, R. L. Myklebust, K. F. J. Heinrich, and H. Yakowitz, *Anal. Chem.* **48**, 172 (1976).
13. E. Waldl, H. Wolfermann, N. Rusovic, and H. Warlimont, *Anal. Chem.* **47**, 1017 (1975).
14. R. J. Gehrke and R. C. Davies, *Anal. Chem.* **47**, 1537 (1975).
15. K. F. J. Heinrich, in *Advances in X-Ray Analysis*, Vol. 7 (W. M. Mueller, G. Mallett, and M. Fay, eds.), p. 382, Plenum Press, New York (1964).
16. K. F. J. Heinrich, in *Advances in X-Ray Analysis*, Vol. 6 (W. M. Mueller and M. Fay, eds.), p. 291, Plenum Press, New York (1963).
17. K. F. J. Heinrich, *Technical Note 278*, National Bureau of Standards, Washington, D.C. (1967).
18. K. F. J. Heinrich, D. Vieth, and H. Yakowitz, in *Advances in X-Ray Analysis*, Vol. 9 (G. R. Mallett, M. J. Fay, and W. M. Mueller, eds.), p. 208, Plenum Press, New York (1966).
19. D. R. Beaman, R. Lewis, and J. A. Isasi, in *Fifth International Congress on X-Ray Optics and Microanalysis*. (G. Möllenstedt and K. H. Gaukler, eds.), p. 84, Springer, Berlin (1969).
20. D. R. Beaman, J. A. Isasi, H. K. Birnbaum, and R. Lewis, *J. Phys. E* **5**, 767 (1972).
21. D. B. Wittry, in *Advances in X-Ray Analysis*, Vol. 7 (W. Mueller, G. Mallett, and M. Fay, eds.), p. 395, Plenum Press, New York, (1964).
22. K. F. J. Heinrich, in *Advances in X-Ray Analysis*, Vol. 11 (J. Newkirk, G. Mallett, and H. Pfeiffer, eds.), p. 40, Plenum Press, New York (1968).
23. R. C. Stanley, *Brit. J. Appl. Phys.* **12**, 503 (1961).
24. D. R. Beaman, *Anal. Chem.* **38**, 599 (1966).
25. N. Spielberg, *Rev. Sci. Instr.* **37**, 1268 (1966).
26. N. Spielberg, *Rev. Sci. Instr.* **38**, 291 (1967).
27. A. K. Varshneya, A. R. Cooper, and M. Cable, *J. Appl. Phys.* **37**, 2199 (1966).
28. M. P. Borom and R. E. Hanneman, *J. Appl. Phys.* **38**, 2406 (1967).
29. L. F. Vassamillet and V. E. Caldwell, *J. Appl. Phys.* **40**, 1637 (1969).
30. V. E. Cosslett and R. N. Thomas, *Brit. J. Appl. Phys.* **15**, 235 (1964); **15**, 883 (1964); **15**, 1283 (1964); **16**, 779 (1965).
31. H. E. Bishop, Ph.D. thesis, Cambridge University, Cambridge, England (1966).
32. D. B. Brown, D. B. Wittry, and D. F. Kyser, *J. Appl. Phys.* **40**, 1627 (1969).
33. M. J. Berger and S. M. Seltzer, in *Studies in Penetration of Charged Particles in Matter*, Publication 1133, p. 205, National Academy of Sciences, Washington, D.C. (1964).
34. P. Duncumb and S. J. B. Reed, in *Quantitative Electron Probe Microanalysis* (K. F. J. Heinrich, ed.), Special Publication 298, p. 133, National Bureau of Standards, Washington, D.C. (1968).
35. L. V. Spencer and U. Fano, *Phys. Rev.* **93**, 1172 (1954).
36. H. E. Bishop, in *Optique des Rayons X et Microanalyse* (R. Castaing, P. Deschamps, and J. Philibert, eds.), p. 153, Hermann, Paris (1966).
37. K. F. J. Heinrich, in *Optique des Rayons X et Microanalyse* (R. Castaing, P. Deschamps, and J. Philibert, eds.), p. 159, Hermann, Paris (1966).
38. E. H. Darlington, *J. Phys. D* **8**, 85 (1975).
39. H. Drescher, L. Reimer, and H. Seidel, *Z. Angew. Phys.* **29**(6), 331 (1970).
40. E. H. Darlington and V. E. Cosslett, *J. Phys. D* **5**, 1969 (1972).
41. H. Kulenkampff and W. Spyra, *Z. Phys.* **137**, 416 (1954).
42. M. Green and V. E. Cosslett, *J. Phys. D* **1**, 425 (1968).
43. J. C. Clark, *Phys. Rev.* **48**, 30 (1935).
44. D. L. Webster, W. W. Hansen, and F. B. Duveneck, *Phys. Rev.* **43**, 839 (1933).
45. L. T. Pockman, D. L. Webster, P. Kirkpatrick, and K. Harworth, *Phys. Rev.* **71**, 330 (1947).
46. M. Green, Ph.D. thesis, Cambridge University, Cambridge, England (1962).

47. C. R. Worthington and S. G. Tomlin, *Proc. Phys. Soc.* **69**, 401 (1956).
48. R. W. Fink, R. C. Jopson, H. Mark, and C. D. Swift, *Rev. Mod. Phys.* **38**, 513 (1966).
49. W. Bambynek, B. Crasemann, R. W. Fink, H. U. Freund, H. Mark, C. D. Swift, R. E. Price, and P. V. Rao, *Rev. Mod. Phys.* **44**, 716 (1972).
50. A. E. Sandström, in *Handbuch der Physik*, Band XXX Röntgenstrahlen, p. 197, Springer, Berlin (1957).
51. W. L. Baun, in *Electron Probe Microanalysis* (A. J. Tousimis and L. Marton, eds.), p. 155, Academic Press, New York (1969).
52. W. L. Baun, in *Characterization of Solid Surfaces* (P. F. Kane and G. R. Larrabee, eds.), p. 485, Plenum Press, New York (1974).
53. D. J. Fabian (ed.), *Soft X-Ray Band Spectra*, Academic Press, London (1968).
54. D. B. Brown, in *Quantitative Electron Probe Microanalysis* (K. F. J. Heinrich, ed.), Special Publication 298, p. 63, National Bureau of Standards, Washington, D.C. (1968).
55. H. E. Bishop, *Proc. Phys. Soc.* **85**, 855 (1965).
56. S. Goudsmit and J. L. Saunderson, *Phys. Rev.* **57**, 24 (1940).
57. H. E. Bishop, *Brit. J. Appl. Phys.* **18**, 703 (1967).
58. G. Shinoda, K. Murata, and R. Shimizu, in *Quantitative Electron Probe Microanalysis* (K. F. J. Heinrich, ed.), Special Publication 298, p. 155, National Bureau of Standards, Washington, D.C. (1968).
59. M. J. Berger, in *Methods in Computational Physics* (B. Alder, S. Fernbach, and M. Rotenberg, eds.), Vol. 1, p. 135, Academic Press, New York (1963).
60. R. Shimizu, Y. Kataoka, T. Ikuta, T. Koshikawa, and H. Hashimoto, *J. Phys. D* **9**, 101 (1976).
61. K. Murata, T. Matsukawa and R. Shimizu, in *Sixth International Conference on X-Ray Optics and Microanalysis* (G. Shinoda, K. Kohra, and T. Ichinokawa, eds.), p. 105, University of Tokyo Press, Tokyo (1972).
62. R. Shimizu, N. Nishigori, and K. Murata, in *Sixth International Conference on X-Ray Optics and Microanalysis* (G. Shinoda, K. Kohra, and T. Ichinokawa, eds.), p. 95, University of Tokyo Press, Tokyo (1972).
63. D. B. Brown and R. E. Ogilvie, *J. Appl. Phys.* **37**, 4429 (1966).
64. T. O. Ziebold and R. E. Ogilvie, *Anal. Chem.* **36**, 322 (1964).
65. T. O. Ziebold and R. E. Ogilvie, in *The Electron Microprobe* (T. D. McKinley, K. F. J. Heinrich, and D. B. Wittry, eds.), p. 378, Wiley, New York (1966).
66. T. O. Ziebold and R. E. Ogilvie, *Trans. Met. Soc. AIME* **239**, 942 (1967).
67. A. E. Bence and A. L. Albee, *J. Geol.* **76**, 382 (1968).
68. A. L. Albee and L. Ray, *Anal. Chem.* **42**, 1408 (1970).
69. K. F. J. Heinrich, *Technical Note 542*, National Bureau of Standards, Washington, D.C. (1970).
70. D. R. Beaman and J. A. Isasi, *Anal. Chem.* **42**, 1540 (1970).
71. G. Springer, *Neues Jahrb. Mineral. Monatsch.*, **1967**, 304 (1967).
72. J. Philibert and R. Tixier, *J. Phys. D* **1**, 685 (1968).
73. P. M. Thomas, *Brit. J. Appl. Phys.* **14**, 397 (1963).
74. P. Duncumb, P. K. Shields-Mason, and C. daCasa, in *Fifth International Congress on X-Ray Optics and Microanalysis* (G. Möllenstedt and K. H. Gaukler, eds.), p. 146, Springer, Berlin (1969).
75. K. F. J. Heinrich, in *The Electron Microprobe* (T. D. McKinley, K. F. J. Heinrich, and D. B. Wittry, eds.), p. 296, Wiley, New York (1966).
76. R. Castaing and J. Descamps, *J. Phys. Rad.* **16**, 304 (1955).
77. R. Castaing, in *Advances in Electronics and Electron Physics*, Vol. 13 (L. Marton, ed.), p. 317, Academic Press, New York (1960).
78. A. Vignes and G. Dez, *J. Phys. D* **1**, 1309 (1968).

79. R. Castaing and J. Hénoc, in *Optique des Rayons X et Microanalyse* (R. Castaing, P. Deschamps, and J. Philibert, eds.), p. 120, Hermann, Paris (1966).
80. J. D. Brown, in *Electron Probe Microanalysis* (A. J. Tousimis and L. Marton, eds.), p. 45, Academic Press, New York (1969).
81. J. Philibert, in *X-Ray Optics and X-Ray Microanalysis* (H. H. Pattee, V. E. Cosslett, and A. Engström, eds.), p. 379, Academic Press, New York (1966).
82. P. Duncumb and P. K. Shields, in *The Electron Microprobe* (T. D. McKinley, K. F. J. Heinrich, and D. B. Wittry, eds.), p. 284, Wiley, New York (1966).
83. K. F. J. Heinrich, in *Second National Conference on Electron Probe Analysis*, Paper 7 (1967). Extended abstracts are published by and available from Microbeam Analysis Society (MAS).
84. G. Love, M. G. C. Cox, and V. D. Scott, *J. Phys. D* **8**, 1686 (1975).
85. G. Love, M. G. C. Cox, and V. D. Scott, *J. Phys. D* **9**, 7 (1976).
86. H. E. Bishop, *J. Phys. D* **7**, 2009 (1974).
87. K. F. J. Heinrich and H. Yakowitz, *Anal. Chem* **47**, 2408 (1975).
88. M. J. Hénoc, F. Maurice, and A. Zemskoff, in *Fifth International Congress on X-Ray Optics and Microanalysis* (G. Möllenstedt and K. H. Gaukler, eds.), p. 187, Springer, Berlin (1969).
89. K. F. J. Heinrich, *Anal. Chem.* **44**, 350 (1972).
90. R. Castaing, Ph.D. thesis, University of Paris, Paris (1951).
91. S. J. B. Reed, *Brit. J. Appl. Phys.* **16**, 913 (1965).
92. C. Springer and B. Rosner, in *Fifth International Congress on X-Ray Optics and Microanalysis* (G. Möllenstedt and K. H. Gaukler, eds.), p. 170, Springer, Berlin (1969).
93. J. Hénoc, in *Quantitative Electron Probe Microanalysis* (K. F. J. Heinrich, ed.), Special Publication 298, p. 197, National Bureau of Standards, Washington, D.C. (1968).
94. T. S. Rao-Sahib, and D. B. Wittry, in *Sixth International Conference on X-Ray Optics and Microanalysis* (G. Shinoda, K. Kohra, and T. Ichinokawa, eds.), p. 131, University of Tokyo Press, Tokyo (1972).
95. T. Shiraiwa and N. Fujino, *Japan J. Appl. Phys.* **9**, 976 (1970).
96. T. Shiraiwa, N. Fujino, and J. Murayama, in *Sixth International Conference on X-Ray Optics and Microanalysis* (G. Shinoda, K. Kohra, and T. Ichinokawa, eds.), p. 213, University of Tokyo Press, Tokyo (1972).
97. I. K. Openshaw, R. J. Pearce, and B. M. Jeffery, *Brit. J. Appl. Phys.* **18**, 1004 (1967).
98. C. A. Andersen, *Brit. J. Appl. Phys.* **18**, 1033 (1967).
99. P. Duncumb and D. A. Melford, in *Optique des Rayons X et Microanalyse* (R. Castaing, P. Deschamps, and J. Philibert, eds.), p. 240, Hermann, Paris (1966).
100. G. Love, M. G. C. Cox, and V. D. Scott, *J. Phys. D* **7**, 2131 (1974).
101. G. Love, M. G. C. Cox, and V. D. Scott, *J. Phys. D* **7**, 2142 (1974).
102. J. Ruste and M. Gantois, *J. Phys. D* **8**, 872 (1975).
103. S. J. Ingrey and W. D. Westwood, *J. Phys. D* **6**, 896 (1973).
104. D. F. Kyser, in *Sixth International Conference on X-Ray Optics Microanalysis* (G. Shinoda, K. Kohra, and T. Ichinokawa, eds.), p. 147, University of Tokyo Press, Tokyo (1972).
105. K. F. J. Heinrich and H. Yakowitz, in *Fifth International Congress on X-Ray Optics and Microanalysis* (G. Möllenstedt and K. H. Gaukler, eds.), p. 151, Springer, Berlin (1969).
106. K. F. J. Heinrich and H. Yakowitz, *Mikrochim. Acta* **1970**, 123 (1970).
107. H. Yakowitz and K. F. J. Heinrich, *Mikrochim. Acta* **1968**, 182 (1968).
108. K. F. J. Heinrich and H. Yakowitz, *Mikrochim. Acta* **1968**, 905 (1968).
109. J. Hénoc, K. F. J. Heinrich, and R. L. Myklebust, *Technical Note 769*, National Bureau of Standards, Washington, D.C. (1973).

110. H. Yakowitz, C. E. Fiori, and R. E. Michaelis, *Special Publication 260-22*, National Bureau of Standards, Washington, D.C. (1971). The Fe-3Si alloy is designated as SRM 483.

111. K. F. J. Heinrich, R. L. Myklebust, S. D. Rasberry, and R. E. Michaelis, *Special Publication 260-28*, National Bureau of Standards, Washington, D.C. (1971). The Au-Ag alloy series and the Au-Cu alloy series are designated as SRM 481 and 482, respectively.

112. J. I. Goldstein, in *Electron Probe Microanalysis* (A. J. Tousimis and L. Marton eds.), p. 245, Academic Press, New York (1969).

113. K. Keil, *Fortschr. Mineral.* **44**(1), 4 (1967).

114. K. Keil, in *Microprobe Analysis* (C. A. Andersen, ed.), p. 189, Wiley Interscience, New York (1973).

115. T. R. Sweatman and J. V. P. Long, *J. Petrol.* **10**, 332 (1969).

116. C. A. Andersen, in *Methods of Biochemical Analysis*, Vol. 15 (D. Glick, ed.), p. 147, Wiley Interscience, New York (1967).

117. T. Hall, in *Quantitative Electron Probe Microanalysis* (K. F. J. Heinrich, ed.), Special Publication 298, p. 269, National Bureau of Standards, Washington, D.C. (1968).

118. W. L. Robinson, in *Microprobe Analysis* (C. A. Andersen, ed.), p. 271, Wiley Interscience, New York (1973).

119. W. T. Kane, in *Microprobe Analysis* (C. A. Andersen, ed.), p. 241, Wiley Interscience, New York (1973).

120. E. W. White, in *Microprobe Analysis* (C. A. Andersen, ed.), p. 349, Wiley Interscience, New York (1973).

121. H. H. Pattee, V. E. Cosslett, and A. Engström, eds., *X-Ray Optics and Microanalysis* (Stanford Conference, 1962), Academic Press, New York (1963).

122. R. Castaing, P. Deschamps, and J. Philibert (eds.), *Optique des Rayons X et Microanalyse* (Orsay Conference 1965), Hermann, Paris (1966).

123. G. Möllenstedt and K. H. Gaukler (eds), *Fifth International Congress on X-Ray Optics and Microanalysis* (Tübingen Conference 1968), Springer, Berlin (1969).

124. G. Shinoda, K. Kohra, and T. Ichinokawa (eds.), *Proceedings of the Sixth International Conference on X-Ray Optics and Microanalysis* (Osaka Conference, 1971), University of Tokyo Press, Tokyo (1972).

125. MAS Microbeam Analysis Society, formerly EPASA Electron Probe Analysis Society of America. Yearly national meeting since 1966.

126. *Advances in X-Ray Analysis*, Plenum Press, New York. Yearly volume of papers given at x-ray conference sponsored by University of Denver Research Institute.

127. K. F. J. Heinrich (ed.), *Quantitative Electron Probe Microanalysis*, Special Publication 298, National Bureau of Standards, Washington, D.C. (1968).

128. A. J. Tousimis and L. Marton (eds.), *Electron Probe Microanalysis*, Academic Press, New York (1969).

129. T. D. McKinley, K. F. J. Heinrich, and D. B. Wittry (eds.), *The Electron Microprobe*, Wiley, New York (1966).

130. C. A. Andersen (ed.), *Microprobe Analysis*, Wiley, New York (1973).

131. G. Hallerman and M. L. Picklesimer, in *Electron Probe Microanalysis* (A. J. Tousimis and L. Marton, eds.), p. 197, Academic Press, New York (1969).

132. G. A. Hutchins, in *Advances in Applied Spectroscopy*, Vol. 7A (E. L. Grove and A. J. Perkins, eds.), p. 80, Plenum Press, New York (1969).

133. G. V. Kidson, *J. Nuclear Materials* **3**, 21 (1961).

134. Y. Adda, M. Beyeler, and A. Kirianenko. *C.R. Acad. Sci. Paris* **250**, 115 (1960).

135. J. I. Goldstein and R. E. Ogilvie, *Trans. Met. Soc., AIMA* **233**, 2083 (1965).

136. J. I. Goldstein and R. E. Ogilvie, in *Optique des Rayons X et Microanalyse* (R. Castaing, P. Deschamps, and J. Philibert, eds.), p. 594, Hermann, Paris (1966).
137. A. E. Austin and N. A. Richard, *J. Appl. Phys.* **32**, 1462 (1961).
138. D. M. Koffman, Sc.D. thesis, Massachusetts Institute of Technology, Cambridge, Massachusetts (1964).
139. J. C. Fisher, *J. Appl. Phys.* **22**, 74 (1951).
140. R. T. P. Whipple, *Phil. Mag.* **45**, 1225 (1954).
141. A. Vignes and J. P. Sabatier, *Trans. Met. Soc.*, *AIME* **245**, 1795 (1969).
142. M. Tenenbaum, *Trans. Met. Soc.*, *AIME* **245**, 1675 (1969).
143. E. A. Starke, Jr., *J. Metals* **22**, 54 (1970).
144. P. N. T. Unwin, G. W. Lorimer, and R. B. Nicholson, *Acta Metall.* **17**, 1363 (1969).
145. W. L. Patterson and G. A. Shirn, *J. Vacuum Sci. Technol.* **4**, 343 (1967).
146. A. P. Schneider, in *Sixth International Conference on X-Ray Optics and Microanalysis* (G. Shinoda, K. Kohra, and T. Ichinokawa, eds.), p. 211, University of Tokyo Press, Tokyo (1972).
147. F. W. Perry, G. A. Hutchins, and L. E. Cross, *Materials Res. Bull.* **2**, 409 (1967).
148. C. W. Mead, in *Electron Probe Microanalysis* (A. J. Tousimis and L. Marton, eds.), p. 227, Academic Press, New York (1969).
149. T. A. Hall and H. J. Höhling, in *Fifth International Congress on X-Ray Optics and Microanalysis* (G. Möllenstedt and K. H. Gaukler, eds.), p. 582, Springer, Berlin (1969).
150. D. J. Marshall and T. A. Hall, *J. Phys. D* **1**, 1651 (1968).
151. W. E. Sweeney, R. E. Seebold, and L. S. Birks, *J. Appl. Phys.* **31**, 1061 (1960).
152. G. H. Cockett and C. D. Davis, *Brit. J. Appl. Phys.* **14**, 813 (1963).
153. G. A. Hutchins, in *The Electron Microprobe* (T. D. McKinley, K. F. J. Heinrich, and D. B. Wittry, eds.), p. 390, Wiley, New York (1966).
154. B. Djurić and D. Cerović, in *Fifth International Congress on X-Ray Optics and Microanalysis* (G. Möllenstedt and K. H. Gaukler, eds.), p. 99, Springer, Berlin (1969).
155. J. W. Colby, in *Advances in X-Ray Analysis*, Vol. 11 (J. Newkirk, G. Mallett, and H. Pfeiffer, eds.), p. 287, Plenum Press, New York (1968).
156. R. Tixier and J. Philibert, in *Fifth International Congress on X-Ray Optics and Microanalysis* (G. Möllenstedt and K. H. Gaukler, eds.), p. 180, Springer, Berlin (1969).
157. W. Reuter, *Surface Sci.* **25**, 80 (1971).
158. W. Reuter, in *Sixth International Conference on X-Ray Optics and Microanalysis* (G. Shinoda, K. Kohra, and T. Ichinokawa, eds.), p. 121, University of Tokyo Press, Tokyo (1972).
159. H. E. Bishop and D. M. Poole, *J. Phys. D* **6**, 1142 (1973).
160. E. Preuss, in *Quantitative Analysis with Electron Microprobes and Secondary Ion Mass Spectrometry* (E. Preuss, ed.), p. 80, Kernforschungsanlage Jülich GmbH, Jülich, Germany (1973).
161. J. A. Chandler and M. S. Morton, *Anal. Chem.* **48**, 1316 (1976).
162. P. Duncumb and D. A. Melford, in *First National Conference on Electron Probe Analysis*, paper 12 (1966).
163. G. A. Hutchins and R. D. Wantman, in *Second National Conference on Electron Probe Microanalysis*, paper 2 (1967).
164. J. D. Brown and L. Parobek, in *Seventh National Conference on Electron Probe Analysis*, paper 5 (1972).
165. G. H. Maher, Ph.D. thesis, Rensselaer Polytechnic Institute, Troy, New York (1971).
166. G. A. Hutchins and A. Shepela, *Thin Solid Films* **18**, 343 (1973).
167. A. R. Janus and G. A. Shirn, *J. Vacuum Sci. Technol.* **4**, 37 (1967).

# 7

# Research Techniques in Detergency

## Anthony M. Schwartz

## 1. Introduction

For the purposes of this discussion, "detergency" will be defined as the removal of an unwanted soil from a substrate by the physicochemical action of an aqueous bath. The bath could consist for reference purposes of pure water or of water of specified hardness, salt content, pH, etc. In the practical situation, however, the bath will contain one or more solutes that greatly increase both the rate and the degree of soil removal over that achievable by water alone. These solutes include surfactants, builders (materials that by various mechanisms promote the soil-removing action of surfactants), and antiredeposition agents (materials that inhibit the recombination of soil with previously separated substrate). Substrates can vary quite widely in both chemical composition and physical form. In general, they must be solid, with a melting or softening point well above the washing temperature, and they must be insoluble and resistant to attack by the bath; in short, the type of material that is normally cleaned with "soap and water." Substrates are usually considered, for practical purposes, in two categories:—fibrous, comprising textile fabrics, and hard surface, comprising tile, painted surfaces, metal, glass, ceramic, and similar materials. Soil has broadly and jokingly been defined as "matter out of place." In the study of detergency, however, certain types of "matter out of place" are usually excluded from consideration. Among these are substances easily removed by dissolution or rinsing in water alone, and at the other extreme, stains of dyelike materials that are molecularly dispersed below the surface of the substrate and can be removed in reasonably short time only by chemical action. Soils are generally grouped as solid, liquid (or "oily"), and mixed. They adhere to the

---

*Anthony M. Schwartz* • 2260 Glenmore Terrace, Rockville, Maryland 20850.

substrate but do not exhibit strong chemical interaction with it. The detersive process consists essentially of a breaking or weakening of the soil–substrate adhesive bond so that the soil can be separated and carried away easily by the hydraulic action of the bath. The literature on surfactants and other detergent components and their behavior and properties is exceedingly extensive and will not be considered here. Theoretical discussions of the physical chemistry of detergency are also available elsewhere.[1] The present discussion will be concerned primarily with the methodology and techniques used in studying the detersive process.

Technical studies of detergency are generally undertaken for either of two broad purposes. The first purpose, and by far the commonest, is immediately practical. Manufacturers of detergents and of cleaning equipment are interested in the effectiveness of their products and in the controllable parameters that influence it, as measured by the ease and extent of soil removal. The factors that affect soil removal in practical cleaning systems are well recognized.[2] They include the chemical nature of the substrate and the soil, composition of the bath (a factor most frequently studied by detergent manufacturers), physical and geometrical character of the system, and the mechanical conditions obtaining during the cleaning operation (a factor of major importance to washing machine manufacturers). A very large segment of the technical literature on detergency consists of studies relating variations in gross soil removal to variations in one or more of these factors. Such studies are obviously important and applicable in advancing the technology of cleaning, and the techniques used in performing them will be the subject of the first part of this discussion. They will be referred to for brevity as "practical techniques."

The second broad class of detergency studies is concerned with the physicochemical mechanisms occurring in detersive systems, i.e., with the science of detergency. From this point of view, the focus of attention is the process by which the substrate–soil phase interface is disrupted and replaced by substrate–bath and soil–bath interfaces. For purposes of scientific study, "soil," which in actuality consists of many phases, is advantageously resolved into its constituents and considered one phase at a time. These individual soil phases may be liquid or solid. We are therefore ultimately concerned with molecular effects at the various solid–liquid and solid–solid interfaces in the system and at the corresponding three-phase boundary lines where these interfaces meet the bath phase. The techniques used in studying these effects are, as might be expected, selected from the total investigative armory of colloid and surface chemistry. Those that have proved applicable will be considered in the second part of this discussion and will be referred to as "research techniques."

Historically, the first controlled studies of detergency were of the eminently practical type. A piece of fabric (or other substrate of interest) was soiled in some hopefully reproducible manner and washed under controlled conditions, and the soil removal was measured.[3] These early investigations, however, were aimed as much at understanding the phenomenon of detergency as at making better detergents, and the methods we now regard as "practical" were originally conceived as being research methods.[4] There is accordingly no sharp conceptual dividing line between the practical approach and the research approach. The difference is largely in the degree of sophistication. Both approaches, however, should be clearly distinguished from the development approach used in the industrial development of new cleaning compositions, devices, and processes. The development approach focuses not only on soil removal but also on a large number of other factors that are of at least equal importance. In some instances, actual soil removal may be less important than the appearance of soil removal, as provided, for example, by fluorescent whitening agents, or than imparting a desirable finish to the substrate. To be commercially acceptable, a cleaning product must be innocuous to the user and the cleaning equipment, harmless to the substrate, convenient, economical, etc. In short, it must have a whole host of properties quite unrelated to its soil-removing power. Aside from certain effects of cleaning compositions on the substrate, none of these properties will be considered explicitly in this discussion.

Finally, it should be noted that a large number of separately recognized colloidal and interfacial phenomena are generally concomitant with the detersive effect. They include deflocculation and dispersion of particulate soil, emulsification and/or solubilization of oily soil, wetting, and foaming. These concomitant effects are for the most part better understood scientifically than detergency itself. Furthermore, each of them can play an indispensible role in detergency; so much so that the detersive process can be regarded as always comprising and in some cases consisting of a combination of these simpler effects. The techniques utilized for their study could logically have been included in our discussion, but they are fully covered elsewhere in this treatise. Where appropriate, they will be alluded to but not discussed in detail.

## 2. Study of Practical Detergency

Reduced to its simplest terms, a typical practical detergency study includes the following steps: (1) A soil of known composition is applied to the substrate of interest to form the soil–substrate complex. This soiling step must be sufficiently well controlled to insure a uniform, reproducible, and well-characterized soiled substrate system. The characterization usually includes some measure of quantity of soil on the substrate, although in certain types of comparative tests this is not necessary. (2) The soiled substrate (which, if a fabric, is usually called "soil cloth") is washed under thoroughly controlled conditions in the bath of interest, whose composition is known. The washing conditions of importance include time, temperature, bath ratio (weight ratio of bath to soiled substrate), and type and degree of mechanical energy input. (3) The washed substrate is removed from the bath, rinsed or otherwise freed of excess clinging bath in a reproducible and controlled manner, and (usually) dried. It may also be "finished" in some specified manner, for example, ironed if it is a fabric. (4) The amount of soil that has been removed in the washing process is determined, either on an absolute basis or by comparison with soil removed under similar conditions in a reference system. (5) The soil removal data are interpreted, i.e., related to changes in the parameters being studied.

It is evident that specialized investigative techniques and methods are required in three major areas if the study of practical detergency is to rest on a technically sound basis. These areas are the preparation and characterization of soiled substrate systems; the development of controlled, reproducible washing procedures; and measurement of the soil content of the substrate at all stages of the study.

The preparation of a soiled substrate system suitable for practical detergency investigations involves a thorough knowledge of the unsoiled substrate of interest and the soil to be applied. It also requires a knowledge of how to apply soil to substrate in such a manner as to insure uniform and reproducible washability of the soiled specimens, i.e., a knowledge of the various factors that must be controlled in the soiling procedure and how to control them. The technologist studying practical detergency wants, of course, to use soiled substrate systems that are "realistic," i.e., representative of those encountered in the field. Accordingly, certain soil–substrate combinations that are perfectly possible to prepare and might even be of great theoretical interest are seldom if ever used. There would be little interest, for example, in tile floors soiled with sebum or lingerie soiled with crankcase oil. It is convenient, therefore, to consider the various common types of substrate and, coupled with each type, the various soils that are commonly used with it. Substrates are broadly divided into two categories— textile, or fabric, and hard surface. This division is obviously made on the basis of geometric form and morphology rather than on the basis of chemical composition. Morphology is a more important factor than chemical composition, because the form of the substrate dictates the manner in which the goods will be manipulated during the cleaning operation. Chemical composition of the substrate, and of course of the soil, too, influence the ease of soil removal, but even in this regard, it often is less important than the morphology. This is not surprising, since all materials commonly cleaned with aqueous detergents have high resistance to water and mildly alkaline or acid solutions, and they share many other solubility properties.

In the development or selection of a controlled, reproducible washing procedure, the obvious factors of time, temperature, bath ratio, and bath compositions are always involved. In many instances, a special device or apparatus must be designed or adopted in which the cleaning operation is performed. Such detergency testing devices, to be useful, should be much smaller in size than the practical device they are intended to simulate. They should also allow easy variation and control of the basic detergency factors mentioned above. Otherwise, it would be just as easy, and certainly more realistic, to carry out the experimental detergency studies in full-scale equipment under full-scale conditions. The major problem in designing small-scale experimental washing devices is to provide for the correct energy and power input. Degree and rate of mechanical action in the substrate–soil–bath system is one of the major factors in detergency. For a scaled-down detergency test to be realistic, i.e., to be predictive of results that will be obtained in full-scale operation, the energy and power input must be appropriately adjusted. Theoretically, if the bath ratios in the small-scale and full-scale systems are identical, the power input per unit mass of substrate–soil–bath system should be adjusted to be the same in both systems. This is a fair rule of thumb, but design of the agitation device as well as power input has a pronounced effect on soil separation, and in actual practice, both the power input and the design of the machine are adjusted empirically to simulate full-scale results as closely as possible. Many practical detergency studies are carried out in full-scale equipment when it is not too cumbersome to do so and when the equipment already is provided with controls for the essential detergency variables.[5]

Methods for estimating soil content of the substrate and soil removal in practical detergency studies generally simulate the methods used in full-scale practice. They may involve visual estimates, optical measurements, mass measurements by physical means, or mass measurements via chemical analysis.

## 2.1. Textile Substrates

### 2.1.1. Detergency in Textile Manufacture[6]

Cleaning operations are an important aspect of textile manufacture, being required at almost all stages from fiber preparation to final finishing. Much practical study of detergency is accordingly carried out in textile mill laboratories and in the laboratories of textile chemical supply houses.

The soil–substrate systems in almost all such studies are actual samples of the material being processed in full scale. In studying raw wool scouring, for example, the soil–substrate is the raw wool received at the mill. Similarly, in studying the washing of printed goods or the scouring of desized goods, the soil–substrate is a sample cut from the material in process. The purposes of the studies are usually to ascertain if some new detergent formulation will do a more economical job than the one being used. Sometimes, if the "soils" (sizings, lubricants, etc.) or the substrates (new fiber blends or fabric constructions) are being changed, the detergency studies are aimed at developing an optimum detergent fomulation.

The washing devices used in these studies vary with the specific situation. In studying raw wool scouring, a set of ordinary beakers adequately simulate the bowls of a full-scale wool scouring train, and the gentle agitation of the full-scale process is easily simulated in the laboratory. At the other extreme, the type of mechanical action afforded by a continuous rope or open width scouring train running at high speed cannot easily be duplicated in laboratory-size equipment. Detergency studies in this area are frequently done in the full-scale equipment itself. There are, of course, many intermediate situations, where the results of studies made in such standard devices as the Tergotometer (discussed in a subsequent section) can be confidently translated into the results of full-scale operation.

Measurements of original soil content and soil removal in textile mill detergency studies are almost always patterned after the process control methods used in full-scale operations. In raw wool or wool piece goods scouring, for example, the residual soil is measured by solvent extraction and weighing. In desizing and boil-off tests, the scoured goods are tested for starch residues by staining with iodine. In the after-washing of dyed and printed goods, crocking and bleeding tests are used to ascertain if a sufficient degree of cleanness has been achieved.

## 2.1.2. Detergency in Household and Commercial Laundering[7,9]

The major portion of all practical detergency research is directed toward the laundering of household fabrics (apparel and nonapparel) and corresponding commercial items such as hotel and restaurant linens and the uniforms of service personnel. The three problem areas, as in all detergency research, are the preparation of a suitable soil–substrate system, development of a washing procedure, and methods for measuring soil removal. The choice and preparation of a soil–substrate system is in itself a complicated task, involving fabric selection, soil selection, and suitable methods for applying the soil to the fabric.

*2.1.2.1. Fabric Selection.* The fabric used in preparing soil cloth for a specific detergency investigation should be either the cloth of practical interest or the closest simulant available. Fiber, construction, and finish are all of equal importance. The fibers commonly used include wool, cotton, viscose and acetate rayons, nylon, and polyester. Blends of these fibers are also encountered, the most widely used being the polyester–cotton blends. Construction of the fabric, especially yarn twist and thread count in woven goods, greatly affects the ease of soil removal. Dense fabrics with tightly packed fibers tenaciously retain both solid and liquid soils. Furthermore, they are inherently stiffer than the more open fabrics and are not flexed as vigorously in the washing machine. Just as loose weaves of lightly twisted yarns are easier to clean than tightly spun tight weaves, so knitted goods are generally easier to clean than weaves.

Most modern household fabrics, especially the cellulosic fabrics and their blends, bear a permanent finish. This finish, regardless of its purpose, has a strong effect on the soil-releasing properties of the fabric and must be considered in selecting the soil cloth substrate. The "wash–wear," "crease-proof," or "permanent press" finish is applied to cottons and rayons as well as to their blends with polyester. This finish is essentially an aminoplast resin generated within the interior of the fiber from the monomeric precursors (formaldehyde and an amide) that have previously been made to diffuse in. The finish, if properly applied, alters both the bulk and surface properties of the cellulosic fiber. It has less effect, however, on polyester that may be blended with the cellulosic, because the monomers do no diffuse into the polyester, and any resin formed on the surface tends to be nonadherent and easily rinsed away. The permanent press finish profoundly alters the soil-release properties of cotton and rayon, generally making them more difficult to clean.[10] Polyester fabrics, particularly those of 100% polyester content, are notably easy to soil and difficult to clean. They are often given a more or less durable "soil-release" finish by the manufacturer. These are adherent topical finishes that contain hydrophilic moieties such as ether oxygen, hydroxyl, or carboxyl. As intended, these materials make the fiber less prone to soiling in use and easier to clean once they are soiled. Bed linens and table linens usually bear a permanent or semipermanent sizing or bodying finish that has soiling characteristics different from that of the unfinished fabric. An increasing percentage of cellulosic fabrics intended for sleepwear and housewear bear durable flameproof finishes. These generally contain phosphorus and nitrogen and constitute up to 18% of the fabric weight, greatly altering the soil-release properties. Aside from these and other permanent finishes, the fabrics may contain fugitive finishes (for softening or bodying) that are easily removed by one or two launderings. Many fabrics as purchased contain both a durable and a fugitive finish. It is good practice in detergency studies to wash them thoroughly before applying the soil. The washing should be carried out with a relatively nonadsorbing built surfactant (all surfactants adsorb to a certain extent), and the fabric should be rinsed thoroughly to remove residual detergent as completely as possible.

*2.1.2.2. Soil Selection.* A soil to be used in making soil cloth for practical detergency studies should have the following characteristics: (1) It should be "realistic," i.e., it must simulate in composition and in ease of

removal the soils that will actually be picked up in use by the substrate fabric.
(2) The composition should be known at least well enough to be reproducible within the limits of assay. (3) It should be realistic with regard to
measurement. The housewife judges cleanliness largely by visual effect, and
the experimental soils should be chosen to contain enough colored
component to enable making an estimate of cleanness either by subjective
judgment or by a suitable photometric instrument. In most instances, the
color or removal of visible soil is of more practical interest than the removal
of invisible soil (such as white particulate matter or colorless oily matter). (4)
It should be easily applicable to the fabric to form a soiled swatch that is
reasonably uniform in appearance (not splotchy) and releases the soil
uniformly during washing, so that the washed swatch is also reasonably
uniform in appearance.

The problem of formulating a realistic soil, representative of the soil
that accumulates on household fabrics, has long been a goal of detergency
research, although in the opinion of many investigators it is unsolvable and
falsely conceived. Relatively early in the history of detergency research,
many studies were made of the composition of these so-called natural
soils.[11] As might be expected, the soil found on dirty laundry varies greatly,
differing from locality to locality, user to user, and item to item. In almost all
situations, the soil is quite complex, containing a large number of
components. To duplicate a natural soil from any single source is difficult
although not impossible, but such a soil would not necessarily behave like
other equally common soils and therefore would hardly be worth the
necessary effort. It is a fact, however, that although the variation in soils is
real, it is frequently not great. If detergent A outperforms detergent B on
one soil, it will also, more often than not, be superior on other soils.
Attempts were therefore made to prepare model soils of relatively simple
composition that could be used as substitutes for natural soils in studying the
relative effectiveness of detergents. The natural soils of household fabrics
contain four different types of components: (1) Water-insoluble solid particulate matter, with particles small enough to be adherent to the fibers.
Particles larger than 10 $\mu$m are usually easy to remove by rinsing alone and
are not considered in the soil range. Particles smaller than about 100 nm
tend to be very difficult to remove from most fibers by ordinary detergent
solutions.[12] (2) Water-insoluble nonvolatile liquids or "oils." This so-
called oily soil usually exists on the fiber as a single phase containing several
components. (3) Water-soluble nonadsorbed material that can be removed
by simple water rinsing and is not considered a cleaning problem.
(4) Water-soluble strongly adsorbed material, usually colored, generally
referred to as "stains." Most stains are not removed completely by ordinary
detergents and are considered a separate cleaning problem. A model soil
should therefore contain solid particulate soil and/or oily soil.

The oldest and simplest artificial soil to be used in making soil cloth for detergency research is a mixture of finely divided carbon with an oil. The commonly used oil component of this mixture is heavy white mineral oil of medicinal grade and/or refined glyceride oil. The ratio of solid to oil can vary quite widely, depending on how dark the experimenter wants his soil cloth to be. Many variations of this carbon–oil soil have been described and are used. The carbon may be in the form of carbon black, lampblack or graphite, and may be purchased for use in the form of aqueous or nonaqueous dispersions (e.g., Aquablack, Oildag, Aquadag, etc.). The oils may have varying degrees of unsaturation and may contain free fatty acids as well as glycerides, waxes (nonglyceride fatty esters), and hydrocarbons. Soil cloths bearing a carbon–oil type of soil on a stipulated type of fabric substrate have long been available commercially, and the various suppliers furnish full information on the composition of their soils. Despite the fact that they have been justifiably criticized as unrealistic (i.e., not able to reliably predict how effectively a test detergent will clean an actual bundle of laundry), the commercial carbon–oil soil cloths are very widely used in detergent evaluation.[13a,c] They are effective in preliminary screening tests, and they do give a reasonably good indication of the minimum effective concentration (MEC) of the test detergent.[14a] The MCE is an important characteristic of detergents and is determined by plotting a curve of soil removal versus detergent concentration under fixed washing conditions (water hardness, time, temperature, degree of mechanical action, etc.). These detergency versus concentration curves are usually sigmoid in form, i.e., below a certain concentration (the MEC), the soil removal is no better than that of water alone, but in the MEC range, it rises rapidly and soon reaches a plateau above which further increases in concentration cause no increase in soil removal. The MEC of a detergent does not in general vary with the soil on which the detergent is being tested, but the height of the detergency plateau does. Carbon–oil soils are therefore not a reliable means for ranking a series of detergents with regard to soil-removing power on actual household laundry.

Many attempts have been made to formulate model soils that would simulate natural soil in the ranking of test detergents.[14b] Some of these attempts have been based on analyses of natural soil, others on the results of empirical test programs, in most of which the solid and oily components have been investigated separately. One of the most popular substitutes for carbon in the newer model soils is clay.[15a–c] Commercial clays of reasonably constant tint (gray or brown) and stipulated origin are available from laboratory supply houses for detergency testing. Siliceous mineral soils have also been used but are not as yet standardized or accepted to the same extent as the clays. Iron oxides of different shades varying from black through brown and red to yellow are well standardized as pigments. Many of these, both individually and in combination, have been used as the solid component of model soils.[15c] One of the difficulties in using iron oxide is that certain forms can become so finely divided as to penetrate into the fiber substance and deposit as a stain rather than as a soil removable in the ordinary detersive process.

The attempt to find more realistic soils has led quite logically to the use of natural soil components themselves as components of model artificial soil. The clays are, of course, a move in this direction. Two other types of solid particulate matter from natural soil have been extensively used, one from air conditioner dirt,[16] the other from vacuum cleaner dirt.[17] The dirt that collects on air conditioner filters within any reasonably restricted area is remarkably homogeneous. It contains a certain amount of oily material, water-soluble material, and water-insoluble particulate solid that is usually very dark in color. This material can easily be isolated by extraction with water and polar or nonpolar solvents, and forms a good, realistic, solid component for model soils. Vacuum cleaner dirt varies from batch to batch much more than air conditioner dirt, although the dirt that can be obtained from individual carpet-cleaning establishments or hotel or office-cleaning services is reasonably constant in composition. It is more difficult to process than air conditioner dirt, containing more staining substances and much more material too coarse to act as soil. Classification by means of water and solvents is not unduly arduous, however, and the yield of material suitable for a model solid soil component is generally quite satisfactory.

The development of realistic oily components for model soils has followed a relatively smooth course. It was early recognized that free fatty acids and other highly polar fatty compounds as well as the neutral oils were components of natural household soil. One of the commonest oily compounds of household soil is sebum, the secretion of the human skin. Sebum has been analyzed and found to comprise a large number of fatty components, including hydrocarbons, esters, acids, and alcohols.[18a] "Artificial sebum," made up from these components in the proper proportions, is frequently used as the oily component of model soils.[18] Artificial sebum combined with air conditioner dirt solids, vacuum cleaner dirt solids, clay, or mixed iron oxides is a favored model soil for testing and ranking heavy-duty household laundry detergents.[19] Since oils, i.e., water-insoluble nonvolatile liquids, are almost always mutually soluble so that several components can form a single phase, it is easy to formulate them to simulate or even duplicate the practical soil of interest. Thus, the types of oily soil that collect on garage floors, kitchen ranges, aircraft surfaces, etc., differ from one another, but they can all be readily simulated for laboratory-scale experiments. Sebum and the various cooking and food oils are of most interest in household soiling studies, although hydrocarbon oils are also encountered.

One of the most widely accepted test procedures for household laundry detergency involves the use of natural facial soil. The test swatches are soiled by rubbing them on the foreheads and faces of a group of panelists, one set of swatches being assigned to each panelist. The swatches are then washed, each set under its own test conditions (the difference between sets usually lying in the batch composition), and after a series of soilings and washings the swatches are compared for cleanness. By suitable statistical treatment of the data, and suitable design of the experiment, a remarkably valid estimate of the relative effectiveness of various detergents can be obtained.[20a] The soil in this case is a mixture of natural skin secretions (sebum and perspiration), skin detritus, cosmetic residues, and airborne soil deposited on the panelists face during the course of the day.[20b]

    *2.1.2.3. Application of Soil to Fabric.* The manner in which soils of any type, solid, oily, or composite, artificial or natural, are applied to the substrate greatly influences the ease with which they can be removed in a washing bath. In comparing detergents, it is therefore necessary that they be tested on soil cloths that are equally refractory and have been prepared in an identical manner. There are two approaches to this problem. One, associated with single cycle testing (described in the following section) is to prepare a set of swatches all bearing the same soil load. These are known as "standard soil cloth swatches." The second approach, associated with multicycle testing, is to soil all swatches at one time in a single bath. The soil is thereby randomly (but not necessarily uniformly) distributed over the swatches. This is referred to as the "random soiling approach."

    Standard soil cloth swatches to be used in detergency tests should be uniform and reproducible in the following aspects: (1) The ease of soil removal should not vary from swatch to swatch nor from area to area on an individual swatch. (2) The soil should be evenly distributed from area to area on the swatch, so that the swatch not only presents a uniform appearance but actually bears the same quantity of soil on each square centimeter. This property is of greater importance in single-cycle than in multiple-cycle testing. (3) The soil should be uniformly distributed through the thickness of the fabric and through the individual yarns, i.e., it should constitute "ground in" or "soaked in" soil, rather than superficially "laid on" soil. This feature contributes to uniform ease (or difficulty) of removal. (4) All swatches in any batch should have the same or nearly the same reflectance values.

    Soils that consist entirely of particulate solid material or contain only a small amount of oily material are sometimes applied dry by rubbing or brushing them into the fabric. Soil cloth made this way is generally unsatisfactory by the above criteria and is seldom used except for special studies. Dry-soiled swatches of this type are usually much easier to clean than wet-soiled swatches. In wet soiling, the usual soiling procedure, the soil is diluted with a volatile diluent, which may be water or a nonaqueous solvent, and applied by a dipping and squeezing procedure. The volatile diluent is then removed by evaporation. Water is the easiest and generally most satisfactory diluent for soiling. In using water for an ordinary oil–solid soil mixture it is advisable to use an emulsifying agent to help disperse the oily component and maintain the soiling bath homogeneous. Nonaqueous diluents are less favored in modern practice than they were formerly. They have the advantages of dissolving the oily component of the soil and of evaporating rapidly. This is offset by two serious disadvantages. First, in the padding operation (using rubber squeeze rolls), there is a large buildup of static electricity. Unless special precautions are taken this causes the solid portion of the soil to be unevenly and loosely deposited. Second, during drying, the oily soil migrates in the fabric and becomes unevenly deposited. Despite these disadvantages many commercial soil cloths are prepared in a single operation using solvents of the type used in drycleaning. A two-stage soiling procedure, which generally gives better results, consists first in applying the solid soil from aqueous suspension; the goods are then dried, and the oily soil is sprayed or padded on in a volatile solvent. The soiled swatches obtained by this procedure are quite uniform. A third way of applying soil to a test fabric is to print it on. In this procedure, the soil is made up to a thick paste, similar in consistency to the ordinary dye pastes used in textile printing, and is applied by the same intaglio printing process. Printed soil cloth tends to be poorly reproducible from batch to batch. The soil lies mostly on the surface of the fabric and does not penetrate the yarns evenly. This type of soil cloth, developed during World War II, is still used to a certain extent but is not highly regarded by experienced investigators in this field.

In multicycle wash testing, the swatches are put through a series of alternate soilings and washings, during which a small but increasing amount of residual unremoved soil accumulates on them.[21] At the end of the last washing (anywhere from three to ten or more cycles are used), the appearance or soil content is compared with that of the original swatches before the first soiling to give a "difference value." The soilings in this procedure need not be "standard," i.e., all the swatches need not be uniformly soiled nor match one another in soil content. The only requirements are that all swatches in the test be soiled in a single operation and that a sufficient number of swatches be used to make the final difference values statistically significant. This usually consists in immersing the swatches in a large agitated soiling bath, withdrawing, squeezing out the excess liquor, and drying. Other procedures, including dry soiling procedures, may be used as appropriate, provided only that the randomness requirement be observed.

*2.1.2.4. Washing Procedures.* Although a washing process may be defined broadly as the treatment of a soiled substrate with an aqueous bath, with the expected result that soil will become separated from substrate, practical processes differ greatly in detail. As mentioned previously, the factors that affect soil removal from a soiled specimen of stipulated character are: (1) weight ratio of bath to specimen, usually referred to as the bath ratio;

(2) concentration and composition of the bath (e.g., how much and what type of detergent is present); (3) Temperature and duration of the operation (i.e., the kinetics of the situation); (4) type and extent of mechanical energy input into the system; (5) provisions for removing separated soil so that it does not redeposit on the substrate. Any washing procedure used for comparative or investigative purposes must provide for keeping all of these factors under control. On this basis, there has been described a wide variety of devices intended to perform a standardized washing operation. The one factor that is most difficult to simulate realistically in a laboratory washing device is the mechanical energy input. Consider, for example, a tumble-type laundering machine ("laundry wheel") such as is commonly used for commercial and institutional laundering. A typical machine consists essentially of a cylindrical drum about 4 ft in diameter rotating on a horizontal axis and containing about 100 lb of laundry and about 100 gal of detergent solution. The mechanical action consists of the heavy load being alternately lifted and dropped an average equivalent distance of about 2 ft every revolution of the drum. This heavy pounding and rubbing cannot be duplicated in a miniature drum, even if the rate of rotation and oscillation is greatly increased. The degree of soil removal obtained in laboratory washing devices cannot, therefore, be used to predict quantitatively the results in full-scale equipment. Laboratory devices are useful and reliable, however, in assessing and comparing the effects on soil removal of several very important factors. These include composition of the detergent bath and composition of the substrate–soil system, the two factors of greatest interest to detergent producers and fabric manufacturers.

By far the most popular laboratory device for studying fabric detergency is the Tergotometer.[®(22)] The working unit in the instrument consists essentially of a stainless steel beaker immersed in a thermostatted bath and fitted with a reciprocating agitator of fixed design simulating the design of a standard upright washing machine agitator. The agitation rate can be controlled, as can the temperature. The unit is essentially a miniature (about 1-liter working capacity) upright home laundering machine. Four units are provided in a single bank, all immersed in the same thermostatted bath, with their agitators all driven off a single shaft. Standard soiled swatches are washed in the Tergotometer at a fixed temperature and rate of agitation for all four units. The instrument is thus most satisfactory for comparing the effects of different detergent baths with regard to composition, concentration of different ingredients, pH, etc. In normal laundering, the bath ratio is so high that rinsing becomes relatively unimportant. Most users of the Tergotometer rinse the washed swatches under the tap or in a relatively large container of water.

*2.1.2.5. Measurement of Soil Removal.* Since detergency is concerned with the removal of soil from a substrate, it is necessary in any study of detergency to have a method for measuring the soil content of the substrate, both before and after treatment with the detersive bath. Depending on the system, this is done in most cases either by mass measurements or by optical measurements. Mass measurements are, of course, the more fundamental and important in the scientific study of detergency as a physicochemical phenomenon. In many practical situations, however, the prime requirement is a removal of *visible* soil, and within reasonable limits, a considerable mass of nonfunctional foreign material can be tolerated on the substrate provided it is not visible or otherwise obtrusive.

As mentioned above, the soil most commonly used in studies of laundering is a mixture of oily constituents (invisible in the concentrations normally used) and visible soiled particulate matter. This solid soil is invariably chosen to have some color, which may range from a pale gray or tan to solid black. The soil of the test swatches at all stages is conventionally expressed in terms of their whiteness. Although a measurement of whiteness says little or nothing about the oil content of the fabric, and except under special conditions cannot be related quantitatively to the mass of solid soil present on the swatch, it is the most "realistic" measure of cleanness. Householders and laundry operators judge the cleanness of their goods visually, and unless the fabric has an unusual concentration or a concentrated spot of invisible soil they are unaware of it.

Comparative judgments of whiteness (i.e., swatch A is whiter or darker than swatch B) are easily made by eye, and in fact, visual comparisons were used exclusively in detergency studies until the advent of reliable inexpensive commercial photometers. The human eye can tell differences in whiteness or reflectance of any color about as well as any but the most sophisticated instruments. It cannot, however, assign a numerical value to the reflectance. Essentially all modern detergency studies use a photoelectric reflectance meter of some type to measure the whiteness of test swatches. Most commercial reflectance meters (photometers) have a readout which expresses the reflectance of the test material as a percentage of the reflectance of a block of pure magnesium oxide taken as 100. A typical clean bleached cotton print cloth will have a reflectance in the neighborhood of 85. Typical test swatches soiled evenly with carbon black as the solid particulate constituent will have reflectances that might range from 20 (dark soil) to 60 (light soil). The most useful photometers are equipped with color filters, so that the test specimen can be illuminated as desired with blue, green, or amber light and the reflectance values for each of these colors read. The green reading is often taken as the index of whiteness, but a more realistic whiteness value in terms of both the blue and green readings can be calculated.[23a,b,c] By including the amber reading a value of yellowness can also be calculated. This is of considerable practical importance, because certain detergents, particularly when interacting with certain types of naturally occurring soils, tend to impart a yellow cast to the originally white fabric.

The reflectance measurement is by no means linearly related to the amount of soil on the fabric. In fact, the relationship between mass of superficial particulate soil and reflectance is very complex, depending on particle size range and size distribution, degree of agglomeration of the individual particles, distribution of particles and agglomerates on the surface of the fibers, and several other factors, as well as on the reflectance spectrum (color) of the individual soil particles. An approximate correlation between mass of soil and reflectance of soiled fabric is given by the Kubelka–Munk equation, originally developed to express the hiding power of paints in terms of their pigment content.[24a,b] There is, however, among detergent researchers little real interest in the mass of visible soil on a test swatch. Mass as such is not the important factor in visible soil–substrate relationships. A substrate can be judged dirty when it bears adherent and visible particles of total mass so small as to be negligible. The adhesion of the particles, which in general is in inverse proportion to their individual masses, is the key factor in soil removal. The function of the detergent is to reduce this adhesive force, thus enabling removal of the soil particles by the mechanical and hydraulic forces operative during the washing process. It is sometimes of interest to study the removal of solid soils that are not visible, for example, white clays, silicas, starches, etc. In these cases, mass is usually important, since the undesirable effects of these soils are proportional to the quantity present and are generally not related to appearance but rather to stiffness, abrasiveness, handle, or some other preceptible characteristic. Determination of such soils can often be done by chemical methods. A nitrogen determination, for example, can be used for estimating proteinaceous soil on a nonnitrogenous substrate. More often, invisible solid soils of this type are determined by tracer methods, outlined later in this discussion.

Oily soil is of considerable importance in household laundry, since most of the soil on underapparel consists of oily sebum and perspiration residues. In studying detergency in relation to household laundering, it is customary to apply, together with the solid particulate soil, a considerable quantity of oily soil to the test swatches. This may vary from less than 0.5% to as much as 5%, based on the weight of the fabric swatch. About 2% is considered a satisfactory oily soil level for test purposes, although it is far higher than the level actually encountered in dirty laundry. An oil or grease stain on a shirt, however, may contain within its area this percentage of oil. The residual oil content of the swatches after washing usually is determined by extraction.[24b] If the washing is very effective and/or the investigator is especially interested in the presence of very small amounts of residual oil, tracer methods are used.

The introduction of radiotracer methods of analysis revolutionized the study of detergency from both the technological and the scientific point of view. Not only radiotagged oils but also radiotagged solid particulate soils have been used extensively in detergency studies. $^{14}C$ and $^{3}H$ are the tags most commonly used for oily soils. Radioactive calcium is used as a tag in clay soils and $^{14}C$ in carbonaceous solid soils. The usual methods for measuring radioactivity are used in quantitative work. By using appropriate counters, the radioactive content of both the washed substrate and the spent bath can be measured, and a material balance on the soil can be established. Autoradiography is often used as a rapid semiquantitative or demonstration method for estimating the quantity of soil on a substrate. Radiotracer procedures are of great value in situations where the soil content of the test pieces is small or where it is not easily measured by conventional means. They are of greatest value in the theoretical study of detergency, to be discussed in a subsequent section.

## 2.2. Hard Surface Substrates

There is no single device for measuring hard surface cleaning in the laboratory that has won wide acceptance, although several have been described. For the most part, these devices have been developed for use in purchase specifications rather than in research and development. The types adapted for panel and floor cleaners exert a controlled scrubbing or squeegeeing action on the test specimen, while a specified quantity (in terms of ounces per square foot) of detergent bath is applied to the soiled test panel.[25] Rinsing is a very important step in panel cleaning, since it is the main way for separating fouled bath from the substrate. Rinsing procedures to be used after instrumental cleaning are usually specified in detail. Estimation of cleanness is made visually.

Dishwashing resembles fabric washing in that the soiled substrate is treated with a "long" bath (high bath ratio). Manual dishwashing is conventionally studied by washing standard soiled dishes by hand in baths of known concentration and temperature. The standard soiling consists in spreading onto the dish a prescribed weight per unit area of a prescribed soiling mixture, which usually consists of fat, a proteinaceous foodstuff such as peanut butter, and a farinaceous foodstuff such as cooked starch.[26] The data obtained can hardly be used as a model of scientific precision, but procedures of this type easily separate satisfactory from unsatisfactory performance and provide a realistic ranking of dishwashing detergents.

The study of machine dishwashing is almost always carried out in household or commercial size machines. Most of the technical literature in this area describes studies on drinking glasses. Glassware shows even slight amounts of undesirable soil, and the relative efficacy of different detergents on cleaning procedures is more easily judged on glassware than on porcelain or china.[27] Quantitative estimates of detergent effectiveness can be obtained by using soiled microscope slides, washing in a standard commercial machine or a specially designed small machine, and measuring the clarity of the washed slides with a suitable photometer.[28]

Stamped and machined metal parts are cleaned to remove the fabricating lubricants. Studies of the effect of washing conditions on the degree of cleaning achieved are usually performed in full-scale washing equipment rather than in laboratory devices.[29] If the metal is to be finished with an organic coating or if it is to be electroplated, it should be completely free of fatty contamination. The water break test is commonly used to estimate if a sufficient degree of cleanness has been achieved. In this test, the metal piece is sprayed or dipped into water. If the water forms a continuous layer and drains evenly without withdrawing from the metal surface at any point (i.e., "showing a break"), the cleaning is judged satisfactory. For more quantitative data, radiotagged soils are usually used.

Food-handling equipment, including dairy equipment, must be kept clean continuously, and cost effectiveness of the detergent and the cleaning procedure can be an important factor in profitabilty of the total operations. Studies of these factors are usually made in full-scale equipment. Assessment of cleanness is made in most cases by visual inspection. Sometimes microbiological tests are made to estimate the residual bacterial count on the cleaned surfaces.[30a,b]

### 2.3. Special Considerations in the Study of Practical Detergency

### 2.3.1. Assessment of Cleanness

The primary purpose of any practical cleaning operation is to remove undesirable contamination from the substrate without undesirably changing the character of the substrate itself. Both the degree to which the surface is cleaned and the degree to which the substrate is affected are essential parameters of practical detergency and must be carefully defined.

A truly "clean" solid surface may be regarded as one that has the same molecular composition as the interior. An an example, we might cite a crystal surface freshly cleaved in a hard vacuum. Since such a surface immediately adsorbs a certain proportion of any molecules or multimolecular particles that come into contact with it via the vapor phase, it can be regarded as a laboratory curiosity rather than an object of any practical significance. "Clean" in the practical sense means "free of *undesirable* foreign matter" or, to be even more precise, "free of perceptible amounts of undesirable foreign matter." The cleaning process is accordingly intended to reduce the amount of undesirable foreign matter on the substrate surface to the point where it is no longer perceptible to the user by any test he may wish to employ. It follows that whether a surface is clean in the practical sense depends on what substances the user considers to be undesirable. A white dress shirt, for example, may be deemed clean even though it contains a heavy surface coating of the starch–oil–wax mixture normally used as a finish after laundering. On the other hand, if it contains enough waxy material to have an undesirable feel (to touch) even though the waxy material is quite invisible, the fabric will be judged dirty. All solid surfaces adsorb, and they adsorb preferentially. This is one of the basic principles underlying the whole detersive process. Most materials are cleanable by aqueous surfactant solutions simply because they adsorb aqueous surfactant in preference to the soil with which they are burdened. Freshly laundered fabrics invariably bear an adsorbed layer of surfactant, the amount generally being in inverse proportion to the thoroughness of the rinsing operation. It is accordingly evident that the composition of a substrate surface depends on its history, especially on its recent history. These facts must be implicit in any discussion of how clean a particular item might be.

## 2.3.2. Effect on Substrate

Another ubiquitous consideration in practical detergency is the effect of the cleaning process on the subsurface or bulk of the substrate. This point is so obvious that it hardly needs elaboration, but it is sometimes overlooked in considering the effectiveness of a detergent. It is fruitless to wash soil from a printed garment if the dyestuff is also removed or to clean a painted wall if the paint becomes marred. In studying practical detergency, the changes suffered by the substrate during cleaning are fully as important as the extent of soil removal. In more theoretical studies subsurface effects are equally important, because cleaning that is due to subsurface effects must be distinguished from true detergency. Detergency is by definition a surface effect. It is easy to remove dirt from any material by paring away mechanically or chemically the outer layer that bears the dirt. This is not true detergency, even though such techniques are frequently used in practical cleaning, for example, in cleaning metals with acid.

It should be emphasized that the above remarks apply only to *permanent* alterations in the subsurface. Many substrates, notably the natural fibers and the synthetic cellulosics, are significantly swollen and weakened by water, but regain their original dimensions and strength on drying. Insofar as such substrates are not affected any more by the warm agitated detergent solution than by plain water, they suffer no permanent attrition or alteration due to the cleaning process.

### 2.3.3. Effects of Substrate Geometry

The configuration of the substrate surface plays as important a role as any other single factor in determining the ease of soil removal. This is true for both solid and oily soils. The macroscopic shape of a surface has, of course, no effect on the specific adhesion of the surface to either a solid or a liquid material that may be in contact with it, the specific adhesion being defined as the free energy of adhesion per unit area of contact. Many objects that require cleaning, however, have tortuous irregular surfaces that can trap or enmesh a soil particle so that it cannot easily be removed even when the specific adhesion has been reduced to zero. This effect is most pronounced with fabrics, especially with tightly constructed fabrics made from swellable fibers. A finely divided siliceous soil, for example, which has very low specific adhesion for cellulose, may be impossible to wash out of a tightly woven rayon fabric into which it has been rubbed while dry. On wetting with the aqueous wash liquor the rayon fibers swell and press against one another, holding in a viselike grip any soil particles that may be trapped between them. Similar effects can occur on nonfabric substrates. A matte painted surface or a concrete floor are both rough enough to hold fine dirt particles in reentrant or overhanging cavities, from which they are unlikely to be removed in an ordinary cleaning process, even though their specific adhesion energy is nil. Oily soils on a geometrically complicated object such as a fabric can be trapped even though the aqueous detergent bath has completely displaced them from the solid surfaces.[31] In a liquid–liquid–solid capillary system, the displaced liquid tends to move toward the flatter or more convex surfaces. In doing so, however, portions of liquid can be separated from the main mass and can be retained in reentrant spaces among the fibers.[32a,b]

The possibility of mechanical retention of soil must be taken into account when comparing various detergent baths for their soil-removing effectiveness on fabrics. The tests must be made on swatches that are identical in construction and in soil load. If any substantial proportion of the original soil load is retained affer washing, the tests should be replicated. They should also be repeated using increased agitation, which tends to release mechanically held soil. The factor of mechanical retention is minimized by using substrates that are as simple as possible in configuration. For fibrous materials, this translates to open weaves, low-twist yarns, and long uncrimped fibers. For nonfibrous materials, it means smooth surfaces that are flat or smoothly curved.

## 3. The Study of Detergency as a Physicochemical Phenomenon

### 3.1. General Considerations

Detergency is essentially the *disjoining* of a solid phase (the substrate) from a second solid phase (a solid soil particle) and/or a liquid phase (oily soil) that are adherent to it. This disjoining is accomplished by introducing a third phase, the bath. The action is a replacement of the substrate–soil interfaces by substrate–bath and soil–bath interfaces. It can be studied from two different points of view—the phenomenological, or descriptive, and the mechanistic, or causative. Since this chapter is concerned primarily with experimental techniques, we shall consider the macroscopic observables of detersive systems rather than the underlying molecular mechanisms.

In practical cleaning operations, as we have seen above, there are as many separate solid–solid phase interfaces present as there are species of solid soil particles. This assumes that the substrate surface is chemically homogeneous; it often consists of two or more phases (as in the case of blended fabrics or composite nonfibrous surfaces), in which case the number of solid–solid phase interfaces is correspondingly increased. There are also present the substrate–liquid soil interfaces and the numerous solid soil–liquid soil interfaces, which may or may not become disjoined during the cleaning operation. The major objective of a practically oriented study is to find out how much undesirable contamination (regardless of its composition) remains on the substrate and the manner in which removal of this contamination or soil is influenced by the variables of the system. The experimental techniques are aimed at control and manipulation of these variables, and at measuring the quantity of soil adherent to the substrate at all stages of the cleaning operation. This is, of course, a straightforward and technically sound approach. In practical studies, however, at least four of the most important factors influencing soil removal are either not varied or not controlled. These are composition and geometry of the substrate and composition and geometry of the soil. Control and systematic variation of these factors distinguish the more sophisticated studies of detergency, which we shall refer to for brevity as "basic" or "theoretical." Another distinctive feature of these scientifically oriented detergency studies is an attempt to observe the physical mechanics and kinetics (i.e., the phenomenology) of soil separation. We are interested not only in the amount of soil removed but also in *how* it is removed; in the case of an oily soil, for example, is the removal accomplished by simple displacement, by emulsification, or by solubilization? Techniques are therefore required for observing and following the removal process. The basic experiment is the same in both practical and theoretical detergency studies and is ostensibly simple. A solid object bearing adherent water-insoluble soil is acted upon by an aqueous bath, as a result of which some or all of the soil is separated.

## 3.2. Model Systems

To study the weakening or breaking of an adhesive bond, it is advantageous to start with a simple and well-defined system. The adherent phases should be homogeneous and well characterized, preferably consisting of one component. If more than one component is present, the composition should be well controlled. The geometry of the substrate phase and of the substrate–soil phase interface should be well characterized and p.eferably simple enough so that it does not interfere with the separation of soil from substrate. Finally, the composition of the aqueous bath phase must be known, and means for determining changes in its composition, i.e., adequate analytical techniques, must be available. Such a system may be referred to as a "model detersive system." Model detersive systems may obviously be divided into three broad classes—liquid soil systems, solid (particulate) soil systems, and mixed soil systems.

The mixed systems most closely approximate the detersive systems encountered in practice and are therefore of special interest. They generally exhibit interaction effects between the two soil phases, the liquid soil either helping or hindering the removal of solid soil. Furthermore, the liquid soil–solid soil complex in the presence of the bath is in itself the physicochemical equivalent of a detersive system. If it becomes disjoined, the solid component may to a certain extent redeposit on the substrate and form a system that is not as easily disjoined as the original system. This redeposition effect is of great importance in practical detergency, especially in the laundering of textile fabrics, and one might expect model redeposition systems to have been set up and studied from the basic point of view. A real distinction must be made, however, between the redeposition described above and a "deposition" system in which the soil is not initially attached to the substrate. Many deposition studies have been published in connection with practical fabric detergency. A typical experiment consists in making a suspension of the desired soil (usually carbon black) in the desired detergent bath, agitating a clean piece of fabric in the soil suspension, withdrawing the fabric, and after light rinsing, measuring the soil pickup by reflectance.[32b] It has recently been shown, however, that this system behaves quite differently from one in which a carbon-soiled fabric and a clean fabric are agitated together in the same detergent bath.[33] The deposition system really belongs to the study of soiling, the process by which a substrate acquires adherent foreign matter. Soiling may be regarded as the inverse of detergency, and knowledge of soiling obviously must contribute to knowledge of detergency. Far fewer studies have been made of soiling, however, than of detergency, and soiling as such will not be considered at any length in this discussion.[34]

The essence of a model detergency system is simplicity—simplicity with regard to geometry and with regard to the phases and components of substrate, soil, and bath. Considering first the geometry, it is evident that some practical systems are in themselves quite close to model systems. Flat glass or plastic surfaces or polished metal surfaces are ideal substrates from the standpoint of geometry. Glazed porcelain plates are in the same category. Some of these materials can be used as valid model substrates for practical systems of more complex geometry. For example, basic detergency data obtained using polyester film as the substrate can be translated directly to polyester fabric, provided film and fabric have the same composition and finish. Differences in cleanability between the film and the fabric in such a situation can confidently be ascribed to the complex geometry of the fabric.[35]

Materials that cannot be obtained in a form presenting extended smooth surfaces can frequently be obtained in the form of fibers. In fact, many of the most important substrates, notably the natural fibers hair, cotton, and wool, are obtainable only in the form of fibers. These fibrous substrates do not have smooth surfaces. If the fibers are loose, unspun, and not tightly packed they can be used *en masse* as substrate material in basic detergency studies.[36] One complicating factor in the use of relatively long fibers is that they can become matted or tangled as they are being agitated in the bath. Furthermore, a mass of loose fibers is not easy to soil uniformly, although by careful manipulation with oily or mixed oily particulate soils, reasonable uniformity can be achieved. One way of overcoming the disadvantages of a mass of loose fibers is to stretch a suitable number of individual fibers, deployed at a considerable space, across a frame, forming a harplike or zitherlike array. This technique has been used in studying the effects of various detergents in cleaning human hair. The array can be soiled uniformly, and the hairs cannot interfere with one another during washing. Single fibers have frequently been used in basic detergency studies. Some of the earliest work demonstrating the rollback mechanism of oily soil removal (discussed in a subsequent section) was done on single fibers.[37,38] The single fibers in these studies are usually stretched across a frame for easy handling and observation. It should be noted that most of the common natural fibers do not have smooth surfaces. The rugosities are capable of holding a certain amount of either solid or oily soil mechanically rather than by pure forces of adhesion.[39] Many synthetic fibers are manufactured with smooth surfaces and cylindrical cross sections and make ideal substrates for detergency studies.

A very effective method for studying detergency on masses of fiber but avoiding the problems of matting and tangling is to chop the fiber into very short lengths, preferably of 1 mm or less. Textile fibers of this length have little tendency to form mats. They can be soiled uniformly and easily separated from the wash liquor on a suction funnel, and the filter mat can be read directly for reflectance.[40,41]

Applying soil to the substrate offers no special problems, and control of both soil composition and bath composition is no more difficult in basic than in practical detergency studies. The problem of controlling energy and power input, however, and of applying a controlled mechanical force to help separate soil from substrate is no easier to solve with a model substrate than with a realistic substrate.[42] The washing procedures that have been used vary from simple immersion in the bath through all types and levels of mechanical agitation, including ultrasonic vibration. There is usually no difficulty in reproducing the overall soil removal result with any of the common methods of agitation. The difficulty from the scientific standpoint lies in not being able to measure quantitatively the force necessary to separate a solid soil particle from the substrate. Attempts have been made to do this indirectly by hydraulic techniques, essentially by directing water jets of known force at a soiled flat plate substrate.[43] One of the most interesting attacks on this problem grew out of some basic studies on adhesion in which a high-speed centrifuge was used as the device for applying a separating force. The model soil particles, spheres of about 10-$\mu$m diameter, were deposited on the model substrate, which formed the outer surface of the centrifuge's rotor. The force necessary to throw off the adherent particles was measured in air, water, and detergent solution.[44,45] Although successful in showing that the soil–substrate bond was actually weakened by the presence of detergent, this experiment served to emphasize one of the most intractable problems in the study of solid soil detergency and indeed in the closely related study of adhesion—determining and controlling the area of contact between soil particle and substrate. This is closely related to the problem of controlling size and shape of the particles. Size and shape are prime factors in determining the traction that can be exerted on the soil–substrate interface, i.e., the disjoining force; and area of true contact determines the adhesive force for any given value of specific adhesion. One of the classic studies in detergency showed quantitatively the effect that particle size has on solid soil detergency.[46] A graded series of carbon blacks was used to soil cotton swatches, which were then washed under identical conditions. The results, plotted as particle size versus soil removal showed that a critical particle size exists above which the soil is easily detached and below which it remains adherent to the substrate. Ideally, studies of solid soil detergency would be made with soil particles all of the same size and shape. Except in very special cases, few if any of which could be related to practical laundering, this is not feasible.

## 3.3. Experimental Methods

### 3.3.1. Visual and Microscopic Observation

As in many other aspects of applied colloid science, visual observation is the basic technique for ascertaining the various phenomena that occur in detergency. The primary tool is the ordinary light microscope, but both the transmission and scanning electron microscopes have proved invaluable in modern detergency studies.[47,48] Present-day knowledge of the various modes in which oily soil is separated from a fiber substrate has been gained largely by microscopic observation. These modes include rollback, emulsification, and solubilization, all of which can be observed directly and can be assessed as function of phase composition, temperature, etc. The usual technique is to stretch the fiber across a frame, apply a droplet of the model oily soil, and immerse the setup in a container of the bath solution mounted on the microscope stage. Suitable manipulators may be used to stir the bath, and the action may be recorded by cinema as desired. This is the simple procedure that was used to elucidate the imbibition effect observed with water-swellable fibrous substrates[49a,b] and to show the importance of mesophase formation in removing oily soil at low bath ratios and moderate surfactant concentrations.[50a,b] Spontaneous or easily induced emulsification is also observable and identifiable in the light microscope. Oily soil removal in the rollback mode has been observed on very fine fibers by means of the scanning electron microscope.

Solid soil particles of practical interest are seldom larger than 5–10 $\mu$m and are often in the submicron range. They are therefore not easily observed in the light microscope under ordinary conditions. The separation behavior of the larger particles, however, has been studied by this technique.[51] One of the more interesting studies showed the role of foam laminae in segregating separated solid soil particles from the bulk of the bath, an effect of considerable importance in rug shampooing, hair shampooing, and other short-bath cleaning operations.[52] Fine solid soil particles are best observed by the electron microscope. This instrument has been used advantageously in studying the disposition and attachment of soil particles on the natural fibers wool and cotton before and after washing.[53]

### 3.3.2. Mass Measurements of Soil Removal

Mass measurements of the soil content of a substrate either before or after washing are no different in basic detergency studies than in practical studies. With oily soils they can be made by solvent extraction and weighing. With certain solid soils (iron oxide, for example), they can be made by wet chemical methods or by spectroscopic or microprobe analysis. Chemical analysis can be used in most cases where the soil contains an element (e.g., nitrogen, sulfur, aluminum, silicon) not contained in the substrate. The disadvantage of chemical methods, aside from being tedious, is they usually require destruction of the substrate. This is less of a disadvantage in basic than in practical studies.

By far the most effective technique for measuring soil content and soil removal is to use radiotagged soil. One of the earliest uses of radiotagged oily soil was in a study of metal cleaning made by Harris and co-workers.[54] These investigators found that even after the metal coupons had been washed to the stage where they showed no water break, they still contained an appreciable quantity of residual oil, presumably adsorbed as a fractional monolayer. Studies of oily soil removal from fabrics under simulated laundering conditions have been published by serveral different groups of investigators.[55–57] Among other important observations it has been demonstrated that polar and nonpolar oily components are differentially removed from a single oily soil phase, presumably due to differential solubilization in the aqueous surfactant bath.[58] This was done by tagging one of the components with tritium and the other with $^{14}$C. Numerous other studies of oily soil removal have been made in model detergency systems as well as in the more practical systems.

Valuable as tracer methods have been in studying oily soil removal, they have been even more advantageous in studying the removal of particulate solid soils.[59,60] They are not only much faster than chemical methods but are also nondestructive of the substrate. Elemental $^{14}$C has been used in estimating carbon soil removal. The calcium in clay can be activated by irradiation, and the behavior of clay soils in detergency has been followed in this manner. Tracer methods are also accurate and revealing in studying the removal of mixed soils.

### 3.3.3. Optical Measurements of Soil Removal

Optical measurements of soil removal include colorimetric and turbidimetric measurements and measurements of whiteness. They are used to a much lesser extent in basic studies than they are in the practical studies discussed previously. Oil-soluble, water-insoluble dyes can be used to tag oily soils, provided the dyes are not adsorbed by the substrate. Turbidimetric methods can be used to estimate the quantity of soil in a used bath, provided that the soil is well and reproducibly dispersed and that no other factors are contributing to the turbidity.[61a–d] Whiteness measurements on fabric soiled with carbon black, as mentioned in the discussion of practical detergency, can be translated to mass of carbon per unit area of fabric by use of the Kubelka–Munk equation. This is seldom done in modern basic studies, however, and in practical studies the whiteness in itself is of greater interest than the mass of soil removed.

## 3.4. Ancillary Studies

It is evident that the detersive process can be regarded as a conglomerate of simpler processes, many of which have been well studied and have become recognized as discrete areas of endeavor in colloid science. An understanding of these simpler processes is necessary if their resultant, the detersive process, is to be understood and adequately described. Oily soil is removed from a substrate by any of three different mechanisms, or any combination or variation thereof. They are rollback, emulsification, and solubilization. The separation of oily soil from solid soil after a mass of mixed soil has been removed from the substrate can also occur by any of these three mechanisms. Separation of a homogeneous solid soil from a substrate is equivalent to the deflocculation of a mixed agglomerate consisting of two solid phases. In practice, solid soil, even though it may be homogeneous, adheres to the substrate in the form of agglomerates. If these homogeneous agglomerates become dispersed in the bath, the individual particles that remain adherent to the substrate may be so unnoticeable as to be unobjectionable. In such a case, the whole detersive process is simply the deflocculation of a homogeneous agglomerate, a phenomenon that has been exhaustively studied and is as well understood as any in the realm of colloid chemistry.[61a] From this point of view, we can list the component phenomena of the detersive process and the ancillary phenomena that accompany it. The techniques used in studying these effects, although they are beyond the scope of this discussion, must be regarded as essential tools in laying an experimental basis for understanding detergency.

## 3.4.1. Rollback

"Rollback" is the name given to the effect in which a mass of adherent oily soil is displaced from a substrate by an aqueous bath without becoming either emulsified or dissolved. In cases where the substrate is a flat polished surface, the action is a classic example of a liquid–liquid–solid system coming to its equilibrium contact angle.[61b–d] For complete displacement of the oil to occur without external mechanical aid, the equilibrium contact angle of the system, measured in the water, must be zero. If external agitation is used, displacement can be achieved at contact angles greater than zero, provided they are substantially less than 90°. This is due to the fluid nature of the oil, which requires that a single mass of it, forced into a sufficiently elongated shape, spontaneously separates into two masses. If the substrate surface is heterogeneous, problems of contact angle hysteresis and dynamic contact angle enter the picture.[62,63]

In geometrically complex substrates such as ordinary textile fabrics, the rollback effect becomes a problem in capillarity. Modern mathematical and computational techniques enable the nonexperimental solution of the simpler problems in capillarity.[64] Experimental approaches are the same whether the system under investigation is regarded as capillary or detersive. The capillary behaviour of multicomponent liquid phases, i.e., solutions, is complicated by the changes in composition that occur progressively as the result of selective adsorption. The study of selective adsorption from the liquid phase is accordingly of direct applicability to detergency in the rollback mode.

### 3.4.2. Emulsification

The study of emulsions and emulsification is generally recognized as a full-fledged subdiscipline of colloid science with a well-developed specialized methodology of its own. In many cleaning systems, emulsification is the main process by which excess liquid soil is detached from the soiled substrate system. It follows that many of the methods used in studying the formation and stability of emulsions can also be used in detergency studies. Systems of greatest interest in laundering are the dilute oil–water emulsions of low viscosity that are formed from a small quantity of oil and a large quantity of water. More concentrated systems, as well as self-emulsifying systems, however, are frequently encountered in other cleaning operations.

### 3.4.3. Solubilization and Mesophase Formation

Solubilization, the process whereby an oil is taken up within the micelles of an aqueous surfactant solution, is probably operative in removing the thin layers and pools of oil left after rollback and emulsification are complete. The study of solubilization as a physicochemical phenomenon should accordingly be considered an integral part of the study of detergency.[65]

Study of the phase behavior of relatively concentrated surfactant solutions in contact with oils of various types, particularly polar oils, has revealed that relatively well-defined mesophases containing both oil and water can be formed by some systems. There is little doubt that such mesophase formation can play an important role in separating oily soil from a solid substrate, and a knowledge of the phase diagrams of water–polar oil–surfactant systems is directly applicable to certain detersive systems.[66] In this connection it should be noted that even the dilute surfactant solutions used in laundering can become quite concentrated in the vicinity of the substrate. It would in theory be possible for mesophases comprising the oily soil to form at the substrate surface and be carried into the bulk of the bath by hydraulic currents. Although this detersive mechanism has not been demonstrated, removal of heavy oil deposits by mesophase formation with concentrated surfactant solutions is a well-established method of cleaning metals and other hard-surface substrates.[67]

### 3.4.4. Deflocculation

Solid soil is attached to the substrate either in the form of individual particles (usually in the size range less than 5 $\mu$m) or as agglomerates, the agglomerate being attached by relatively few of its component particles. Deflocculation of the agglomerate by the bath accomplishes removal of the major portion of soil. Studies of deflocculation are thus directly related to detergency. Such studies are well advanced in both theory and experiment. Most of them have been made with single-component sols. It is only in recent years that a theory applicable to multicomponent sols has been developed.[68,69a,b] This theory, however, could well prove to be a major breakthrough in the study and understanding of detergency. A single soil particle adherent to a substrate constitutes a two-component agglomerate, and the deflocculation of this agglomerate can be regarded as the basic effect in solid soil detergency.

As mentioned earlier, the study of detergency has consisted largely of variations on a single experiment: forming a soil–substrate complex, treating the complex with a bath, and measuring the extent to which the complex is disrupted. It is true that the variations have been numerous, ingenious, and sophisticated and can profitably be extended in the same vein.[69b] As has been demonstrated, however, many other major areas of investigation in colloid science are directly applicable to detergency, and the phenomena being studied are integrally involved in the detersive process. The various techniques and experimental designs used in their study should therefore be regarded as belonging to the study of detergency. There is little doubt that they can contribute importantly to the understanding of this familiar but complex effect.

## References

1. A. M. Schwartz, in *Surface and Colloid Science* (E. Matijevic, ed.), Vol. 5, p. 195ff, John Wiley and Sons, New York (1972).
2. A. M. Schwartz, in *Encyclopedia of Chemical Technology* (R. E. Kirk and D. F. Othmer, eds.) (2nd ed.), Vol. 6, p. 853, Interscience, New York (1965).
3. F. H. Rhodes and S. W. Brainerd, *Ind. Eng. Chem.* **21**, 60 (1929).
4. W. Spring, *Kolloid Z.* **4**, 161 (1909); **6**, 11, 109, 164 (1910).
5. ASTM Standard D2960-76. *ASTM 1976 Book of Standards*, p. 437, ASTM, Philadelphia (1976).

6. A. M. Schwartz, J. W. Perry, and J. Berch, *Surface Active Agents and Detergents*, Vol. II, p. 566ff, Interscience, New York (1958).
7. J. C. Harris, *Detergency Evaluation and Testing*, Interscience, New York (1954).
8. K. Durham (ed.), *Surface Activity and Detergency*, Macmillan, London (1961).
9. J. J. Cramer, in *Detergency, Theory and Test Methods* (W. G. Cutler and R. C. Davis, eds.), p. 323ff., Marcel Dekker, New York (1972).
10. C. E. Warburton, Jr. and A. T. Schindler, *J. Appl. Polymer Sci.* **15**, 2794 (1971).
11. H. L. Sanders and J. L. Lambert, *J. Amer. Oil. Chemists' Soc.* **27**, 153 (1950).
12. J. Compton and W. J. Hart, *Textile Res. J.* **23**, 158, 418 (1953).
13a. F. L. Diehl and J. B. Crowe. *J. Amer. Oil Chemists' Soc.* **31**, 404 (1954).
13b. E. Goette, *Melliand Textilber.* **35**, 534 (1954).
13c. W. M. Linfield *et al.*, *J. Amer. Oil Chemists' Soc.* **53**, 60–76 (1976).
14a. M. E. Ginn and J. C. Harris, *J. Amer. Oil Chemists' Soc.* **38**, 605 (1961).
14b. L. I. Bavika *et al.*, USSR patent 525, 765 (1976); *Chem. Abstr.* **85**, 179418x.
15a. A. R. Martin and R. C. Davis, *Soap Chem. Specialties* **36** (4), 49 (1960); (5), 73 (1960).
15b. W. C. Powe, *Textile Res. J.* **29**, 879 (1959).
15c. T. P. Matson and M. A. Johnson. *J. Amer. Oil Chem. Soc.* **53**, 218 (1976).
16. S. V. Vaeck and E. Maes, *Tenside* **5**, 4 (1968).
17. R. C. Ferris and L. O. Leenerk, *Soap Chem. Specialties*, **32**(7), 37 (1956).
18a. E. Walter, *Fette Seifen Anstrichmittel* **61**, 188 (1959).
18b. L. Kravetz, D. H. Scharer, and H. Stupel, *Household Personal Prods. Inds.* **14**(1), 55 (1977).
19. W. G. Spangler, H. D. Cross III, and B. R. Schaefoma, *J. Amer. Oil Chemists' Soc.* **42**, 723 (1965).
20a. J. R. Trowbridge and J. Rubinfeld, *Proc. 4th Int. Congr. Surface Activity, 1964*, Vol. 3, p. 221, Gordon and Breach, London (1967).
20b. R. G. Anderson, U.S. patent 3,714,076 (assigned to Chevron Research Co.) (1973).
21. A. M. Schwartz and J. Berch, *Soap Chem. Specialties* **39**, 5, 78 (1963).
22. U. S. Testing Co., Hoboken, New Jersey.
23a. R. S. Hunter, *J. Opt. Soc. Am.* **50** 44 (1960).
23b. R. S. Hunter, National Bureau of Standards Circular C-429 (1942).
23c. R. S. Hunter, *J. Amer. Oil Chemists' Soc.* **45**, 362 (1968).
24a. I. Reich, F. D. Snell, and L. Osipow, *Ind. Eng. Chem.* **45**, 137 (1953).
24b. P. A. Carson and P. Tissington, Brit. patent 1,425,343 (1976).
25. U.S. Federal Specification PC-431a. Non-abrasive synthetic detergent cleaning compound.
26. E. H. Mann and C. C. Ruchhoft, *U.S. Public Health Repts.* **61**, 877 (1946).
27. J. L. Wilson and E. E. Mendenhall, *Ind. Eng. Chem., Anal. Ed.* **16**, 251 (1944).
28. A. L. Kimmel, H. M. Gadbury, and D. O. Darby, *Soap Chem. Specialties* **37**(4), 51 (1961).
29. J. C. Harris, W. Stericher, and S. Spring, *Amer. Soc. Testing Mater. Bull.* **204**, 31 (1955).
30a. E. H. Armbruster and G. M. Ridenour, *Soap Chem. Specialties* **28**(6), 83 (1952).
30b. E. H. Armbruster and G. M. Ridenour, *Soap Chem. Specialties* **31**(7), 47 (1955).
31. F. W. Minor, A. M. Schwartz, L. C. Buckles, and E. Wulkow, *Amer. Dyestuff Reptr.* **49**(12), 37 (1960).
32a. S. V. Vaeck and W. Verleye, *Tenside Deterg.* **13**, 216 (1976).
32b. A. M. Schwartz, *Ind. Eng. Chem.* **61**, 10 (1969).
33. J. W. Hensley, *J. Amer. Oil Chemists' Soc.* **42**, 993 (1965).
34. J. Berch, H. Peper, and G. L. Drake, Jr., *Textile Res. J.*, 39 (1964).
35. T. Fort, Jr., H. R. Billica, and T. H. Grindstaff, *J. Amer. Oil Chemists' Soc.* **45**, 354 (1968).
36. G. Barnett and D. H. Powers, *J. Soc. Cosmetic Chemists* **2**, 219 (1951).

37. N. K. Adam, *J. Soc. Dyers Colourists* **53**, 121 (1937).
38. W. Kling, E. Langer, and I. Haussner, *Melliand Textilber.* **25**, 198 (1945).
39. T. H. Grindstaff, H. T. Patterson, and H. R. Billica, *Textile Res. J.* **37**, 564 (1967).
40. J. Powney and A. J. Fenell, *Research* **2**, 331 (1949).
41. W. J. Hart and J. Compton, *Ind. Eng. Chem.* **43**, 1564 (1951).
42. J. Tuzson and B. A. Short, *Textile Res. J.* **30**, 989 (1960).
43. A. D. Zimon, *Adhesion of Dust and Powder*, Plenum Press, New York (1969).
44. H. Krupp, *Advan. Colloid Interface Sci.* **1**, 208 (1967).
45. W. Kling, *Fette, Seifen, Anstrichmittel* **69**, 676 (1967).
46. J. Compton and W. J. Hart, *Textile Res. J.* **24**, 263 (1954).
47. S. Simauchi and H. Mizushima, *Amer. Dyestuff Reptr.* **57**, 462 (1968).
48. W. Kling and H. Mahl, *Melliand Textilber.* **35**, 640 (1954).
49a. D. G. Stevenson, *J. Textile Inst.* **42**, T194 (1951).
49b. D. G. Stevenson, *J. Textile Inst.* **43**, T112 (1952).
50a. A. S. C. Laurence, in *Surface Activity and Detergency* (K. Durham, ed.), p. 180ff, Macmillan, London (1961).
50b. *Discuss. Faraday Soc.* **25**, 51 (1958).
51. I. J. Gruntfest and E. M. Young, *J. Amer. Oil Chemists' Soc.* **26**, 236 (1949).
52. D. G. Stevenson, *J. Soc. Dyers Colourists* **68**, 57 (1952).
53. V. W. Tripp, A. T. Moore, B. R. Porter, and M. L. Rollins, *Textile Res. J.* **28**, 447 (1958).
54. J. C. Harris, R. E. Kamp, and W. H. Yanko, *ASTM Bull.* no. 158, 49 (1949).
55. T. Fort, Jr., H. R. Billica, and T. H. Grindstaff, *Textile Res. J.* **36**, 99 (1966).
56. M. S. Sontag, M. E. Purchase, and B. F. Smith, *Textile Res. J.* **40**, 529 (1970).
57. B. A. Scott, *J. Appl. Chem.* (*London*) **13**, 133 (1963).
58. B. E. Gordon, J. Roddewig, and W. T. Shebs, *J. Amer. Oil Chemists' Soc.* **44**, 289 (1967).
59. B. J. Rutkowski and A. R. Martin, *Textile Res. J.* **31**, 892 (1961).
60. J. W. Hensley and C. G. Inks, *ASTM Spec. Tech. Publ.* no. 268 (1959).
61a. T. H. Vaughn, E. F. Hill, C. E. Smith, and L. R. McCoy, *Ind. Eng. Chem.* **41**, 112 (1949).
61b. T. Imamura and F. Tokiwa, *Nippon Kagaku Kaishi* **1976**, 869 (1976); *Chem. Abstr.* **85**, 80054p (1976).
61c. K. Ogino and W. Agui, *Bull. Chem. Soc. Japan* **49**, 1703 (1976); *Chem. Abstr.* **85**, 80042h (1976).
61d. K. Ogino and K. Shigemura, *Bull. Chem. Soc. Japan* **49**, 3236 (1976); *Chem. Abstr.* **86**, 31251n (1977).
62. T. D. Blake and J. M. Haynes, *Progr. Surface Membrane Sci.* **6**, 125 (1973).
63. R. Shuttleworth and G. L. J. Bailey, *Discuss. Faraday Soc.* **3**, 16 (1948).
64. R. J. Hansen and T. Y. Toong, *J. Colloid Interface Sci.* **37**, 196 (1971).
65. K. Shinoda (ed.), *Solvent Properties of Surfactant Solutions*, Marcel Dekker, New York (1967).
66. D. G. Stevenson, *J. Textile Inst.* **44**, T12 (1953).
67. *Metals Handbook* (8th ed.), Vol. II, Amer. Soc. for Metals, Metals Park, Ohio (1964).
68. E. M. Lifshitz, *Sov. Phys.* **2**, 73 (1956).
69a. L. D. Landau and E. M. Lifshitz, *Electrodynamics of Continuous Media*, Pergamon Press, Oxford (1960).
69b. H. Schleussler, *Brauwissenschaft* **29**, 263 (1976); *Chem. Abstr.* **86**, 57119a (1977).

# Author Index

# Subject Index